P9-DGF-554

Solutions Manual
for
Introduction to Genetic Analysis
Eighth Edition

WILLIAM D. FIXSEN

Interactive Genetics:
A Step-by-Step Approach to Problem Solving
CD-ROM

LIANNA JOHNSON
JOHN MERRIAM

Exploring Genomes:
Web-Based Bioinformatics Tutorials

PAUL G. YOUNG

W. H. FREEMAN AND COMPANY
NEW YORK

ISBN: 0-7167-8665-6
EAN: 978071678665-8

Copyright © 2005 by W. H. Freeman and Company

All rights reserved.

Printed in the United States of America

First printing

W. H. Freeman and Company
41 Madison Avenue
New York, NY 10010
Houndmills, Basingstoke RG21 6XS, England

www.whfreeman.com

Contents

1

Genetics and the Organism

BASIC PROBLEMS

1. *Genetics* is the study of genes and genomes: their biochemical basis; how they function; how they are controlled; how they are organized; how they replicate; how they change; how they can be manipulated; and how they are transmitted from cell to cell and generation to generation.

    The ancient Egyptian racehorse breeders can be only loosely classified as geneticists because their interests were highly focused on producing fast horses, rather than on attempting to understand the mechanisms of heredity. Their understanding of the processes involved in producing fast horses was very incorrect and their methods were not analytic in the modern sense of the word. Nevertheless, they did produce very fast horses through a combination of observation, trial and error, and artificial selection.

2. DNA determines all the specific attributes of a species (shape, size, form, behavioral characteristics, biochemical processes, etc.) and sets the limits for possible variation that is environmentally induced.

3. Properties of DNA that are vital to its being the hereditary molecule are: its ability to replicate; its informational content; and its relative stability while still retaining the ability to change or mutate. Alien life forms might utilize RNA, just as some viruses do, as a hereditary molecule. However, of the types of molecules that can exist on earth, only the nucleic acids possess the necessary characteristics.

4. There are four possible nucleotide pairs at each position (A–T, T–A, G–C, or C–G). Therefore, the general formula for calculating the different possibilities is 4^n (where n = the number of pairs). In this case, there are 4^{10} or 1,048,576 possible DNA molecules!

5. If the DNA is double stranded, A = T and G = C and A + T + C + G = 100%. If T = 15% then C = [100 − 15(2)]/2 = 35%.

6. If the DNA is double stranded, G = C = 24% and A = T = 26%.

7. Human somatic cells are diploid ($2n$) and the haploid number is 23 ($n = 23$). Each chromosome is a double-stranded DNA molecule, so there are 46 molecules of DNA. It is generally stated that the haploid number represents the number of different types of DNA molecules but that does not take into account the difference between the X and Y chromosome in males. So females have 23 different types (called 1–22 and X) and males have 24 (1–22, X, and Y).

8. Keeping this strand antiparallel to the strand given, the sequence would be

    T A A C C A C G T A A T G A A G T C C G A G A

9. Yes. There are no sequence restrictions in single-stranded DNA. The percentage of A must equal the percentage of T only in double-stranded DNA.

10. a. Yes. Because A = T and G = C, the equation A + C = G + T can be rewritten as T + C = C + T by substituting the equal terms.

    b. Yes. The percentage of purines will equal the percentage of pyrimidines in double-stranded DNA.

11. For the following, normal typeface represents previously polymerized nucleotides and *italic* typeface represents newly polymerized nucleotides.

    T T G G C A C G T C G T A A T
    A A C C G T G C A G C A T T A

    T T G G C A C G T C G T A A T
    A A C C G T G C A G C A T T A

12. Remember, the transcript will be antiparallel to the DNA template.
    3′ — U U G G C A C G U C G U A A U — 5′

13. There are several outcomes possible depending on the exact cause of the null allele. For the representation below, it is assumed that the null allele is still transcribed and translated but that the protein product of the allele is altered in a way that affects its mobility within the gel. (For example, it is slightly smaller and migrates to a position that is lower.) For both gels, lane 1 represents two normal alleles, lane 2 represents one normal and one null allele, and lane 3 represents two null alleles. One other possibility is that the mutant allele is neither transcribed nor translated, in which case lane 3 of both blots would be "blank."

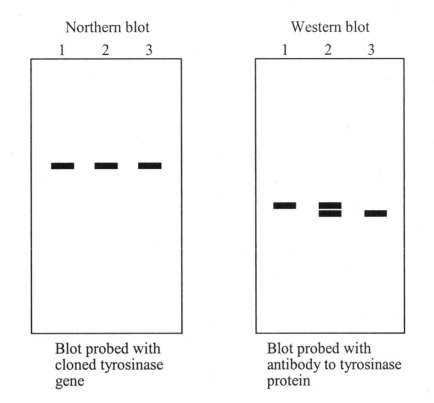

Northern blot

1   2   3

Blot probed with
cloned tyrosinase
gene

Western blot

1   2   3

Blot probed with
antibody to tyrosinase
protein

14. The simplest definition is that a *gene* is a chromosomal region capable of making a functional transcript. However, this does not take into account the regulatory regions near the gene necessary for the proper expression of the gene nor the regions that help control transcription that can be quite distant. Also, many eukaryotes have large regions of noncoding sequences (introns) interspersed within the regions that encode the product (exons.)

15. Many eukaryotes have intervening sequences called *introns*, which are transcribed but then spliced out (removed) prior to translation. In this case, the 25-kb primary transcript has all but 2.1 kb removed to generate the proper mRNA necessary for translation.

16. Electrophoresis separates DNA molecules by size. When DNA is carefully isolated from *Neurospora* (which has seven different chromosomes) seven bands should be produced using this technique. Similarly, the pea has seven different chromosomes and will produce seven bands (homologous chromosomes will co-migrate as a single band). The housefly has six different chromosomes and should produce six bands.

17. mRNA size = gene size − (number of introns × average size of introns)

18. The anticodon associates with the codon in an antiparallel orientation. Thus, to pair with the codon UUA, the anticodon must be AAU.

19. a. For the fungus to be orange (as in mutant 1), orange pigment would have to accumulate and then not be converted into red pigment. This would occur if enzyme C was defective.

   b. Mutant 2 is yellow. Using similar logic as in part (a), enzyme B must be defective.

   c. An organism defective in both enzyme B and C would still have the phenotype associated with the block at the earlier step of this biochemical pathway. In this case, it would be yellow due to the lack of enzyme B.

CHALLENGING PROBLEMS

20. *Unpacking the problem*

   1. *Lathyrus odoratus* (sweet peas) are grown for their fragrant, colorful flowers and in fact, their seeds are poisonous. The edible pea belongs to a different but related genus (*Pisum*).

   2. A pathway is a set of biochemical reactions that begins with a precursor molecule and through a series of steps, converts it into some product. In this case, a precursor molecule is converted into an anthocyanin pigment.

   3. There are two pathways described.

   4. Yes. As described, the two pathways are independent.

   5. In biology, a pigment is any molecule that reflects or transmits specific wavelengths of light — its color depends on which wavelengths of light are absorbed or reflected.

   6. In this case, a plant unable to synthesize pigment would be "colorless" or white. Many common solutes are colorless, such as salt (NaCl) and sucrose (table sugar).

   7. As stated above, white.

   8. Yes. The almost infinite variety of paint colors are obtained by mixing a very few primary pigments into a white base. In sweet peas, purple is achieved through the mixing of red and blue pigments.

   9. For now, a mutation is any alteration of the DNA that leads to a different and discontinuous variant, compared to the more common "normal" phenotype.

10. A null mutation is a change in the DNA that prevents the production of a functional product (in this case, enzyme).

11. Null mutations may arise through many types of changes: small deletions or duplications leading to frameshift mutations; nucleotide-pair substitutions that destroy the active site, lead to premature stop codons, or affect proper splicing (in eukaryotes); insertions of DNA; etc.

12. The organism is diploid. It has two copies of each of its chromosomes and therefore two copies of each of its genes.

13. The enzymes encoded by the genes are proteins. Whether or not the proteins are properly encoded (and the enzymes active) is the basis of this analysis.

14. No. The genetic locations of these genes are not relevant to this problem.

    In the following schematics, the "nucleus" separates the DNA and its transcription from translation. Also, the "defect" within the gene that results in the production of non-functional enzyme is marked by an "x."

15.

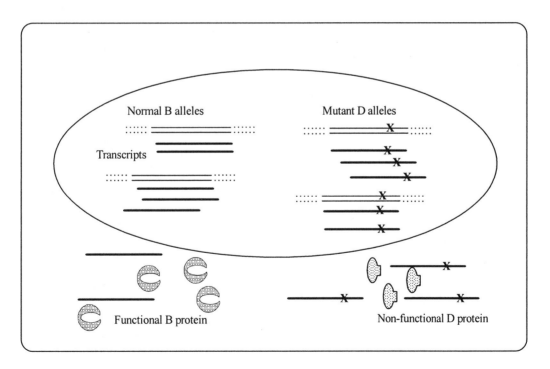

16.

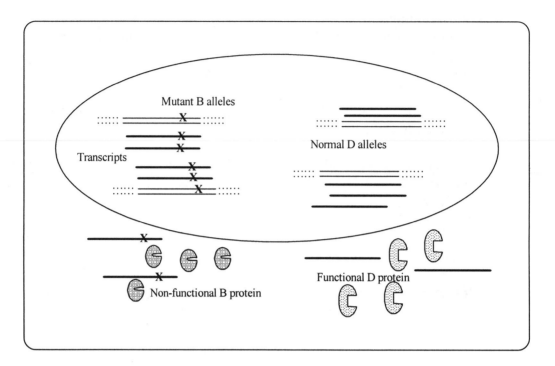

17.

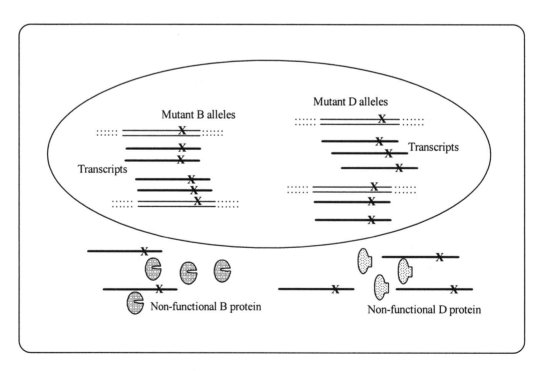

18. Sweet peas are able to produce two different pigments or colors. Just like mixing paints or crayons, these two colors can be blended to produce a

third. Only those plants that inherited the ability to make both blue and red pigments have purple flowers. Some plants, however, do not inherit this ability. Some have lost the ability to produce red and therefore have blue flowers, and conversely, some have lost the ability to produce blue and have red flowers. If both of these abilities are lost (or not inherited), the plant has white flowers.

Solving the problem

a.   A plant unable to synthesize red pigment would be blue.

b.   A plant unable to synthesize blue pigment would be red.

c.   A plant unable to synthesize both blue and red pigments would be white.

21.   There are a number of explanations that might explain the positional clustering of null mutants in this *Neurospora* gene. However, based on the material covered in this chapter, it is possible that the gene codes for an enzyme that uses its central amino acids (encoded by the central region of the gene) to form its active site. Mutations that alter the active site amino acids are most likely to destroy enzymatic function and would be classified as null (loss-of-activity) mutations.

22.   Protein function can be destroyed by a mutation that causes the substitution of a single amino acid, even though the protein has the same immunological properties. For example, enzymes require very specific amino acids in exact positions within their active site. A substitution of one of these key amino acids might have no effect on overall size and shape of the protein while completely destroying its enzymatic activity.

23.   **a.**   The mothers had an excess of phenylalanine in their blood, and that excess was passed through the placenta into the fetal circulatory system, where it caused brain damage prior to birth.

b.   The diet had no effect because the neurological damage has already happened *in utero* prior to birth.

c.   A fetus with two mutant copies of the allele that causes PKU makes no functional enzyme. However, the mother of such a child is heterozygous and makes enough enzyme to block any brain damage; the excess phenylalanine in the fetal circulatory system enters the maternal circulatory system and is processed by the maternal gene product. After birth, which is when PKU damage occurs in a PKU child, dietary restrictions block a buildup of phenylalanine in the circulatory system until brain development is completed.

The fetus of a mother with PKU is exposed to the very high level of phenylalanine in its circulatory system during the time of major brain development. Therefore, brain damage occurs before birth, and no dietary restrictions after birth can repair that damage.

d. The obvious solution to the brain damage seen in the babies of PKU mothers is to return the mother to a restricted diet during pregnancy in order to block high levels of exposure to her child.

e. PKU is characterized as a rare recessive disorder. A child with PKU has two parents that carry a mutant allele for the metabolism of phenylalanine. When two individuals who are heterozygous for PKU have a child, the risk that the child will have PKU is 25 percent. A PKU child is unable to make a functional enzyme that converts phenylalanine to tyrosine. As a result, an excess level of phenylalanine is found in the blood, and the excess is detected as an increase in phenylpyruvic acid in both the blood and the urine. The excess phenylpyruvic acid blocks normal development of the brain, resulting in retardation.

24. a. Recessive. The one normal allele provides enough enzyme to be sufficient for normal function (the definition of haplosufficient).

b. There are many ways to mutate a gene to destroy enzyme function. One possible mutation might be a frameshift mutation within an exon of the gene. Assuming that a single basepair was deleted, the mutation would completely alter the translational product 3′ to the mutation.

c. Hormone replacement could be given to the patient.

d. If the hormone is required before birth, it would be supplied by the mother.

25. DNA has often been called the "blueprint of life" but how does it actually compare to a blueprint used for house construction? Both are abstract representations of instructions for building three-dimensional forms and both require interpretation for their information to be useful. But real blueprints are two-dimensional renderings of the various views of the final structure drawn to scale. There is one-to-one correlation between the lines on the drawing and the real form. The information in the DNA is encoded in a linear array — a one dimensional set of instructions which only becomes three-dimensional as the encoded linear array of amino acids fold into their many forms. Also included in the informational content of the DNA are all the directions required for its "house" to maintain and repair itself, respond to change, and replicate. Let's see a real blueprint do that!

26. Other than the human ability to synthesize and wear synthetic materials, living systems are "proteins or something that has been made by a protein."

27. The norm of reaction is the phenotypic variation that exists for a species of a fixed genotype within a varying environment. The variability itself can become vital to a species with a change in environment, for it allows for the possibility that some individuals may be able to survive under the new environmental conditions long enough to reproduce.

28. Phenotypic variation within a species can be due to genotype, environmental effects, and pure chance (random noise). Showing that variation of a particular trait has a genetic basis, especially for traits that show continuous variation, therefore requires very carefully controlled analyses. This is discussed in detail in Chapter 20 of the companion text.

29. The formula *genotype + environment = phenotype* is not quite accurate. While phenotype is a product of the genotype and the environment, there is also an inherent variation due to random noise. Given complete information regarding both genotype and the environment, it is still impossible to specify the phenotype completely, although a close approximation can be achieved.

2

Patterns of Inheritance

BASIC PROBLEMS

1. Mendel's first law states that alleles are in pairs and segregate during meiosis, and his second law states that genes assort independently during meiosis.

2. Do a testcross (cross to a/a). If the fly was A/A, all the progeny will be phenotypically A; if the fly was A/a, half the progeny will be A and half will be a.

3. The progeny ratio is approximately 3:1, indicating classic heterozygous-by-heterozygous mating. Since Black (B) is dominant to white (b):
    Parents: B/b × B/b
    Progeny: 3 black:1 white (1 B/B : 2 B/b : 1 b/b)

4. a. This is simply a matter of counting genotypes; there are nine genotypes in the Punnett square. Alternatively, you know there are three genotypes possible per gene, for example R/R, R/r, and r/r, and since both genes assort independently, there are 3 × 3 = 9 total genotypes.

    b. Again, simply count. The genotypes are
1 R/R ; Y/Y	1 r/r ; Y/Y	1 R/R ; y/y	1 r/r ; y/y
2 R/r ; Y/Y	2 r/r ; Y/y	2 R/r ; y/y	
2 R/R ; Y/y			
4 R/r ; Y/y			

    c. To find a formula for the number of genotypes, first consider the following:

Number of genes	Number of genotypes	Number of phenotypes
1	$3 = 3^1$	$2 = 2^1$
2	$9 = 3^2$	$4 = 2^2$
3	$27 = 3^3$	$8 = 2^3$

Note that the number of genotypes is 3 raised to some power in each case. In other words, a general formula for the number of genotypes is 3^n, where n equals the number of genes.

For allelic relationships that show complete dominance, the number of phenotypes is 2 raised to some power. The general formula for the number of phenotypes observed is 2^n, where n equals the number of genes.

d. The round, yellow phenotype is $R/-$; $Y/-$. Two ways to determine the exact genotype of a specific plant are through selfing or conducting a testcross.

With selfing, complete heterozygosity will yield a 9:3:3:1 phenotypic ratio. Homozygosity at one locus will yield a 3:1 phenotypic ratio, while homozygosity at both loci will yield only one phenotypic class.

With a testcross, complete heterozygosity will yield a 1:1:1:1 phenotypic ratio. Homozygosity at one locus will yield a 1:1 phenotypic ratio, while homozygosity at both loci will yield only one phenotypic class.

5. Each die has six sides, so the probability of any one side (number) is $1/6$.
To get specific red, green, and blue numbers involves "and" statements that are independent. So each independent probability is multiplied together.

a. $(1/6)(1/6)(1/6) = (1/6)^3 = 1/216$

b. $(1/6)(1/6)(1/6) = (1/6)^3 = 1/216$

c. $(1/6)(1/6)(1/6) = (1/6)^3 = 1/216$

d. To not roll any sixes is the same as getting anything but sixes:
$(1 - 1/6)(1 - 1/6)(1 - 1/6) = (5/6)^3 = 125/216$.

e. There are three ways to get two sixes and one five:

6R, 6G, 5B	$(1/6)(1/6)(1/6)$
or	+
6R, 5G, 6B	$(1/6)(1/6)(1/6)$
or	+
5R, 6G, 6B	$(1/6)(1/6)(1/6)$

$$= 3(1/6)^3 = 3/216 = 1/72$$

f. Here there are "and" and "or" statements: p(three sixes "or" three fives)
$= p$(6R "and" 6G "and" 6B "or" 5R "and" 5G "and" 5B)
$= (1/6)^3 + (1/6)^3 = 2(1/6)^3 = 2/216 = 1/108$

g. There are six ways to fulfill this:
$$(1/6)^3 + (1/6)^3 + (1/6)^3 + (1/6)^3 + (1/6)^3 + (1/6)^3 = 6(1/6)^3 = 1/36$$

h. The easiest way to approach this problem is to consider each die separately. The first die thrown can be any number. Therefore, the probability for it is 1.

The second die can be any number except the number obtained on the first die. Therefore, the probability of not duplicating the first die is $1 - p$(first die duplicated) $= 1 - 1/6 = 5/6$.

The third die can be any number except the numbers obtained on the first two dice. Therefore, the probability is $1 - p$(first two dice duplicated) $= 1 - 2/6 = 2/3$.

Finally, the probability of all different dice is $(1)(5/6)(2/3) = 10/18 = 5/9$.

6. You are told that the disease being followed in this pedigree is very rare. If the allele that results in this disease is recessive, then the father would have to be homozygous and the mother would have to be heterozygous for this allele. On the other hand, if the trait is dominant, then all that is necessary to explain the pedigree is that the father is heterozygous for the allele that causes the disease. This is the better choice as it is more likely, given the rarity of the disease.

7. a. By considering the pedigree (see below), you will discover that the cross in question is $T/t \times T/t$. Therefore, the probability of being a taster is $3/4$, and the probability of being a nontaster is $1/4$.

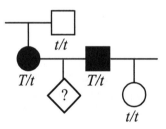

Also, the probability of having a boy equals the probability of having a girl equals $1/2$.

  (1) p(nontaster girl) $= p$(nontaster) $\times p$(girl) $= 1/4 \times 1/2 = 1/8$
  (2) p(taster girl) $= p$(taster) $\times p$(girl) $= 3/4 \times 1/2 = 3/8$
  (3) p(taster boy) $= p$(taster) $\times p$(boy) $= 3/4 \times 1/2 = 3/8$

b. p(taster for first two children) $= p$(taster for first child) $\times p$(taster for second child) $= 3/4 \times 3/4 = 9/16$

8. *Unpacking the Problem*

1.   Yes. The pedigree is given below.

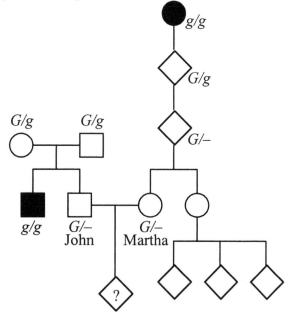

2.   In order to state this problem as a Punnett square, you must first know the genotypes of John and Martha. The genotypes can be determined only through considering the pedigree. Even with the pedigree, however, the genotypes can be stated only as *G/–* for both John and Martha.

The probability that John is carrying the allele for galactosemia is $2/3$, rather than the $1/2$ that you might guess. To understand this, recall that John's parents must be heterozygous in order to have a child with the recessive disorder while still being normal themselves (the assumption of normalcy is based on the information given in the problem). John's parents were both *G/g*. A Punnett square for their mating would be:

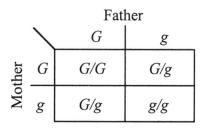

The cross is:

P	$G/g \times G/g$	
F$_1$	*g/g*	John's brother
	G/–	John (either *G/G* or *G/g*)

The expected ratio of the F_1 is 1 *G/G* : 2 *G/g* : 1 *g/g*. Because John does not have galactosemia (an assumption based on the information given in the problem), he can be either *G/G* or *G/g*, which occurs at a ratio of 1:2. Therefore, his probability of carrying the *g* allele is $^2/_3$.

The probability that Martha is carrying the *g* allele is based on the following chain of logic. Her great-grandmother had galactosemia, which means that she had to pass the allele to Martha's grandparent. Because the problem states nothing with regard to the grandparent's phenotype, it must be assumed that the grandparent was normal, or *G/g*. The probability that the grandparent passed it to Martha's parent is $^1/_2$. Next, the probability that Martha's parent passed the allele to Martha is also $^1/_2$, assuming that the parent actually has it. Therefore, the probability that Martha's parent has the allele and passed it to Martha is $^1/_2 \times {}^1/_2$, or $^1/_4$.

In summary:

John $p(G/G) = {}^1/_3$
$p(G/g) = {}^2/_3$

Martha $p(G/G) = {}^3/_4$
$p(G/g) = {}^1/_4$

This information does not fit easily into a Punnett square.

3. While the above information could be put into a branch diagram, it does not easily fit into one and overcomplicates the problem, just as a Punnett square would.

4. The mating between John's parents illustrates Mendel's first law.

5. The scientific words in this problem are *galactosemia*, *autosomal*, and *recessive*.

    Galactosemia is a metabolic disorder characterized by the absence of the enzyme galactose-1-phosphate uridyl transferase, which results in an accumulation of galactose. In the vast majority of cases, galactosemia results in an enlarged liver, jaundice, vomiting, anorexia, lethargy, and very early death if galactose is not omitted from the diet (initially, the child obtains galactose from milk).

    Autosomal refers to genes that are on the autosomes.

    Recessive means that in order for an allele to be expressed, it must be the only form of the gene present in the organism.

6. The major assumption is that if nothing is stated about a person's phenotype, the person is of normal phenotype. Another assumption that may be of value, but is not actually needed, is that all people marrying into these two families are normal and do not carry the allele for galactosemia.

7. The people not mentioned in the problem, but who must be considered, are John's parents and Martha's grandparent and parent descended from her affected great-grandmother.

8. The major statistical rule needed to solve the problem is the product rule (the "and" rule).

9. Autosomal recessive disorders are assumed to be rare and to occur equally frequently in males and females. They are also assumed to be expressed if the person is homozygous for the recessive genotype.

10. Rareness leads to the assumption that people who marry into a family that is being studied do not carry the allele, which was assumed in entry (6) above.

11. The only certain genotypes in the pedigree are John's parents, John's brother, and Martha's great-grandmother and grandmother. All other individuals have uncertain genotypes.

12. John's family can be treated simply as a heterozygous-by-heterozygous cross, with John having a $2/3$ probability of being a carrier, while it is unknown if either of Martha's parents carry the allele.

13. The information regarding Martha's sister and her children turns out to be irrelevant to the problem.

14. The problem contains a number of assumptions that have not been necessary in problem solving until now.

15. I can think of a number. Can you?

Solution to the Problem

p(child has galactosemia) = p(John is G/g) × p(Martha is G/g) × p(both parents passed g to the child) = $(2/3)(1/4)(1/4) = 2/48 = 1/24$

9. Charlie, his mate, or both, obviously were not pure-breeding, because his F_2 progeny were of two phenotypes. Let A = black and white, and a = red and white. If both parents were heterozygous, then red and white would have been

expected in the F_1 generation. Red and white were not observed in the F_1 generation, so only one of the parents was heterozygous. The cross is:

$$P \qquad A/a \times A/A$$
$$F_1 \qquad 1\ A/a\ :\ 1\ A/A$$

Two F_1 heterozygotes (A/a) when crossed would give 1 A/A (black and white) : 2 A/a (black and white) : 1 a/a (red and white). If the red and white F_2 progeny were from more than one mate of Charlie's, then the farmer acted correctly. However, if the F_2 progeny came only from one mate, the farmer may have acted too quickly.

10. Because the parents are heterozygous, both are A/a. Both twins could be albino or both twins could be normal (and, or, and = multiply, add, multiply). The probability of being normal ($A/-$) is $3/4$, and the probability of being albino (a/a) is $1/4$.

$$p(\text{both normal}) + p(\text{both albino})$$
$$p(\text{first normal}) \times p(\text{second normal}) + p(\text{first albino}) \times p(\text{second albino})$$
$$(3/4)(3/4) + (1/4)(1/4) = 9/16 + 1/16 = 5/8$$

11. The plants are approximately 3 blotched : 1 unblotched. This suggests that blotched is dominant to unblotched and that the original plant which was selfed was a heterozygote.

    a. Let A = blotched, a = unblotched.
        $$P \qquad A/a\ (\text{blotched}) \times A/a\ (\text{blotched})$$
        $$F_1 \qquad 1\ A/A : 2\ A/a : 1\ a/a$$
        $$3\ A/-\ (\text{blotched}) : 1\ a/a\ (\text{unblotched})$$

    b. All unblotched plants should be pure-breeding in a testcross with an unblotched plant (a/a), and one-third of the blotched plants should be pure-breeding.

12. In theory, it cannot be proved that an animal is not a carrier for a recessive allele. However, in an $A/- \times a/a$ cross, the more dominant-phenotype progeny produced, the less likely it is that the parent is A/a. In such a cross, half the progeny would be a/a and half would be A/a. With n dominant phenotype progeny, the probability that the parent is A/a is $(1/2)^n$.

13. The results suggest that winged ($A/-$) is dominant to wingless (a/a) (cross 2 gives a 3 : 1 ratio). If that is correct, the crosses become

		Number of progeny plants	
Pollination	Genotypes	Winged	Wingless
winged (selfed)	$A/A \times A/A$	91	1*
winged (selfed)	$A/a \times A/a$	90	30
wingless (selfed)	$a/a \times a/a$	4*	80
winged × wingless	$A/A \times a/a$	161	0
winged × wingless	$A/a \times a/a$	29	31
winged × wingless	$A/A \times a/a$	46	0
winged × winged	$A/A \times A/-$	44	0
winged × winged	$A/A \times A/-$	24	0

The five unusual plants are most likely due either to human error in classification or to contamination. Alternatively, they could result from environmental effects on development. For example, too little water may have prevented the seed pods from becoming winged even though they are genetically winged.

14. **a.** The disorder appears to be dominant because all affected individuals have an affected parent. If the trait was recessive, then I-1, II-2, III-1, and III-8 would all have to be carriers (heterozygous for the rare allele).

   b. Assuming dominance, the genotypes are
   I : *d/d, D/d*
   II : *D/d, d/d, D/d, d/d*
   III : *d/d, D/d, d/d, D/d, d/d, d/d, D/d, d/d*
   IV : *D/d, d/d, D/d, d/d, d/d, d/d, d/d, D/d, d/d*

   c. The mating is *D/d × d/d*. The probability of an affected child (*D/d*) equals $1/2$, and the probability of an unaffected child (*d/d*) equals $1/2$. Therefore, the chance of having four unaffected children (since each is an independent event) is: $1/2 \times 1/2 \times 1/2 \times 1/2 = 1/16$.

15. **a.** *Pedigree 1*: The best answer is recessive, because two unaffected individuals had affected progeny. Also, the disorder skips generations and appears in a mating between two related individuals.

   Pedigree 2: The best answer is dominant, because two affected parents have an unaffected child. Also, it appears in each generation, roughly half the progeny are affected, and all affected individuals have an affected parent.

   Pedigree 3: The best answer is dominant, for many of the reasons stated for pedigree 2. Inbreeding, while present in the pedigree, does not allow an explanation of recessive because it cannot account for individuals in the second generation.

Pedigree 4: The best answer is recessive. Two unaffected individuals had affected progeny.

b.   Genotypes of pedigree 1:
Generation I: *A/–, a/a*
Generation II: *A/a, A/a, A/a, A/–, A/–, A/a*
Generation III: *A/a, A/a*
Generation IV: *a/a*

Genotypes of pedigree 2:
Generation I: *A/a, a/a, A/a, a/a*
Generation II: *a/a, a/a, A/a, A/a, a/a, a/a, A/a, A/a, a/a*
Generation III: *a/a, a/a, a/a, a/a, a/a, A/–, A/–, A/–, A/a, a/a*
Generation IV. *a/a, a/a, a/a*

Genotypes of pedigree 3:
Generation I: *A/–, a/a*
Generation II: *A/a, a/a, a/a, A/a*
Generation III: *a/a, A/a, a/a, a/a, A/a, a/a*
Generation IV. *a/a, A/a, A/a, A/a, a/a, a/a*

Genotypes of pedigree 4:
Generation I: *a/a, A/–, A/a, A/a*
Generation II: *A/a, A/a, A/a, a/a, A/–, a/a, A/–, A/–, A/–, A/–, A/–*
Generation III: *A/a, a/a, A/a, A/a, a/a, A/a*

16.   **a.**   The pedigree is

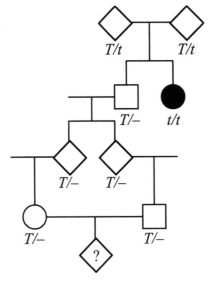

b.   The probability that the child of the two first cousins will have Tay-Sachs disease is a function of three probabilities: *p*(the woman is *T/t*) × *p*(the man is *T/t*) × *p*(both donate *t*);
$$= (^2/_3)(^1/_2)(^1/_2) \times (^2/_3)(^1/_2)(^1/_2) \times ^1/_4 = ^1/_{144}$$

To understand the probabilities of the first two events, see the discussion for problem 8 part (2) of this chapter.

17. **a.** Autosomal recessive: affected individuals inherited the trait from unaffected parents and a daughter inherited the trait from an unaffected father.

    b. Both parents must be heterozygous to have a $^1/_4$ chance of having an affected child. Parent 2 is heterozygous, since her father is homozygous for the recessive allele and parent 1 has a $^1/_2$ chance of being heterozygous, since his father is heterozygous because 1's paternal grandmother was affected. Overall, $1 \times {}^1/_2 \times {}^1/_4 = {}^1/_8$.

18. **a.** Yes. It is inherited as an autosomal dominant trait.

    b. Susan is highly unlikely to have Huntingtons disease. Her great-grandmother (individual II-2) is 75 years old and has yet to develop it, when nearly 100 percent of people carrying the allele will have developed the disease by that age. If her great-grandmother does not have it, Susan cannot inherit it.

Alan is somewhat more likely than Susan to develop Huntingtons disease. His grandfather (individual III-7) is only 50 years old, and approximately 20 percent of the people with the allele have yet to develop the disease by that age. Therefore it can be estimated that the grandfather has a 10 percent chance of being a carrier (50 percent chance he inherited the allele from his father × 20 percent chance he has not yet developed symptoms). If Alan's grandfather eventually develops Huntingtons disease, then there is a probability of 50 percent that Alan's father inherited it from him, and a probability of 50 percent that Alan received that allele from his father. Therefore, Alan has a $^1/_{10} \times {}^1/_2 \times {}^1/_2 = {}^1/_{40}$ current probability of developing Huntingtons disease and $^1/_2 \times {}^1/_2 = {}^1/_4$ probability, if his grandfather eventually develops it.

19. **a.** Assuming the trait is rare, expect that all individuals marrying into the pedigree do not carry the disease-causing allele.

$$\text{I}: P/P, p/p, p/p, P/P$$
$$\text{II}: P/P, P/p, P/p, P/p, P/p$$
$$\text{III}: P/P, P/-, P/-, P/P$$
$$\text{IV}: P/-, P/-$$

    b. For their child to have PKU, both A and B must be carriers and both must donate the recessive allele.

The probability that individual A has the PKU allele is derived from individual II-2. II-2 must be P/p since her father must be p/p. Therefore, the probability that II-2 passed the PKU allele to individual III-2 is $1/2$. If III-2 received the allele, the probability that he passed it to individual IV-1 (A) is $1/2$. Therefore, the probability that A is a carrier is $1/2 \times 1/2 = 1/4$.

The probability that individual B has the allele goes back to the mating of II-3 and II-4, both of whom are heterozygous. Their child, III-3, has a $2/3$ probability of having received the PKU allele and a probability of $1/2$ of passing it to IV-2 (B). Therefore, the probability that B has the PKU allele is $2/3 \times 1/2 = 1/3$.

If both parents are heterozygous, they have a $1/4$ chance of both passing the p allele to their child.

$$p(\text{child has PKU}) = p(\text{A is } P/p) \times p(\text{B is } P/p) \times p(\text{both parents donate } p)$$
$$1/4 \quad \times \quad 1/3 \quad \times \quad 1/4 \quad = 1/48$$

c. If the first child is normal, no additional information has been gained and the probability that the second child will have PKU is the same as the probability that the first child will have PKU, or $1/48$.

d. If the first child has PKU, both parents are heterozygous. The probability of having an affected child is now $1/4$, and the probability of having an unaffected child is $3/4$.

20. a. Sons inherit their X chromosome from their mother. The mother has earlobes, the son does not. If the allele for earlobes is dominant and the allele for lack of earlobes recessive, then the mother could be heterozygous for this trait and the gene could be X-linked.

b. It is not possible from the data given to decide which allele is dominant. If lack of earlobes is dominant, then the father would be heterozygous and the son would have a 50 percent chance of inheriting the dominant "lack-of-earlobes" allele. If lack of earlobes is recessive, then the trait could be autosomal or X-linked but in either case, the mother would be heterozygous.

21. a. Let C stand for the normal allele and c stand for the allele that causes cystic fibrosis.

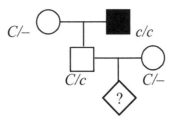

b. The man has a 100 percent probability of having the *c* allele. His wife, who is from the general population, has a $1/50$ chance of having the *c* allele. If both have the allele, then $1/4$ of their children will have cystic fibrosis. The probability that their first child will have cystic fibrosis is

$$p(\text{man has } c) \times p(\text{woman has } c) \times p(\text{both pass } c \text{ to the child})$$
$$1.0 \quad \times \quad 1/50 \quad \times \quad 1/4 = 1/200 = 0.005$$

c. If the first child does have cystic fibrosis, then the woman is a carrier of the *c* allele. Because both parents are *C/c*, the chance that the second child will be normal is the probability of a normal child in a heterozygous × heterozygous mating, or $3/4$.

22. The cross is *C/c* × *c/c* so there is a $1/2$ chance that a progeny would be black (*C/c*). Because each progeny's genotype is independent of the others, the chance that all 10 progeny are black is $(1/2)^{10}$.

23. **a.** *C/c* ; *S/s* × *C/c* ; *S/s* — There are 3 short:1 long, and 3 dark:1 albino. Therefore, each gene is heterozygous in the parents.

  b. *C/C* ; *S/s* × *C/C* ; *s/s* — There are no albino, and there are 1 long:1 short indicating a testcross for this trait.

  c. *C/c* ; *S/S* × *c/c* ; *S/S* — There are no long, and there are 1 dark:1 albino.

  d. *c/c* ; *S/s* × *c/c* ; *S/s* — All are albino, and there are 3 short:1 long.

  e. *C/c* ; *s/s* × *C/c* ; *s/s* — All are long, and there are 3 dark:1 albino.

  f. *C/C* ; *S/s* × *C/C* ; *S/s* — There are no albino, and there are 3 short:1 long.

  g. *C/c* ; *S/s* × *C/c* ; *s/s* — There are 3 dark:1 albino, and 1 short:1 long.

24. **a. and b.** Cross 2 indicates that purple (*G*) is dominant to green (*g*), and cross 1 indicates cut (*P*) is dominant to potato (*p*).

  Cross 1 : *G/g* ; *P/p* × *g/g* ; *P/p* — There are 3 cut : 1 potato, and 1 purple : 1 green.

  Cross 2 : *G/g* ; *P/p* × *G/g* ; *p/p* — There are 3 purple : 1 green, and 1 cut : 1 potato.

  Cross 3: *G/G* ; *P/p* × *g/g* ; *P/p* — There are no green, and there are 3 cut:1 potato.

  Cross 4: *G/g* ; *P/P* × *g/g* ; *p/p* — There are no potato, and there are

Cross 5: G/g ; p/p × g/g ; P/p 1 purple:1 green.
There are 1 cut:1 potato, and there are
1 purple:1 green.

25. P s^+/s^+ × s/Y
        ↓

F$_1$ $1/2\ s^+/s$ normal female

      $1/2\ s^+/Y$ normal male

      s^+/s × s^+/Y
        ↓

F$_2$ $1/4\ s^+/s^+$ normal female

      $1/4\ s^+/s$ normal female

      $1/4\ s^+/Y$ normal male

      $1/4\ s/Y$ small male

P s^+/s × s/Y
        ↓

Progeny $1/4\ s^+/s$ normal female

      $1/4\ s/s$ small female

      $1/4\ s^+/Y$ normal male

      $1/4\ s/Y$ small male

26. Let H = hypophosphatemia and h = normal. The cross is $H/Y \times h/h$, yielding H/h (females) and h/Y (males). The answer is 0%.

27. **a.** You should draw pedigrees for this question.

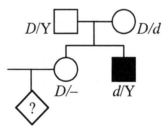

The "maternal grandmother" had to be a carrier, D/d. The probability that the woman inherited the d allele from her is $1/2$. The probability that she passes it to her child is $1/2$. The probability that the child is male is $1/2$. The total probability of the woman having an affected child is $1/2 \times 1/2 \times 1/2 = 1/8$.

b. The same pedigree as part (a) applies. The "maternal grandmother" had to be a carrier, D/d. The probability that your mother received the allele is $1/2$.

The probability that your mother passed it to you is $^1/_2$. The total probability is $^1/_2 \times {}^1/_2 = {}^1/_4$.

c.

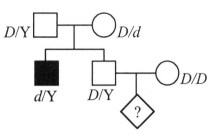

Because your father does not have the disease, you cannot inherit the allele from him. Therefore, the probability of inheriting an allele will be based on the chance that your mother is heterozygous. Since she is "unrelated" to the pedigree, assume that this is zero.

28. a. Because none of the parents are affected, the disease must be recessive. Because the inheritance of this trait appears to be sex-specific, it is most likely X-linked. If it were autosomal, all three parents would have to be carriers and by chance, only sons and none of the daughters inherited the trait (which is quite unlikely).

b. I $A/Y, A/a, A/Y$

II $A/Y, A/-, a/Y, A/-, A/Y, a/Y, a/Y, A/-, a/Y, A/-$

29. You should draw the pedigree before beginning.

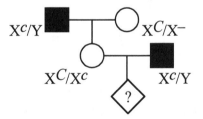

a. $X^C/X^c, X^c/X^c$

b. $p(\text{colorblind}) \times p(\text{male}) = (^1/_2)(^1/_2) = {}^1/_4$

c. The girls will be 1 normal (X^C/X^c) : 1 colorblind (X^c/X^c).

d. The cross is $X^C/X^c \times X^c/Y$, yielding 1 normal : 1 colorblind for both sexes.

30. a. This problem involves X-inactivation. Let B = black and b = *orange*.

	Females	Males
	X^B/X^B = black	X^B/Y = black
	X^b/X^b = orange	X^b/Y = orange
	X^B/X^b = calico	

b. P X^b/X^b (orange) × X^B/Y (black)
     F_1 X^B/X^b (calico female)
          X^b/Y (orange male)

c. P X^B/X^B (black) × X^b/Y (orange)
     F_1 X^B/X^b (calico female)
          X^B/Y (black male)

d. Because the males are black or orange, the mother had to have been calico. Half the daughters are black, indicating that their father was black.

e. Males were orange or black, indicating that the mothers were calico. Orange females indicate that the father was orange.

31. **a.** Recessive (unaffected parents have affected progeny) and X-linked (only assumption is that the grandmother, I-2, is a carrier). If autosomal, then I-1, I-2, and II-6 would all have to be carriers.

b.
Generation I: X^A/Y, X^A/X^a
Generation II: X^A/X^A, X^a/Y, $X^A Y$, X^A/X^-, X^A/X^a, X^A/Y
Generation III: X^A/X^A, X^A/Y, X^A/X^a, X^A/X^a, X^A/Y, $X^A X^A$, X^a/Y, X^A/Y, X^A/X^-

c. Because it is stated that the trait is rare, the assumption is that no one marrying into the pedigree carries the recessive allele. Therefore, the first couple has no chance of an affected child because the son received a Y chromosome from his father. The second couple has a 50 percent chance of having affected sons and no chance of having affected daughters. The third couple has no chance of having an affected child, but all of their daughters will be carriers.

32. **a.** From cross 6, Bent (B) is dominant to normal (b). Both parents are "bent," yet some progeny are "normal."

b. From cross 1, it is X-linked. The trait is inherited in a sex-specific manner — all sons have the mother's phenotype.

c. In the following table, the Y chromosome is stated; the X is implied.

| | Parents | | Progeny | |
Cross	Female	Male	Female	Male
1	b/b	B/Y	B/b	b/Y
2	B/b	b/Y	B/b, b/b	B/Y, b/Y
3	B/B	b/Y	B/b	B/Y
4	b/b	b/Y	b/b	b/Y
5	B/B	B/Y	B/B	B/Y
6	B/b	B/Y	B/B, B/b	B/Y, b/Y

33. *Unpacking the Problem*

1. *Normal* is used to mean wild type, or red eye color and long wings.

2. Both *line* and *strain* are used to denote pure-breeding fly stocks, and the words are interchangeable.

3. Your choice.

4. Three characters are being followed: eye color, wing length, and sex.

5. For eye color, there are two phenotypes: red and brown. For wing length, there are two phenotypes: long and short. For sex, there are two phenotypes: male and female.

6. The F_1 females designated normal have red eyes and long wings.

7. The F_1 males that are called short-winged have red eyes and short wings.

8. The F_2 ratio is

    $3/8$ red eyes, long wings
    $3/8$ red eyes, short wings
    $1/8$ brown eyes, long wings
    $1/8$ brown eyes, short wings

9. Because there is not the expected 9:3:3:1 ratio, one of the factors that distorts the expected dihybrid ratio must be present. Such factors can be sex linkage, epistasis, genes on the same chromosome, environmental effect, reduced penetrance, or a lack of complete dominance in one or both genes.

10. With sex linkage, traits are inherited in a sex-specific way. With autosomal inheritance, males and females have the same probabilities of inheriting the trait.

11. The F_2 does not indicate sex-specific inheritance.

12. The F_1 data does show sex-specific inheritance — all males are brown-eyed, like their mothers, while all females are red-eyed, like their fathers.

13. The F_1 suggests that long is dominant to short and red is dominant to brown. The F_2 data show a 3 red : 1 brown ratio indicating the dominance of red but a 1 : 1 long : short ratio indicative of a test cross. Without the F_1 data, it is not possible to

14. If Mendelian notation is used, then the red and long alleles need to be designated with uppercase letters, for example R and L, while the brown (r) and short (l) alleles need to be designated with lowercase letters. If *Drosophila* notation is used, then the brown allele may be designated with a lowercase b and the wild-type (red) allele with a b^+; the short wing-length gene with a s and the wild-type (long) allele with an s^+. (Genes are often named after their mutant phenotype.)

15. To deduce the inheritance of these phenotypes means to provide all genotypes for all animals in the three generations discussed and account for the ratios observed.

Solution to the Problem

Start this problem by writing the crosses and results so that all the details are clear.

P	brown, short female × red, long male
F_1	red, long females
	red, short males

These results tell you that red-eyed is dominant to brown-eyed and since both females and males are red-eyed, that this gene is autosomal. Since males differ from females in their genotype with regard to wing length, this trait is sex-linked. Knowing that *Drosophila* females are XX and males are XY, the long-winged females tell us that long is dominant to short and that the gene is X-linked. Let B = red, b = brown, S = long, and s = short. The cross can be rewritten as follows:

P	b/b ; s/s × B/B ; S/Y
F_1	$1/2$ B/b ; S/s females
	$1/2$ B/b ; s/Y males

F_2	$1/16$ B/B ; S/s	red, long, female
	$1/16$ B/B ; s/s	red, short, female
	$1/8$ B/b ; S/s	red, long, female

$^1/_8\ B/b\ ;\ s/s$	red, short, female
$^1/_{16}\ b/b\ ;\ S/s$	brown, long, female
$^1/_{16}\ b/b\ ;\ s/s$	brown, short, female
$^1/_{16}\ B/B\ ;\ S/Y$	red, long, male
$^1/_{16}\ B/B\ ;\ s/Y$	red, short, male
$^1/_8\ B/b\ ;\ S/Y$	red, long, male
$^1/_8\ B/b\ ;\ s/Y$	red, short, male
$^1/_{16}\ b/b\ ;\ S/Y$	brown, long, male
$^1/_{16}\ b/b\ ;\ s/Y$	brown, short, male

The final phenotypic ratio is

$^3/_8$ red, long
$^3/_8$ red, short
$^1/_8$ brown, long
$^1/_8$ brown, short

with equal numbers of males and females in all classes.

34. **a.** Because photosynthesis is affected and the plants are yellow, rather than green, it is likely that the chloroplasts are defective. If the defect maps to the DNA of the chloroplast, the trait will be maternally inherited. This fits the data that all progeny have the phenotype of the female parent and not the phenotype of the male (pollen-donor) parent.

   b. If the defect maps to the DNA of the chloroplast, the trait will be maternally inherited. Use pollen from sickly, yellow plants and cross to emasculated flowers of a normal dark green-leaved plant. All progeny should have the normal dark green phenotype.

   c. The chloroplasts contain the green pigment chlorophyll and are the site of photosynthesis. A defect in the production of chlorophyll would give rise to all the stated defects.

35. Maternal inheritance of chloroplasts results in the green-white color variegation observed in *Mirabilis*.
   Cross 1:   variegated female × green male → variegated, green, or white progeny
   Cross 2:   green female × variegated male → green progeny

In both crosses, the pollen (male contribution) contains no chloroplasts and thus does not contribute to the inheritance of this phenotype. Eggs from a variegated female plant can be of three types : contain only "green" chloroplasts, contain only "white" chloroplasts, or contain both (variegated). The offspring will have the phenotype associated with the egg's chloroplasts.

36. The crosses are

Cross 1:	stop-start female × wild-type male → all stop-start progeny
Cross 2:	wild-type female × stop-start male → all wild-type progeny
	mtDNA is inherited only from the "female" in *Neurospora*.

37. The genetic determinants of R and S are showing maternal inheritance and are therefore cytoplasmic. It is possible that the gene that confers resistance maps either to the mtDNA or cpDNA.

38. To use the chi-square test, first state the hypothesis being tested and the expected results. In this case, the hypothesis is that the phenotypic difference is due to two alleles of one gene. The expected results would be 300 purple and 100 blue (or an expected 3:1 ratio in the F_2), The formula to use is

$$\chi^2 = \text{(observed-expected)}^2/\text{expected}$$

Class	Observed (O)	Expected (E)	$(O–E)^2$	$(O–E)^2/E$
purple	320	300	400	1.33
blue	80	100	400	3.00

Total = χ^2 = 4.33

This χ^2 value must now be looked up in a table (Table 2-2 in the companion text) which will give you the probability that these data fit the stated hypothesis. But first, you must also determine the degrees of freedom (df) or the number of independent variables in the data. Generally, the degrees of freedom can be calculated as the number of phenotypes in the problem minus one. In this example, there are two phenotypes (purple and blue) and therefore, one degree of freedom. Using the table, you will find that the data have a p value that is < 0.05 or $< 5\%$. The p value is the probability that the deviation of the observed from that expected is due to chance alone. The relative standard commonly used in biological research for rejecting a hypothesis is $p < 0.05$. In this case, the data do not support the hypothesis.

39. The hypothesis is that the organism being tested is a heterozygote and that the *A/a* and *a/a* progeny are of equal viability. The expected values would be that phenotypes occur with equal frequency. There are two genotypes in each case, so there is one degree of freedom.

$$\chi^2 = \text{(observed-expected)}^2/\text{expected}$$

a. $\chi^2 = [(120–110)^2 + (100–110)^2]/110$
$= 1.818; p > 0.10$, nonsignificant; hypothesis cannot be rejected

b. $\chi^2 = [(5000-5200)^2 + (5400-5200)^2]/5200$
$= 15.385; p < 0.005$, significant; hypothesis must be rejected

c. $\chi^2 = [(500-520)^2 + (540-520)^2]/520$
$= 1.538; p > 0.10$, nonsignificant; hypothesis cannot be rejected

d. $\chi^2 = [(50-52)^2 + (54-52)^2]/52$
$= 0.154; p > 0.50$, nonsignificant; hypothesis cannot be rejected

40. The hypothesis is that the organism being tested is a dihybrid with independently assorting genes and that all progeny are of equal viability. The expected values would be that phenotypes occur with equal frequency. There are four phenotypes so there are 3 degrees of freedom.

$$\chi^2 = \sum(\text{observed-expected})^2/\text{expected}$$

$\chi^2 = [(230-233)^2 + (210-233)^2 + (240-233)^2 + (250-233)^2]/233$
$= 2.215; p > 0.50$, nonsignificant; hypothesis cannot be rejected

41. The hypothesis is that the organism being tested is a dihybrid with independently assorting genes and that all progeny are of equal viability. The expected values would be that phenotypes occur in a $9 : 3 : 3 : 1$ ratio.. There are four phenotypes so there are 3 degrees of freedom.

$$\chi^2 = \sum(\text{observed-expected})^2/\text{expected}$$

$\chi^2 = (178-180)^2/180 + (62-60)^2/60 + (56-60)^2/60 + (24-20)^2/20$
$= 1.156; p > 0.50$, nonsignificant; hypothesis cannot be rejected

CHALLENGING PROBLEMS

42. a. Before beginning the specific problems, calculate the probabilities associated with each jar.

jar 1 $p(R)\ 600/(600 + 400) = 0.6$
$p(W) = 400/(600 + 400) = 0.4$

jar 2 $p(B) = 900/(900 + 100) = 0.9$
$p(W)\ 100/(900 + 100) = 0.1$

jar 3 $p(G)\ 10/(10 + 990) = 0.01$
$p(W)\ 990/(10 + 990) = 0.99$

(1) $p(R, B, G) = (0.6)(0.9)(0.01) = 0.0054$

(2) p(W, W, W) = (0.4)(0.1)(0.99) = 0.0396

(3) Before plugging into the formula, you should realize that, while white can come from any jar, red and green must come from specific jars (jar 1 and jar 3). Therefore, white must come from jar 2:
p(R, W, G) = (0.6)(0.1)(0.01) = 0.0006

(4) p(R, W, W) = (0.6)(0.1)(0.99) = 0.0594

(5) There are three ways to satisfy this:
      R, W, W   or   W, B, W   or   W, W, G
= (0.6)(0.1)(0.99) + (0.4)(0.9)(0.99) + (0.4)(0.1)(0.01)
= 0.0594 + 0.3564 + 0.0004 = 0.4162

(6) At least one white is the same as 1 minus no whites:
p(at least 1 W) = $1 - p$(no W) = $1 - p$(R, B, G)
= $1 - (0.6)(0.9)(0.01) = 1 - 0.0054 = 0.9946$

b. The cross is $R/r \times R/r$. The probability of red ($R/–$) is $3/4$, and the probability of white (r/r) is $1/4$. Because only one white is needed, the only unacceptable result is all red.

In n trials, the probability of all red is $(3/4)^n$. Because the probability of failure must be no greater than 5 percent:
$(3/4)^n < 0.05$
$n > 10.41$, or 11 seeds

c. The p(failure) = 0.8 for each egg. Since all eggs are implanted simultaneously, the p(5 failures) = $(0.8)^5$. The p(at least one success) = $1 - (0.8)^5 = 1 - 0.328 = 0.672$

43. a. Galactosemia pedigree

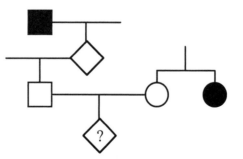

b. Both parents must be heterozygous for this child to have a $1/4$ chance of inheriting the disease. Since the mother's sister is affected with galactosemia, their parents must have both been heterozygous. Since the mother does not have the trait, there is a $^2/_3$ chance that she is a carrier (heterozygous). One of the father's parents must be a carrier since his grandfather had the recessive trait. Thus, the father had a $^1/_2$ chance of inheriting the allele from that parent. Since these are all independent events, the child's risk is

$$^1/_4 \times {}^2/_3 \times {}^1/_2 = {}^1/_{12}$$

    c.   If the child has galactosemia, both parents must be carriers and thus those probabilities become 100 percent. Now all future children have a $^1/_4$ chance of inheriting the disease.

44. **a.** In order to draw this pedigree, you should realize that if an individual's status is not mentioned, then there is no way to assign a genotype to that person. The parents of the boy in question had a phenotype (and genotype) that differed from his. Therefore, both parents were heterozygous and the boy, who is a non-roller, homozygous recessive. Let *R* stand for the ability to roll the tongue and *r* stand for the inability to roll the tongue. The pedigree becomes:

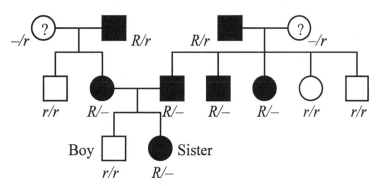

    b. Assuming the twins are identical, there might be either an environmental component to the expression of that gene or developmental noise (see Chapter 1). Another possibility is that the *R* allele is not fully penetrant and that some genotypic "rollers" do not express the phenotype.

45. **a.** The inheritance pattern for red hair suggested by this pedigree is recessive since most red-haired individuals are from parents without this trait.

    b. By observing those around us, the allele appears to be somewhat rare.

46. *Taster by taster cross*: Tasters can be either *T/T* or *T/t*, and the genotypic status cannot be determined until a large number of progeny are observed. A failure to obtain a 3 : 1 ratio in the matings of two tasters would be expected because there are three types of matings:

Mating	Genotypes	Phenotypes
T/T × *T/T*	all *T/T*	all tasters
T/T × *T/t*	1 *T/T* : 1 *T/t*	all tasters
T/t × *T/t*	1 *T/T* : 2 *T/t* : 1 *t/t*	3 tasters : 1 non-tasters

Taster by non-taster cross: There are two types of matings that resulted in the observed progeny:

Mating	Genotypes	Phenotypes
T/T × *t/t*	all *T/t*	all tasters
T/t × *t/t*	1 *T/t* : 1 *t/t*	1 tasters : 1 non-tasters

Again, the failure to obtain either a 1 : 0 ratio or a 1 : 1 ratio would be expected because of the two mating types.

Non-taster by non-taster cross: There is only one mating that is non-taster by non-taster (*t/t* × *t/t*), and 100 percent of the progeny would be expected to be non-tasters. Of 223 children, five were classified as tasters. Some could be the result of mutation (unlikely), some could be the result of misclassification (likely), some could be the result of a second gene that affects the expression of the gene in question (possible), some could be the result of developmental noise (possible), and some could be due to illegitimacy (possible).

47. Use the following symbols:

Gene function	Dominant allele	Recessive allele
color	*R* = red	*r* = yellow
loculed	*L* = two	*l* = many
height	*H* = tall	*h* = dwarf

The starting plants are pure-breeding, so their genotypes are

red, two-loculed, dwarf *R/R* ; *L/L* ; *h/h*
and
yellow, many-loculed, tall *r/r* ; *l/l* ; *H/H*.

The farmer wants to produce a pure-breeding line that is yellow, two-loculed, and tall, which would have the genotype *r/r* ; *L/L* ; *H/H*.

The two pure-breeding starting lines will produce an F$_1$ that will be *R/r* ; *L/l* ; *H/h*. By doing an F$_1$ × F$_1$ cross and selecting yellow, two-loculed, and tall plants, the known genotype will be *r/r* ; *L/–* ; *H/–*. The task then will be to do sequential testcrosses for both the *L/–* and *H/–* genes among these yellow, two-loculed, and tall plants. Because the two genes in question are possibly homozygous recessive in different plants, each plant that is yellow, two-loculed, and tall will have to be testcrossed twice.

For each testcross, the plant will obviously be discarded if the testcross reveals a heterozygous state for the gene in question. If no recessive allele is detected, then the minimum number of progeny that must be examined to be 95 percent confident that the plant is homozygous, is based on the frequency of the dominant phenotype if heterozygous, which is $1/2$. In n progeny, the probability of obtaining all dominant progeny in a testcross, given that the plant is heterozygous, is $(1/2)^n$. To be 95 percent confident of homozygosity, the following formula is used, where 5 percent is the probability that it is not homozygous:

$$(1/2)^n = 0.05$$

$n = 4.3$, or 5 phenotypically dominant progeny must be obtained from each testcross to be 95 percent confident that the plant is homozygous.

48. a. Because each gene assorts independently, each probability should be considered separately and then multiplied together for the answer.

For (1) $1/2$ will be A, $3/4$ will be B, $1/2$ will be C, $3/4$ will be D, and $1/2$ will be E.
$$1/2 \times 3/4 \times 1/2 \times 3/4 \times 1/2 = 9/128$$

For (2) $1/2$ will be a, $3/4$ will be B, $1/2$ will be c, $3/4$ will be D, and $1/2$ will be e.
$$1/2 \times 3/4 \times 1/2 \times 3/4 \times 1/2 = 9/128$$

For (3) it is the sum of (1) and (2) $= 9/128 + 9/128 = 9/64$

For (4) it is $1 - $ (part 3) $= 1 - 9/64 = 55/64$

b. For (1) $1/2$ will be A/a, $1/2$ will be B/b, $1/2$ will be C/c, $1/2$ will be D/d, and $1/2$ will be E/e.
$$1/2 \times 1/2 \times 1/2 \times 1/2 \times 1/2 = 1/32$$

For (2) $1/2$ will be a/a, $1/2$ will be B/b, $1/2$ will be c/c, $1/2$ will be D/d, and $1/2$ will be e/e.
$$1/2 \times 1/2 \times 1/2 \times 1/2 \times 1/2 = 1/32$$

For (3) it is the sum of (1) and (2) $= 1/16$.

For (4) it is $1 - $ (part 3) $= 1 - 1/16 = 15/16$

49. **a.** Cataracts appear to be caused by a dominant allele because affected people have affected parents. Dwarfism appears to be caused by a recessive allele because affected people have unaffected parents.

b. Using A for cataracts, a for no cataracts, B for normal height, and b for dwarfism. The genotypes are:

III : a/a ; B/b, a/a ; B/b, A/a ; $B/-$, a/a ; $B/-$, A/a ; B/b, A/a ; B/b, a/a ; $B/-$, a/a ; $B/-$, A/a ; b/b

c. The mating is a/a ; b/b (IV-1) $\times$ $A/-$; $B/-$ (IV-5). Recall that the probability of a child's being affected by any disease is a function of the probability of each parent carrying the allele in question and the probability that one parent (for a dominant disorder) or both parents (for a recessive disorder) donate it to the child. Individual IV-1 is homozygous for these two genes, therefore, the only task is to determine the probabilities associated with individual IV-5.

The probability that individual IV-5 is heterozygous for dwarfism is $2/3$. Thus the probability that she has the b allele and will pass it to her child is $2/3 \times 1/2 = 1/3$.

The probability that individual IV-5 is homozygous for cataracts is $1/3$; the probability that she is heterozygous is $2/3$. If she is homozygous for the allele that causes cataracts, she must pass it to her child or if she is heterozygous for cataracts, she has a probability of $1/2$ of passing it to her child.

The probability that the first child is a dwarf with cataracts is the probability that the child inherits the A and b alleles from its mother which is $(1/3 \times 1)(2/3 \times 1/2) + (2/3 \times 1/2)(2/3 \times 1/2) = 2/9$. Alternatively, you can calculate the chance of inheriting the b allele ($2/3 \times 1/2$) and not inheriting the a allele ($1 - 1/3$) or $1/3 \times 2/3 = 2/9$.

The probability of having a phenotypically normal child is the probability that the mother donates the a and B (or not b) alleles, which is $(2/3 \times 1/2)(1 - 1/3) = 2/9$.

50. **a. and b.** Begin with any two of the three lines and cross them. If, for example, you began with a/a ; B/B ; $C/C \times A/A$; b/b ; C/C, the progeny would all be A/a ; B/b ; C/C. Crossing two of these would yield:

9	$A/-$; $B/-$; C/C
3	a/a ; $B/-$; C/C
3	$A/-$; b/b ; C/C
1	a/a ; b/b ; C/C

The *a/a* ; *b/b* ; *C/C* genotype has two of the genes in a homozygous recessive state and occurs in $1/16$ of the offspring. If that were crossed with *A/A* ; *B/B* ; *c/c*, the progeny would all be *A/a* ; *B/b* ; *C/c*. Crossing two of them (or "selfing") would lead to a 27:9:9:9:3:3:3:1 ratio, and the plant occurring in $1/64$ of the progeny would be the desired *a/a* ; *b/b* ; *c/c*.

There are several different routes to obtaining *a/a* ; *b/b* ; *c/c*, but the one outlined above requires only four crosses.

51. If the historical record is accurate, the data suggest Y linkage. Another explanation is an autosomal gene that is dominant in males and recessive in females. This has been observed for other genes in both humans and other species.

52. The different sex-specific phenotypes found in the F_1 indicate sex-linkage — the females inherit the trait of their fathers. The first cross also indicates that the wild-type large spots is dominant over the lacticolor small spots. Let A = wild type and a = lacticolor.

Cross 1: If the male is assumed to be the hemizygous sex, then it soon becomes clear that the predictions do not match what was observed:

P	*a/a* female $\times$ *A/Y* male
F_1	*A/a* wild-type females
	a/Y lacticolor males

Therefore, assume that the female is the hemizygous sex. Let Z stand for the sex-determining chromosome in females. The cross becomes:

P	*a/Z* female $\times$ *A/A* male	
F_1	*A/a*	wild-type male
	A/Z	wild-type female
F_2	$1/4$ *A/Z*	wild-type females
	$1/2$ *A/–*	wild-type males
	$1/4$ *a/Z*	lacticolor females

Cross 2:

P	*A/Z* female $\times$ *a/a* male	
F_1	*a/Z*	lacticolor females
	A/a	wild-type males

F₂   ¹/₄   *A/Z*   wild-type females
       ¹/₄   *A/a*   wild-type males
       ¹/₄   *a/Z*   lacticolor females
       ¹/₄   *a/a*   lacticolor males

53. Note that only males are affected and that in all but one case, the trait can be traced through the female side. However, there is one example of an affected male having affected sons. If the trait is X-linked, this male's wife must be a carrier. Depending on how rare this trait is in the general population, this suggests that the disorder is caused by an autosomal dominant with expression limited to males.

54. First, draw the pedigree.

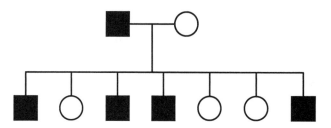

Let the genes be designated by the pigment produced by the normal allele : red pigment, *R;* green pigment, *G;* and blue pigment, *B.*

Recall that the sole X in males comes from the mother, while females obtain an X from each parent. Also recall that a difference in phenotype between sons and daughters is usually due to an X-linked gene. Because all the sons are colorblind and neither the mother nor the daughters are, the mother must carry a different allele for colorblindness on each X chromosome. In other words, she is heterozygous for both X-linked genes, and they are in repulsion: *R g/r G.* With regard to the autosomal gene, she must be *B/–.*

Because all the daughters are normal, the father, who is colorblind, must be able to complement the defects in the mother with regard to his X chromosome. Because he has only one X with which to do so, his genotype must be *R G/Y ; b/b.* Likewise, the mother must be able to complement the father's defect, so she must be *B/B.*

The original cross is therefore

    P   *R g/r G ; B/B* × *R G/Y ; b/b*

    F₁   Females   Males
            R g/R G ; B/b   *R g/Y ; B/b*
            r G/R G ; B/b   *r G/Y ; B/b*

55.   Because the disorder is X-linked recessive, the affected male had to have received the allele, *a*, from the female common ancestor in the first generation. The probability that the affected man's wife also carries the *a* allele is the probability that she also received it from the female common ancestor. That probability is $1/8$.

The probability that the couple will have an affected boy is
$$p(\text{father donates Y}) \times p(\text{the mother has } a) \times p(\text{mother donates } a)$$
$$1/2 \times \quad 1/8 \times \quad 1/2 = 1/32$$

The probability that the couple will have an affected girl is:
$$p(\text{father donates X}^a) \times p(\text{the mother has } a) \times \text{p(mother donates } a)$$
$$1/2 \times \quad 1/8 \times \quad 1/2 = 1/32$$

The probability of normal children:
$$= 1 - p(\text{affected children})$$
$$= 1 - p(\text{affected male}) - p(\text{affected female})$$
$$= 1 - 1/32 - 1/32 = 30/32 = 15/16$$

Half the normal children will be boys, with a probability of $15/32$, and half will be girls, with a probability of $15/32$.

56.   a.   The complete absence of male offspring is the unusual aspect of this pedigree. In addition, all progeny that mate carry the trait for lack of male offspring. If the male lethality factor were nuclear, the male parent would be expected to alter this pattern. Therefore, cytoplasmic inheritance is suggested.

b.   If all females resulted from chance alone, then the probability of this result is $(1/2)^n$, where n = the number of female births. In this case n is 72. Chance is an unlikely explanation for the observations.
The observations can be explained by cytoplasmic factors by assuming that the proposed mutation in mitochondria is lethal only in males.
Mendelian inheritance cannot explain the observations, because all the fathers would have had to carry the male-lethal mutation in order to observe such a pattern. This would be highly unlikely.

57.   a.   The pedigree clearly shows maternal inheritance.

b.   Most likely, the mutant DNA is mitochondrial.

Tips on Problem Solving

Ratios: a 1:2:1 (or 3:1) ratio indicates that one gene is involved (see Problem 3). A 9:3:3:1 ratio, or some modification of it, indicates that two genes are involved (see Problem 4). A testcross results in a 1:1 ratio if the organism being tested is heterozygous and a 1:0 ratio if it is homozygous (see Problem 2).

Pedigrees: normal parents have affected offspring in recessive disorders (see Problem 8). Normal parents have normal offspring and affected parents have affected and/or normal offspring in dominant disorders (see Problems 7). If phenotypically identical parents produce progeny with two phenotypes, the parents were both heterozygous (see Problems 27).

Probability: when dealing with two or more independently assorting genes, consider each gene separately (see Problem 48).

X-linked or autosomal: if the male phenotype is different from the female phenotype, X linkage is involved for the allele carried by the female (see Problems 15, 27).

Inheritance patterns: there are only seven possible inheritance patterns for a gene . Usually only numbers 1 – 4 will be encountered:

1. Autosomal dominant
2. Autosomal recessive
3. X-linked dominant
4. X-linked recessive
5. Autosomal with expression limited to one sex
6. Y-linked
7. X- and Y-linked (pseudoautosomal)

A Systematic Approach to Problem Solving

Now that you have struggled with a number of genetics problems, it may be worthwhile to make some generalizations about problem solving beyond what has been presented for each chapter so far.

The first task always is to determine exactly what information has been presented and what is being asked. Frequently, it is necessary to rewrite the problem or to symbolize the presented information in some way.

The second task is to formulate and test hypotheses. If the results generated by a hypothesis contradict some aspect of the problem, then the hypothesis is rejected. If the hypothesis generates data compatible with the problem, then it is retained.

A systematic approach is the only safe approach in working genetics problems. Shortcuts in thought processes usually lead to an incorrect answer. Consider the following two types of problems.

1. When analyzing pedigrees, there are usually only four possibilities (hypotheses) to be considered: autosomal dominant, autosomal recessive, X-linked dominant, and X-linked recessive. The criteria for each should be checked against the data. Additional factors that should be kept in mind are epistasis, penetrance, expressivity, age of onset, incorrect diagnosis in earlier generations, adultery, adoptions that are not mentioned, and inaccurate information in general. All these factors can be expected in real life, although few will be encountered in the problems presented here.

2. When studying matings, frequently the first task is to decide whether you are dealing with one gene, two genes, or more than two genes (hypotheses). The location of the gene or genes may or may not be important. If location is important, then there are two hypotheses: autosomal and X-linked. If there are two or more genes, then you may have to decide on linkage relationships between them. There are two hypotheses: unlinked and linked.

    If ratios are presented, then 1:2:1 (or some modification signaling dominance) indicates one gene, 9:3:3:1 (or some modification reflecting epistasis) indicates two genes, and 27:9:9:9:3:3:3:1 (or some modification signaling epistasis) indicates three genes. If ratios are presented that bear no relationship to the above, such as 35:35:15:15, then you are dealing with two linked genes (see Chapter 4 for a discussion of linkage).

    If phenotypes rather than ratios are emphasized in the problem, then a cross of two mutants that results in wild type indicates the involvement of two genes rather than alleles of the same gene. Both mutants are recessive to wild type. A correlation of sex with phenotype indicates X linkage for the gene mutant in the female parent, while a lack of correlation indicates autosomal location.

    If the problem involves X linkage, frequently the only way to solve it is to focus on the male progeny.

    Once you determine the number of genes being followed and their location, the problem essentially solves itself if you make a systematic listing of genotype and phenotype.

    Sometimes, the final portion of a problem will give additional information that requires you to adjust all the work that you have done up to that point. As an example, in Problem 64 of Chapter 6, crosses 1 – 3 lead you to assume that you are working with one gene. In cross 4, data incompatible with this assumption are presented. Your initial assumption of one gene is correct for the information given in the first three crosses; it is not a mistake. Other than a lack of systematic thought, the greatest mistake that a student can make is to label a rejected hypothesis an error. This decreases self-confidence and increases anxiety, with the result that real mistakes will likely follow. The beginner needs to keep in mind that science progresses by the rejection of hypotheses. When a hypothesis is rejected, something concrete is known: the proposed hypothesis does not explain the results. An unrejected hypothesis may be right or it may be wrong, and there is no way to know without further experimentation.

A very generalized procedure for problem solving would look like this:

1. Determine what information is being presented and what is being asked.

2. Formulate all possible hypotheses.

3. Check the consequences of each hypothesis against the data (the given information).
    Reject all hypotheses that are incompatible with the data.
    Retain all hypotheses that are compatible with the data.
    If no hypothesis is compatible with the data, return to step 1.

3

The Chromosomal Basis of Inheritance

BASIC PROBLEMS

1. The key function of mitosis is to generate two daughter cells genetically identical to the original parent cell.

2. Two key functions of meiosis are to halve the DNA content and to reshuffle the genetic content of the organism to generate genetic diversity among the progeny.

3. Its pretty hard to beat several billions of years of evolution but it might be simpler if DNA did not replicate prior to meiosis. The same events responsible for halving the DNA and producing genetic diversity could be achieved in a single cell division if homologous chromosomes paired, recombined, randomly aligned during metaphase, and separated during anaphase, etc. However, you would lose the chance to check and repair DNA that replication allows.

4. In large part, this question is asking, why sex? Parthogenesis (the ability to reproduce without fertilization — in essence, cloning) is not common among multicellular organisms. Parthenogenesis occurs in a some species of lizards and fishes, and several kinds of insects but it is tbe only means of reproduction in only a few of these species. In plants, about 400 species can reproduce asexually by a process called apomixis. These plants produce seeds without fertilization. However, the majority of plants and animals reproduce sexually. Sexual reproduction produces a wide variety of different offspring by forming new combinations of traits inherited from both the father and the mother. Despite the numerical advantages of asexual reproduction, most multicellular species that have adopted it as their only method of reproducing have become extinct. However, there is no agreed upon explanation of why the loss of sexual reproduction usually leads to early extinction or conversely, why sexual reproduction is associated with evolutionary success.

    On the other hand, the immediate effects of such a scenario are obvious. All offspring will be genetically identical to their mothers, and males would be extinct within one generation.

5.   As cells divide mitotically, each chromosome consists of identical sister chromatids that are separated to form genetically identical daughter cells. Although the second division of meiosis appears to be a similar process, the "sister" chromatids are likely to be different. Recombination during earlier meiotic stages has swapped regions of DNA between sister and nonsister chromosomes such that the two daughter cells of this division typically are not genetically identical.

6.   a.   Because the sister-chromatids have not separated, the cell is undergoing meiosis I. As the dyads are at the poles, the cell is at the end of anaphase I.

   b.   Each chromosome will have been replicated and each would contain two sister-chromatids. The replicated chromosome becomes the visible bivalent during prophase I of meiosis. So, an organism with 14 chromosomes will have 14 bivalents.

7.

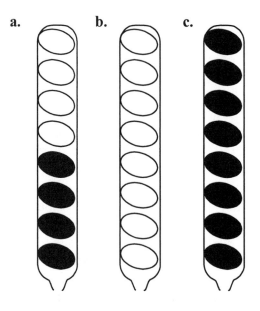

Note: The octad illustrated in (a) happens to have no crossover in the region between the centromere and the gene being followed. A crossover here would change the pattern of ascospores in the ascus but would not effect the 1:1 ratio of alleles in the meiotic products.

8.   The four stages of mitosis are: prophase, metaphase, anaphase, and telophase. The first letters, PMAT, can be remembered by a mnemonic such as: Playful Mice Analyze Twice.

   The five stages of prophase I are: leptotene, zygotene, pachytene, diplotene, and diakinesis. The first letters, LZPDD, can be remembered by a mnemonic such as: Large Zoos Provide Dangerous Distractions.

9. Each diploid meiocyte undergoes meiosis to form four products and then each undergoes a postmeiotic mitotic division to give a total of eight ascospores per meiocyte. In total then, 100 meiocytes will give rise to 800 ascospores.

10. The resulting cells will have the identical genotype as the original cell: *A/a* ; *B/b*.

11. Yes. It could work but certain DNA repair mechanisms (such as postreplication recombination repair) could not be invoked prior to cell division. There would be just two cells as products of this meiosis, rather than four.

12. The general formula for the number of different male/female centromeric combinations possible is 2^n, where n = number of different chromosome pairs. In this case, $2^5 = 32$.

13. Because the DNA levels vary six-fold, the range covers cells that are haploid (spores or cells of the gametophyte stage) to cells that are triploid (the endosperm) and dividing (after DNA has replicated but prior to cell division). The following cells would fit the DNA measurements:

0.7 haploid cells
1.4 diploid cells in G_1 or haploid cells after S but prior to cell division
2.1 triploid cells of the endosperm
2.8 diploid cells after S but prior to cell division
4.2 triploid cells after S but prior to cell division

14.

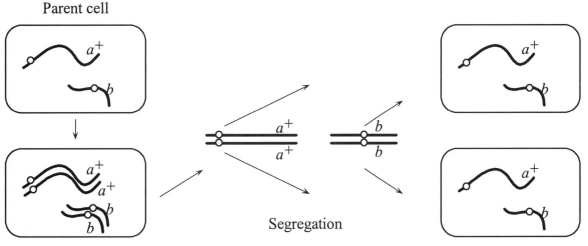

Parent cell

Chromosome duplication

Segregation

Daughter cells

15. PFGE separates DNA molecules by size. When DNA is carefully isolated from *Neurospora* (which has seven different chromosomes) seven bands should be produced using this technique. Similarly, the pea has seven different chromosomes and will produce seven bands (homologous chromosomes will co-migrate as a single band).

16. There is a total of 4 m of DNA and nine chromosomes per haploid set. On average, each is $^4/_9$ m long. At metaphase, their average length is 13 μm, so the average packing ratio is 13×10^{-6} m : 4.4×10^{-1} m or roughly 1:34,000! This remarkable achievement is accomplished through the interaction of the DNA with proteins. At its most basic, eukaryotic DNA is associated with histones in units called nucleosomes and during mitosis, coils into a solenoid. As loops, it associates with and winds into a central core of nonhistone protein called the scaffold.

17. The nucleus contains the genome and separates it from the cytoplasm. However, during cell division, the nuclear envelope dissociates (breaks down). It is the job of the microtubule-based spindle to actually separate the chromosomes (divide the genetic material) around which nuclei reform during telophase. In this sense, it can be viewed as a passive structure that is divided by the cell's cytoskeleton.

18. Yes. Half of our genetic makeup is derived from each parent, each parent's genetic makeup is derived half from each of their parents, etc.

19. Because the "half" inherited is very random, the chances of receiving exactly the same half is vanishingly small. Ignoring recombination and focusing just on which chromosomes are inherited from one parent (for example, the one they inherited from their father or the one from their mother?), there are $2^{23} = 8,388,608$ possible combinations!

20. Remember, the endosperm is formed from two polar nuclei (which are genetically identical) and one sperm nucleus.

Female	Male	Polar nuclei	Sperm	Endosperm
s/s	S/S	s and s	S	S/s/s
S/S	s/s	S and S	s	S/S/s
S/s	S/s	$^1/_2$ S and S	$^1/_2$ S	$^1/_4$ S/S/S
				$^1/_4$ S/S/s
		$^1/_2$ s and s	$^1/_2$ s	$^1/_4$ S/s/s
				$^1/_4$ s/s/s

21. a. A sporophyte of *A/a* ; *B/b* genotype will produce gametophytes in the following proportions:

$^1/_4$ *A* ; *B*
$^1/_4$ *A* ; *b*
$^1/_4$ *a* ; *B*
$^1/_4$ *a* ; *b*

b. Random fertilization of the spores from the above gametophytes can occur $4 \times 4 = 16$ possible ways. Four of these combinations (A ; $B \times a$; b, a ; $b \times A$; B, A ; $b \times a$; B, a ; $B \times A$; b) will result in the desired A/a ; B/b sporophyte genotype. Therefore $1/4$ of next generation should be of this genotype.

22. All daughter cells will still be A/a ; B/b ; C/c. Mitosis generates daughter cells genetically identical with that of the original cell.

23.

P	ad^- ; $a \times ad^+$; α
Transient diploid	ad^+/ad^- ; a/α
F$_1$	$1/4$ ad^+ ; a, white
	$1/4$ ad^- ; a, purple
	$1/4$ ad^+ ; α, white
	$1/4$ ad^- ; α, purple

24.

	Mitosis	Meiosis
fern	sporophyte gametophyte	sporophyte (sporangium)
moss	sporophyte gametophyte	sporophyte (atheridium and archegonium)
plant	sporophyte gametophyte	sporophyte (anther and ovule)
pine tree	sporophyte gametophyte	sporophyte (pine cone)
mushroom	sporophyte gametophyte	sporophyte (ascus or basidium)
frog	somatic cells	gonads
butterfly	somatic cells	gonads
snail	somatic cells	gonads

25. This problem is tricky because the answers depend on how a cell is defined. In general, geneticists consider the transition from one cell to two cells to occur with the onset of anaphase in both mitosis and meiosis, even though cytoplasmic division occurs at a later stage.

a. 46 chromosomes, each with two chromatids = 92 chromatids

    b.   46 chromosomes, each with two chromatids = 92 chromatids

    c.   46 physically separate chromosomes in each of two about-to-be-formed cells

    d.   23 chromosomes in each of two about-to-be-formed cells, each with two chromatids = 46 chromatids

    e.   23 chromosomes in each of two about-to-be-formed cells

26.   (5) chromosome pairing (synapsis)

27.   His children will have to inherit the satellite-containing 4 (probability = $1/2$), the abnormally-staining 7 (probability = $1/2$), and the Y chromosome (probability = $1/2$). To get all three, the probability is $(1/2)(1/2)(1/2) = 1/8$.

28.   The parental set of centromeres can match either parent, which means there are two ways to satisfy the problem. For any one pair, the probability of a centromere from one parent going into a specific gamete is $1/2$. For n pairs, the probability of all the centromeres being from one parent is $(1/2)^n$. Therefore, the total probability of having a haploid complement of centromeres from either parent is $2(1/2)^n = (1/2)^{n-1}$.

29.   Dear Monk Mendel:

I have recently read your most engrossing manuscript detailing the results of your most wise experiments with garden peas. I salute both your curiosity and your ingenuity in conducting said experiments, thereby opening up for scientific exploration an entire new area of our Maker's universe. Dear Sir, your findings are extraordinary!

While I do not pretend to compare myself to you in any fashion, I beg to bring to your attention certain findings I have made with the aid of that most fascinating and revealing instrument, the microscope. I have been turning my attention to the smallest of worlds with an instrument that I myself have built, and I have noticed some structures that may parallel in behavior the factors which you have postulated in the pea.

I have worked with grasshoppers, however, not your garden peas. Although you are a man of the cloth, you are also a man of science, and I pray that you will not be offended when I state that I have specifically studied the reproductive organs of male grasshoppers. Indeed, I did not limit myself to studying the organs themselves; instead, I also studied the smaller units that make up the male organs and have beheld structures most amazing within them.

These structures are contained within numerous small bags within the male organs. Each bag has a number of these structures, which are long and threadlike at some times and short and compact at other times. They come together in the middle of a bag, and then they appear to divide equally. Shortly thereafter, the bag itself divides, and what looks like half of the threadlike structures goes into each new bag. Could it be, Sir, that these threadlike structures are the very same as your factors? I know, of course, that garden peas do not have male organs in the same way that grasshoppers do, but it seems to me that you found it necessary to emasculate the garden peas in order to do some crosses, so I do not think it too far-fetched to postulate a similarity between grasshoppers and garden peas in this respect.

Pray, Sir, do not laugh at me and dismiss my thoughts on this subject even though I have neither your excellent training nor your astounding wisdom in the Sciences. I remain your humble servant to eternity!

30. When results of a cross are sex-specific, sex linkage should be considered. In moths, the heterogametic sex is actually the female while the male is the homogametic sex. Assuming that dark (D) is dominant to light (d), then the data can be explained by the dark male being heterozygous (D/d) and the dark female being hemizygous (D). All male progeny will inherit the D allele from their mother and therefore be dark while half the females will inherit D from their father (and be dark) and half will inherit d (and be light).

31. **a.** Both the gametophyte and the sporophyte are closer in shape to the mother than the father. Note that a size increase occurs in each type of cross.

    b. Gametophyte and sporophyte morphology are affected by extranuclear factors. Leaf size may be a function of the interplay between nuclear genome contributions.

    c. If extranuclear factors are affecting morphology while nuclear factors are affecting leaf size, then repeated backcrosses could be conducted, using the hybrid as the female. This would result in the cytoplasmic information remaining constant while the nuclear information becomes increasingly like that of the backcross parent. Leaf morphology should therefore remain constant while leaf size would decrease toward the size of the backcross parent.

32. The goal here is to generate a plant with the cytoplasm of plant A and the nuclear genome predominantly of plant B. Remember that the cytoplasm is contributed by the egg only. So using plant A as the maternal parent, cross to B (as the paternal parent) and then backcross the progeny of this cross using plant B again as the paternal parent. Repeat for several generations until virtually the entire nuclear genome is from the B parent.

33. If the variegation is due to a chloroplast mutation, then the phenotype of the offspring will be controlled solely by the phenotype of the maternal parent. Look for flowers on white branches and test to see if they produce seeds that grow into all white plants regardless of the source of the pollen. To test whether a dominant nuclear mutation is responsible for the variegation, cross pollen from flowers on white branches to green plants (or flowers on green branches of same plant) and see if half the progeny are white and half are green (assuming the dominant mutation is heterozygous) or all white, if the dominant mutation is homozygous.

34. Progeny plants inherited only normal cpDNA (lane 1); only mutant cpDNA (lane 2); or both (lane 3). In order to get homozygous cpDNA, seen in lanes 1 and 2, segregation of chloroplasts had to occur.

CHALLENGING PROBLEMS

35. In the following schematic drawings, chromosomes (or chromatids) that are radioactive are indicated be the grains that would be observed after radioautography. After the second mitotic division, a number of outcomes are possible due to the random alignment and separation of the radioactive and non-radioactive chromatids.

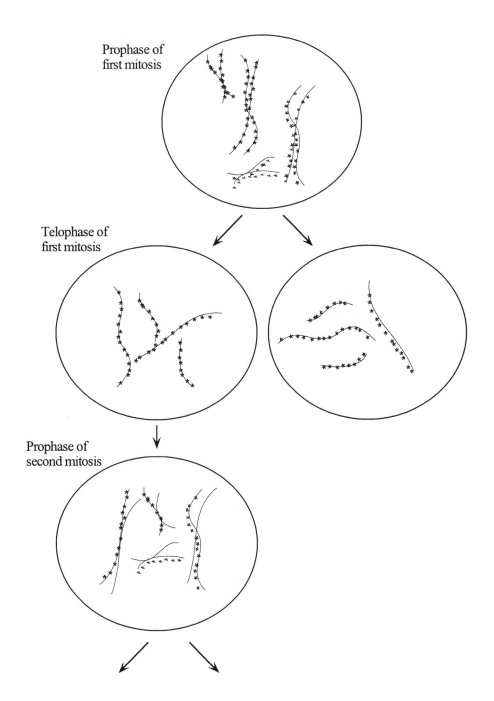

Prophase of first mitosis

Telophase of first mitosis

Prophase of second mitosis

Telophase of
second mitosis

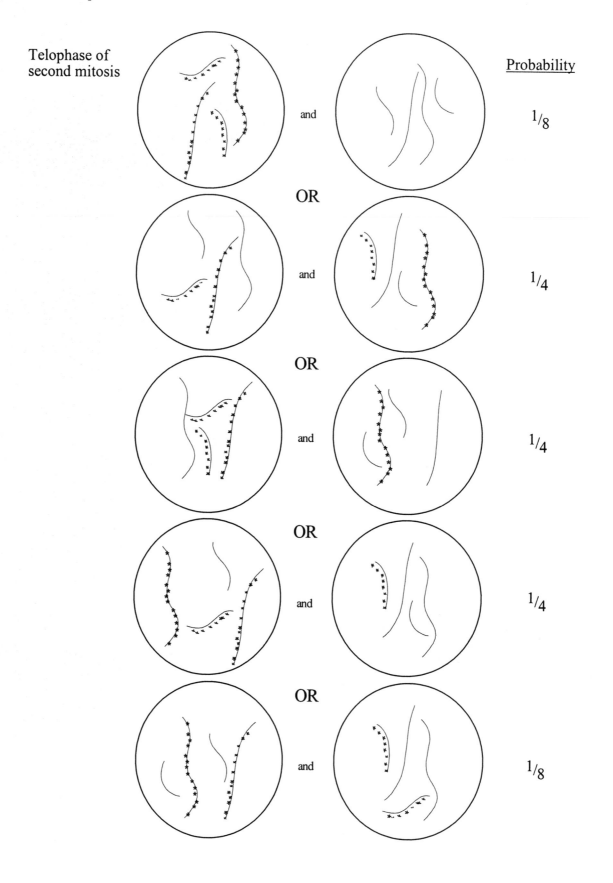

Probability

and   1/8

OR

and   1/4

OR

and   1/4

OR

and   1/4

OR

and   1/8

36.

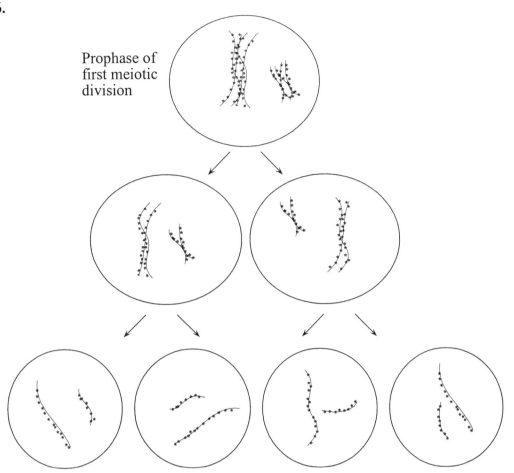

Prophase of
first meiotic
division

37. **a.** In a diploid cell, expect two chromosomes (a pair of homologs) to each have a single locus of radioactivity.

b. Expect many regions of radioactivity scattered throughout the chromosomes. The exact number and pattern would be dependent on the specific sequence in question, and where and how often it is present within the genome.

c. The multiple copies of the genes for ribosomal RNA are organized into large tandem arrays called nucleolar organizers (NO). Therefore, expect broader areas of radioactivity compared to (a). The number of these regions would equal the number of NO present in the organism.

d. Each chromosome end would be labeled by telomeric DNA.

e. The multiple repeats of this heterochromatic DNA are organized into large tandem arrays. Therefore, expect broader areas of radioactivity compared to (a). Also, there may be more than one area in the genome of the same simple repeat.

38. The following is meant to be examples of what is possible. It is also possible, for instance, that more than one band would be present in (a), depending on the position of the restriction sites within the sequence complementary to the probe used.

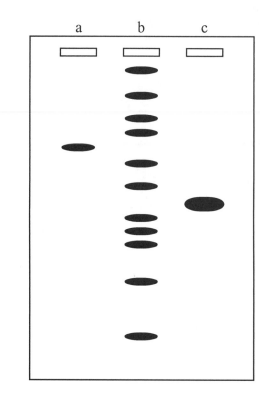

For (d) and (e), the specifics or where the DNA is cut relative to the telomeric DNA or heterochromatic DNA will effect what is observed. Assuming the restriction sites are not within the telomeric or heterochromatic DNA, then (d) will be similar to (b), and (c) will have one or several very large bands.

39. First, examine the crosses and the resulting genotypes of the endosperm:

Female	Male	Polar nuclei	Sperm	Endosperm
f'/f' (floury)	f''/f''	f' and f'	f''/f''	$f'/f'/f''$
f''/f'' (flinty)	f'/f'	f'' and f''	f'/f'	$f''/f''/f'$

As can be seen, the phenotype of the endosperm correlates to the predominant allele present.

40. (1)   Impossible: the alleles of the same genes are on nonhomologous chromosomes
      (2)   Meiosis II

(3)   Meiosis II
(4)   Meiosis II
(5)   Mitosis
(6)   Impossible: appears to be mitotic anaphase but alleles of sister chromatids are not identical
(7)   Impossible: too many chromosomes
(8)   Impossible: too many chromosomes
(9)   Impossible: too many chromosomes
(10)   Meiosis I
(11)   Impossible: appears to be meiosis of homozygous *a/a* ; *B/B*
(12)   Impossible: the alleles of the same genes are on nonhomologous chromosomes

41. Recall that each cell has many mitochondria, each with numerous genomes. Also recall that cytoplasmic segregation is routinely found in mitochondrial mixtures within the same cell.

The best explanation for this pedigree is that the mother in generation I experienced a mutation in a single cell that was a progenitor of her egg cells (primordial germ cell). By chance alone, the two males with the disorder in the second generation were from egg cells that had experienced a great deal of cytoplasmic segregation prior to fertilization, while the two females in that generation received a mixture.

The spontaneous abortions that occurred for the first woman in generation II were the result of extensive cytoplasmic segregation in her primordial germ cells: aberrant mitochondria were retained. The spontaneous abortions of the second woman in generation II also came from such cells. The normal children of this woman were the result of extensive segregation in the opposite direction: normal mitochondria were retained. The affected children of this woman were from egg cells that had undergone less cytoplasmic segregation by the time of fertilization, so that they developed to term but still suffered from the disease.

42. For the following, S will signify cytoplasm of a male-sterile line and N will signify cytoplasm of a non-male-sterile line. *Rf* will signify the dominant nuclear restorer allele and *rf*, the recessive non-restorer allele.

a. If S *rf/rf* (male-sterile plants) are crossed with pollen from N *Rf/Rf* plants, the offspring will all be S *Rf/rf* and male fertile. If these offspring are then crossed with pollen from N *rf/rf* plants, half the offspring will be S *Rf/rf* (male-fertile) and half will be S *rf/rf* (male-sterile). The S cytoplasm will not be altered or affected even though the maternal offspring parent plant was *Rf/rf*.

b. The cross is S *rf/rf* × N *Rf/Rf* so all the progeny will be S *Rf/rf* and male-fertile.

c. The cross is S *Rf/rf* × N *rf/rf* so half the progeny will be S *Rf/rf* (male-fertile) and half will be S *rf/rf* (male-sterile).

d. i. The cross is S *rf-1/rf-1* ; *rf-2/rf-2* × N *Rf-1/rf-1* ; *Rf-2/rf-2*
    The progeny will be: $\frac{1}{4}$ S *Rf-1/rf-1* ; *Rf-2/rf-2* (male-fertile)
                            $\frac{1}{4}$ S *Rf-1/rf-1* ; *rf-2/rf-2* (male-fertile)
                            $\frac{1}{4}$ S *rf-1/rf-1* ; *Rf-2/rf-2* (male-fertile)
                            $\frac{1}{4}$ S *rf-1/rf-1* ; *rf-2/rf-2* (male-sterile)

  ii. The cross is S *rf-1/rf-1* ; *rf-2/rf-2* × N *Rf-1/Rf-1* ; *rf-2/rf-2*
    The progeny will all be: S *Rf-1/rf-1* ; *rf-2/rf-2* (male-fertile)

  iii. The cross is S *rf-1/rf-1* ; *rf-2/rf-2* × N *Rf-1/rf-1* ; *rf-2/rf-2*
    The progeny will be: $\frac{1}{2}$ S *Rf-1/rf-1* ; *rf-2/rf-2* (male-fertile)
                            $\frac{1}{2}$ S *rf-1/rf-1* ; *rf-2/rf-2* (male-sterile)

  vi. The cross is S *rf-1/rf-1* ; *rf-2/rf-2* × N *Rf-1/rf-1* ; *Rf-2/Rf-2*
    The progeny will be: $\frac{1}{2}$ S *Rf-1/rf-1* ; *Rf-2/rf-2* (male-fertile)
                            $\frac{1}{2}$ S *rf-1/rf-1* ; *Rf-2/rf-2* (male-fertile)

4

Eukaryotic Chromosome Mapping by Recombination

BASIC PROBLEMS

1. You perform the following cross and are told that the two genes are 10 m.u. apart.

    $$A\ B/a\ b \ \times a\ b/a\ b$$

    Among their progeny, 10 percent should be recombinant (*A b/a b* and *a B/a b*) and 90 percent should be parental (*A B/a b* and *a b/a b*). Therefore, *A B/a b* should represent $1/2$ of the parentals or 45 percent.

2. P $A\ d\ /\ A\ d \ \ \times \ \ a\ D\ /\ a\ D$

    F_1 $A\ d\ /\ a\ D$

    F_2 1 $A\ d\ /\ A\ d$ phenotype: A d

              2   $A\ d\ /\ a\ D$ phenotype: A D

              1   $a\ D\ /\ a\ D$ phenotype: a D

3. P $R\ S/r\ s \times R\ S/r\ s$

    gametes $1/2\ (1 - 0.35)$ $R\ S$

                     $1/2\ (1 - 0.35)$ $r\ s$

                     $1/2\ (0.35)$ $R\ s$

                     $1/2\ (0.35)$ $r\ S$

    F_1 genotypes   0.1056 *R S/R S*   0.1138 *r s/r S*

                        0.1056 *r s/r s*   0.1138 *r s/R s*

                        0.2113 *R S/r s*   0.0306 *R s/R s*

                        0.1138 *R S/r S*   0.0306 *r S/r S*

                        0.1138 *R S/R s*   0.0613 *R s/r S*

F$_1$ phenotypes   0.6058 R S
0.1056 r s
0.1444 R s
0.1444 r S

4.   The cross is *E/e · F/f* × *e/e · f/f*. If independent assortment exists, the progeny should be in a 1:1:1:1 ratio, which is not observed. Therefore, there is linkage. *E f* and *e F* are recombinants equaling one-third of the progeny. The two genes are 33.3 map units (m.u.) apart.

$$RF = 100\% \times 1/3 = 33.3\%$$

5.   Because only parental types are recovered, the two genes must be tightly linked and recombination must be very rare. Knowing how many progeny were looked at would give an indication of how close the genes are.

6.   The problem states that a female that is *A/a · B/b* is test crossed. If the genes are unlinked, they should assort independently and the four progeny classes should be present in roughly equal proportions. This is clearly not the case. The *A/a · B/b* and *a/a · b/b* classes (the parentals) are much more common than the *A/a · b/b* and *a/a · B/b* classes (the recombinants). The two genes are on the same chromosome and are 10 map units apart.

$$RF = 100\% \times (46 + 54)/1000 = 10\%$$

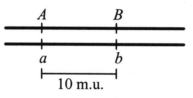

7.   The cross is *A/A · b/b* × *a/a · B/B*. The F$_1$ would be *A/a · B/b*.

     a.   If the genes are unlinked, all four progeny classes from the test cross (including *a/a ; b/b*) would equal 25 percent.

     b.   With completely linked genes, the F$_1$ would produce only *A b* and *a B* gametes. Thus, there would be a 0 percent chance of having *a b/a b* progeny from a test cross of this F$_1$.

     c.   If the two genes are linked and 10 map units apart, 10 percent of the test cross progeny should be recombinants. Since the F$_1$ is *A b/a B*, *a b* is one of the recombinant classes (*A B* being the other) and should equal $1/2$ of the total recombinants or 5 percent.

     d.   12 percent (see part c)

8. Meiosis is occurring in an organism that is *C d/c D*, ultimately producing haploid spores. The parental genotypes are *C d* and *c D*, in equal frequency. The recombinant types are *C D* and *c d*, in equal frequency. Eight map units means 8 percent recombinants. Thus, *C D* and *c d* will each be present at a frequency of 4 percent, and *C d* and *c D* will each be present at a frequency of (100% – 8%)/2 = 46%.

    a.   4 percent

    b.   4 percent

    c.   46 percent

    d.   8 percent

9. To answer this question, you must realize that

(1) One chiasma involves two of the four chromatids of the homologous pair so if 16 percent of the meioses have one chiasma, it will lead to 8 percent recombinants observed in the progeny (one half of the chromosomes of such a meiosis are still parental), and
(2) Half of the recombinants will be *B r,* so the correct answer is 4 percent.

10. *Unpacking the Problem*

    1.   There is no correct drawing; any will do. Pollen from the tassels is placed on the silks of the females. The seeds are the F_1 corn kernels.

    2.   The +'s all look the same because they signify wild type for each gene. The information is given in a specific order, which prevents confusion, at least initially. However, as you work the problem, which may require you to reorder the genes, errors can creep into your work if you do not make sure that you reorder the genes for each genotype in exactly the same way. You may find it easier to write the complete genotype, p^+ instead of +, to avoid confusion.

    3.   The phenotype is purple leaves and brown midriff to seeds. In other words, the two colors refer to different parts of the organism.

    4.   There is no significance in the original sequence of the data.

    5.   A tester is a homozygous recessive for all genes being studied. It is used so that the meiotic products in the organism being tested can be seen directly in the phenotype of the progeny.

6. The progeny phenotypes allow you to infer the genotypes of the plants. For example, *gre* stands for "green," the phenotype of $p^+/-$; *sen* stands for "virus-sensitive," the phenotype of $v^+/-$; and *pla* stands for "plain seed," the phenotype of $b^+/-$. In this test cross, all progeny have at least one recessive allele so the "gre sen pla" progeny are actually $p^+/p \cdot v^+/v \cdot b^+/b$.

7. *Gametes* refers to the gametes of the two pure-breeding parents. F_1 *gametes* refers to the gametes produced by the completely heterozygous F_1 progeny. They indicate whether crossing-over and independent assortment have occurred. In this case, because there is either independent assortment or crossing-over, or both, the data indicate that the three genes are not so tightly linked that zero recombination occurred.

8. The main focus is meiosis occurring in the F_1 parent.

9. The gametes from the tester are not shown because they contribute nothing to the phenotypic differences seen in the progeny.

10. Eight phenotypic classes are expected for three autosomal genes, whether or not they are linked, when all three genes have simple dominant-recessive relationships among their alleles. The general formula for the number of expected phenotypes is 2^n, where n is the number of genes being studied.

11. If the three genes were on separate chromosomes, the expectation is a 1:1:1:1:1:1:1:1 ratio.

12. The four classes of data correspond to the parentals (largest), two groups of single crossovers (intermediate), and double crossovers (smallest).

13. By comparing the parentals with the double crossovers, gene order can be determined. The gene in the middle flips with respect to the two flanking genes in the double-crossover progeny. In this case, one parental is $+++$ and one double crossover is $p++$. This indicates that the gene for leaf color (p) is in the middle.

14. If only two of the three genes are linked, the data can still be grouped, but the grouping will differ from that mentioned in (12) above. In this situation, the unlinked gene will show independent assortment with the two linked genes. There will be one class composed of four phenotypes in approximately equal frequency, which combined will total more than half the progeny. A second class will be composed of four phenotypes in approximately equal frequency, and the combined total will be less than half the progeny. For example, if the cross were $a\ b/+\ + \ ;\ c/+ \ \times \ a\ b/a\ b\ ;\ c/c$, then the parental class (more frequent class) would have four components: $a\ b\ c$, $a\ b\ +$, $+\ +\ c$, and $+\ +\ +$. The recombinant class would be $a\ +\ c$, $a\ +\ +$, $+\ b\ c$, and $+\ b\ +$.

15. *Point* refers to locus. The usage does not imply linkage, but rather a testing for possible linkage. A four-point testcross would look like the following: $a/+ \cdot b/+ \cdot c/+ \cdot d/+ \times a/a \cdot b/b \cdot c/c \cdot d/d$.

16. A *recombinant* refers to an individual who has alleles inherited from two different grandparents, both of whom were the parents of the individual's heterozygous parent. Another way to think about this term is that in the recombinant individual's heterozygous parent, recombination took place among the genes that were inherited from his or her parents. In this case, the recombination took place in the F_1 and the recombinants are among the F_2 progeny.

17. The "recombinant for" columns refer to specific gene pairs and progeny that exhibit recombination between those gene pairs.

18. There are three "recombinant for" columns because three genes can be grouped in three different gene pairs.

19. *R* refers to recombinant progeny, and they are determined by reference back to the parents of their heterozygous parent.

20. Column totals indicate the number of progeny that experience crossing-over between the specific gene pairs. They are used to calculate map units between the two genes.

21. The diagnostic test for linkage is a recombination frequency of less than 50 percent.

22. A map unit represents 1 percent crossing-over and is the same as a centimorgan.

23. In the tester, recombination cannot be detected in the gamete contribution to the progeny because the tester is homozygous. The F_1 individuals have genotypes fixed by their parents' homozygous state and, again, recombination cannot be detected in them, simply because their parents were homozygous.

24. Interference I = 1 − coefficient of coincidence = 1 − (observed double crossovers/expected double crossovers). The expected double crossovers are equal to p(frequency of crossing-over in the first region, in this case between v and p) $\times p$(frequency of crossing-over in the second region, between p and b) $\times$ number of progeny. The probability of crossing-over is equal to map units converted back to percentage.

25. If the three genes are not all linked, then interference cannot be calculated.

26. A great deal of work is required to obtain 10,000 progeny in corn because each seed on a cob represents one progeny. Each cob may contain as many as 200 seeds. While seed characteristics can be assessed at the cob stage, for other characteristics, each seed must be planted separately and assessed after germination and growth. The bookkeeping task is also enormous.

Solution to the Problem

a. The three genes are linked.

b. Comparing the parentals (most frequent) with the double crossovers (least frequent), the gene order is $v\,p\,b$. There were 2200 recombinants between v and p, and 1500 between p and b. The general formula for map units is

$$\text{m.u.} = 100\%(\text{number of recombinants})/\text{total number of progeny}$$

Therefore, the map units between v and p = $100\%(2200)/10,000 = 22$ m.u., and the map units between p and b = $100\%(1500)/10,000 = 15$ m.u. The map is

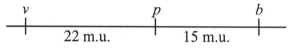

v	p	b
22 m.u.	15 m.u.	

c. I = 1 – observed double crossovers/expected double crossovers
= 1 – 132/(0.22)(0.15)(10,000)
= 1 – 0.4 = 0.6

11. P $\qquad a\,b\,c/a\,b\,c \times a^+\,b^+\,c^+/a^+\,b^+\,c^+$

$F_1 \qquad a^+\,b^+\,c^+/a\,b\,c \times a^+\,b^+\,c^+/a\,b\,c$

F_2	1364	$a^+\,b^+\,c^+$
	365	$a\,b\,c$
	87	$a\,b\,c^+$
	84	$a^+\,b^+\,c$
	47	$a\,b^+\,c^+$
	44	$a^+\,b\,c$
	5	$a\,b^+\,c$
	4	$a^+\,b\,c^+$

This problem is somewhat simplified by the fact that recombination does not occur in male *Drosophila*. Also, only progeny that received the $a\,b\,c$ chromosome from the male will be distinguishable among the F_2 progeny.

a. Because you cannot distinguish between $a^+\,b^+\,c^+/a^+\,b^+\,c^+$ and $a^+\,b^+\,c^+/a\,b\,c$, use the frequency of $a\,b\,c/a\,b\,c$ to estimate the frequency of $a^+\,b^+\,c^+$ (parental) gametes from the female.

parentals   730   (2 × 365)
CO a–b: 91   ($a\ b^+\ c^+$, $a^+\ b\ c = 47 + 44$)
CO b–c: 171   ($a\ b\ c^+$, $a^+\ b^+\ c = 87 + 84$)
DCO: 9   ($a^+\ b\ c^+$, $a\ b^+\ c = 4 + 5$)
    1001

a–b: 100%(91 + 9)/1001 = 10 m.u.
b–c: 100%(171 + 9)/1001 = 18 m.u.

    b.   Coefficient of coincidence = (observed DCO)/(expected DCO)
        = 9/[(0.1)(0.18)(1001)] = 0.5

12.   **a.**   By comparing the two most frequent classes (parentals: *an br$^+$ f$^+$*, *an$^+$ br f*) to the least frequent classes (DCO: *an$^+$ br f$^+$*, *an br$^+$ f*), the gene order can be determined. The gene in the middle switches with respect to the other two (the order is *an f br*). Now the crosses can be written fully.

    P   *an f$^+$ br$^+$/an f$^+$ br$^+$* × *an$^+$ f br/an$^+$ f br*

    F$_1$ *an$^+$ f br/an f$^+$ br$^+$* × *an f br/an f br*

    F$_2$ 355   *an f$^+$ br$^+$/an f br*   parental
        339   *an$^+$ f br/an f br*   parental
        88   *an$^+$ f$^+$ br$^+$/an f br*   CO *an–f*
        55   *an f br/an f br*   CO *an–f*
        21   *an$^+$ f br$^+$/an f br*   CO *f–br*
        17   *an f$^+$ br/an f br*   CO *f–br*
        2   *an$^+$ f$^+$ br/an f br*   DCO
        2   *an f br$^+$/an f br*   DCO
        879

    b.   *an–f*:   100% (88 + 55 + 2 + 2)/879 = 16.72 m.u.
        f–br: 100% (21 + 17 + 2 + 2)/879 = 4.78 m.u.

    c.   Interference = 1 – (observed DCO/expected DCO)
        = 1 – 4/(0.1672)(0.0478)(879)= 0.431

13.   By comparing the most frequent classes (parental: *+ v lg*, *b + +*) with the least frequent classes (DCO: *+ + +*, *b v lg*) the gene order can be determined. The gene in the middle switches with respect to the other two, yielding the following sequence: *v b lg*. Now the cross can be written

P $v\,b^+\,lg/v^+\,b\,lg^+ \times v\,b\,lg/v\,b\,lg$

F$_1$

305	$v\,b^+\,lg/v\,b\,lg$	parental
275	$v^+\,b\,lg^+/v\,b\,lg$	parental
128	$v^+\,b\,lg/v\,b\,lg$	CO b–lg
112	$v\,b^+\,lg^+/v\,b\,lg$	CO b–lg
74	$v^+\,b^+\,lg/v\,b\,lg$	CO v–b
66	$v\,b\,lg^+/v\,b\,lg$	CO v–b
22	$v^+\,b^+\,lg^+/v\,b\,lg$	DCO
18	$v\,b\,lg/v\,b\,lg$	DCO

v–b: $100\%(74 + 66 + 22 + 18)/1000 = 18.0$ m.u.
b–lg: $100\%(128 + 112 + 22 + 18)/1000 = 28.0$ m.u.

c.c.   = observed DCO/expected DCO = $(22 + 18)/(0.28)(0.18)(1000) = 0.79$

14. Let F = fat, L = long tail, and Fl = flagella. The gene sequence is $F\ L\ Fl$ (compare most frequent with least frequent). The cross is

P $F\,L\,Fl/f\,l\,fl \times f\,l\,fl/f\,l\,fl$

F$_1$

398	$F\,L\,Fl/f\,l\,fl$	parental
370	$f\,l\,fl/f\,l\,fl$	parental
72	$F\,L\,fl/f\,l\,fl$	CO L–Fl
67	$f\,l\,Fl/f\,l\,fl$	CO L–Fl
44	$f\,L\,Fl/f\,l\,fl$	CO F–L
35	$F\,l\,fl/f\,l\,fl$	CO F–L
9	$f\,L\,fl/f\,l\,fl$	DCO
5	$F\,l\,Fl/f\,l\,fl$	DCO

L–Fl: $100\%(72 + 67 + 9 + 5)/1000 = 15.3$ m.u.
F–L: $100\%(44 + 35 + 9 + 5)/1000 = 9.3$ m.u.

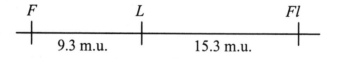

15. a.   The hypothesis is that the genes are not linked. Therefore, a 1:1:1:1 ratio is expected.

b. χ^2 = $(54{-}50)^2/50 + (47{-}50)^2/50 + (52{-}50)^2/50 + (47{-}50)^2/50$
     = $0.32 + 0.18 + 0.08 + 0.18 = 0.76$

c.   With 3 degrees of freedom, the p value is between 0.50 and 0.90.

d. Between 50 percent and 90 percent of the time values this extreme from the prediction would be obtained by chance alone.

e. Accept the initial hypothesis.

f. Because the χ^2 value was insignificant, the two genes are assorting independently. The genotypes of all individuals are

P dp^+/dp^+ ; $e/e \times dp/dp$; e^+/e^+

F_1 dp^+/dp ; e^+/e

tester dp/dp ; e/e

progeny   long, ebony dp^+/dp ; e/e
            long, gray dp^+/dp ; e^+/e
            short, gray dp/dp ; e^+/e
            short, ebony dp/dp ; e/e

16. a.

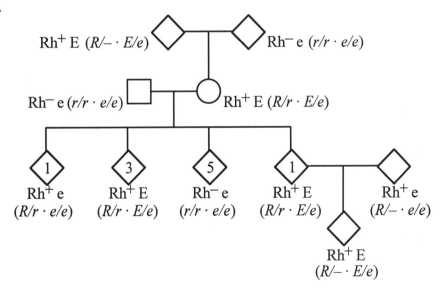

b. Yes

c. Dominant

d. As drawn, the pedigree hints at linkage. If unlinked, expect that the phenotypes of the 10 children should be in a 1:1:1:1 ratio of Rh^+ E, Rh^+ e, Rh^- E, and Rh^- e. There are actually five Rh^- e, four Rh^+ E, and one Rh^+ e. If linked, this last phenotype would represent a recombinant, and the distance between the two genes would be 100%(1/10) = 10 m.u. However, there is just not enough data to strongly support that conclusion.

17. Assume there is no linkage. (This is your hypothesis. If it can be rejected, the genes are linked.) The expected values would be that genotypes occur with equal frequency. There are four genotypes in each case ($n = 4$) so there are 3 degrees of freedom.

$$\chi^2 = \sum (observed - expected)^2/expected$$

Cross 1: $\chi^2 = [(310-300)^2 + (315-300)^2 + (287-300)^2 + (288-300)^2]/300$
$= 2.1266$; $p > 0.50$, nonsignificant; hypothesis cannot be rejected

Cross 2: $\chi^2 = [(36-30)^2 + (38-30)^2 + (23-30)^2 + (23-30)^2]/300$
$= 6.6$; $p > 0.10$, nonsignificant; hypothesis cannot be rejected

Cross 3: $\chi^2 = [(360-300)^2 + (380-300)^2 + (230-300)^2 + (230-300)^2]/300$
$= 66.0$; $p < 0.005$, significant; hypothesis must be rejected

Cross 4: $\chi^2 = [(74-60)^2 + (72-60)^2 + (50-60)^2 + (44-60)^2]/300$
$= 11.60$; $p < 0.01$, significant; hypothesis must be rejected

18. **a.** Both disorders must be recessive to yield the patterns of inheritance that are observed. Notice that only males are affected, strongly suggesting X linkage for both disorders. In the first pedigree there is a 100 percent correlation between the presence or absence of both disorders, indicating close linkage. In the second pedigree, the presence and absence of both disorders are inversely correlated, again indicating linkage. In the first pedigree, the two defective alleles must be cis within the heterozygous females to show 100 percent linkage in the affected males, while in the second pedigree the two defective alleles must be trans within the heterozygous females.

b. and c. Let *a* stand for the allele giving rise to steroid sulfatase deficiency (vertical bar) and *b* stand for the allele giving rise to ornithine transcarbamylase deficiency (horizontal bar). Crossing over cannot be detected without attaching genotypes to the pedigrees. When this is done, it can be seen that crossing-over need not occur in either of the pedigrees to give rise to the observations.

First pedigree:

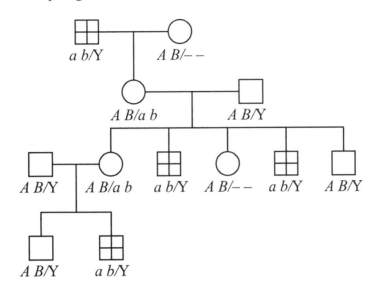

Second pedigree:

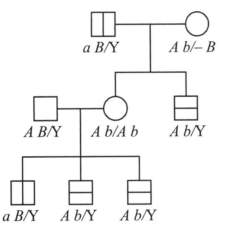

19. **a.** Note that only males are affected by both disorders. This suggests that both are X-linked recessive disorders. Using *p* for protan and *P* for non-protan, and *d* for deutan and *D* for non-deutan, the inferred genotypes are listed on the pedigree below. The Y chromosome is shown, but the X is represented by the alleles carried.

   b. Individual II-2 must have inherited both disorders in the trans configuration (on separate chromosomes). Therefore, individual III-2 inherited both traits as the result of recombination (crossing-over) between his mother's X chromosomes.

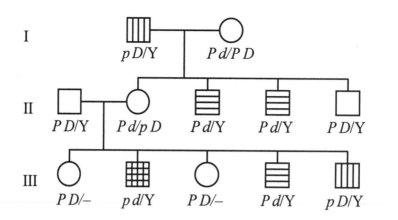

c. Because both genes are X-linked, this represents crossing-over. The progeny size is too small to give a reliable estimate of recombination.

20.   **a. and b.** Again, the best way to determine whether there is linkage is through chi-square analysis, which indicates that it is highly unlikely that the three genes assort independently. To determine linkage by simple inspection, look at gene pairs. Because this is a testcross, independent assortment predicts a 1:1:1:1 ratio.

Comparing shrunken and white, the frequencies are

+	+	(113 + 4)/total
s	wh	(116 + 2)/total
+	wh	(2708 + 626)/total
s	+	(2538 + 601)/total

There is not independent assortment between shrunken and white, which means that there is linkage.

Comparing shrunken and waxy, the frequencies are

+	+	(626 + 4)/total
s	wa	(601 + 2)/total
+	wa	(2708 + 113)/total
s	+	(2538 + 1160/total

There is not independent assortment between shrunken and waxy, which means that there is linkage.
Comparing white and waxy, the frequencies are

+	+	(2538 + 4)/total
wh	wa	(2708 + 2)/total
wh	+	(626 + 116)/total
+	wa	(601 + 113)/total

There is not independent assortment between waxy and white, which means that there is linkage.

Because all three genes are linked, the strains must be $+ s +/wh + wa$ and $wh\ s\ wa/wh\ s\ wa$ (compare most frequent, parentals, to least frequent, double crossovers, to obtain the gene order). The cross can be written as

P $+ s +/wh + wa \times wh\ s\ wa/wh\ s\ wa$

F_1 as in problem

Crossovers between white and shrunken and shrunken and waxy are

113	601
116	626
4	4
2	2
235	1233

Dividing by the total number of progeny and multiplying by 100 percent yields the following map:

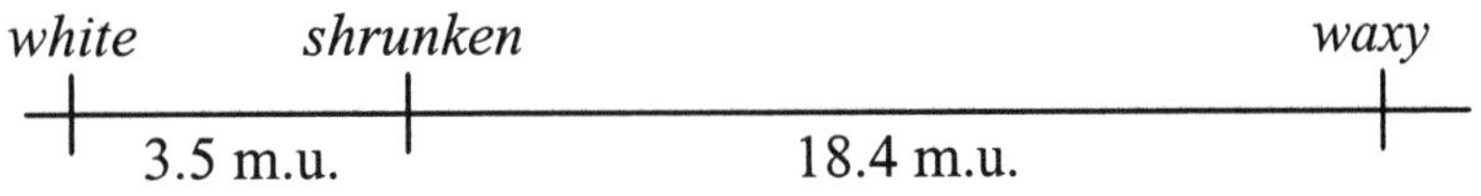

```
white        shrunken                              waxy
  +             +                                    +
      3.5 m.u.            18.4 m.u.
```

c.   Interference $= 1 - $ (observed double crossovers/expected double crossovers)
$= 1 - 6/(0.035)(0.184)(6{,}708) = 0.86$

21. **a.** The results of this cross indicate independent assortment of the two genes. This might be diagrammed as

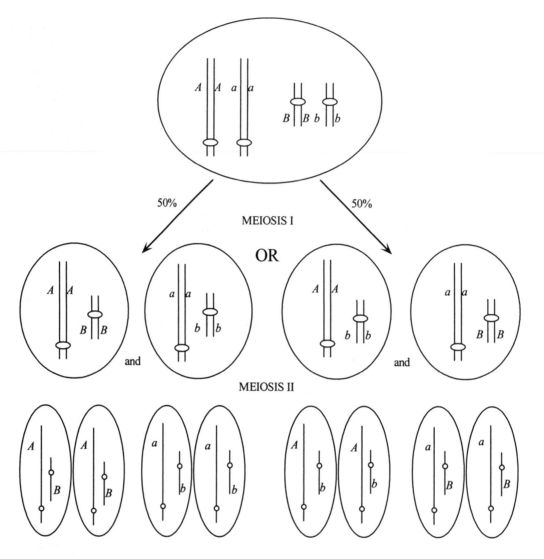

b. The results of this cross indicate that the two genes are linked and 10 m.u. apart. Further, the recessive alleles are in repulsion in the dihybrid (*C d/c D* × *c d/c d*). This might be diagrammed as

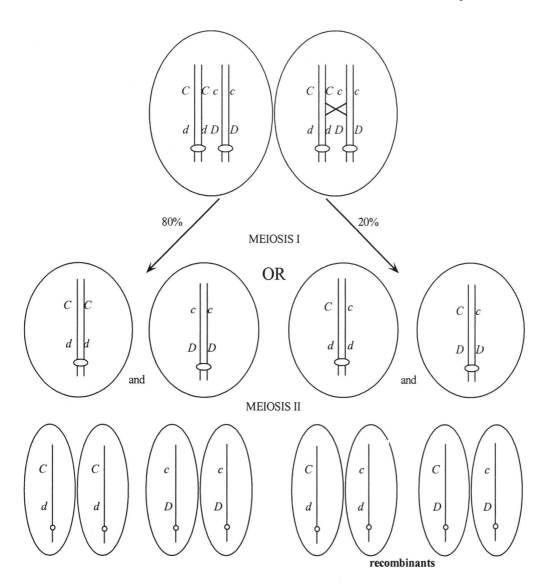

22.   **a.**   If the genes are unlinked, the cross becomes

P   *hyg/hyg* ; *her/her* × *hyg⁺/hyg⁺* ; *her⁺/her⁺*

F₁   *hyg⁺/hyg* ; *her⁺/her* × *hyg⁺/hyg* ; *her⁺/her*

F₂ $^9/_{16}$ *hyg⁺/–* ; *her⁺/–*
                     $^3/_{16}$ *hyg⁺/–* ; *her/her*
                     $^3/_{16}$ *hyg/hyg* ; *her⁺/–*
                     $^1/_{16}$ *hyg/hyg* ; *her/her*

So only $^1/_{16}$ (or 6.25 percent) of the seeds would be expected to germinate.

b. and c. No. More than twice the expected seeds germinated so assume the genes are linked. The cross then becomes

P $\quad$ *hyg her/hyg her* × *hyg⁺ her⁺/hyg⁺ her⁺*

F$_1$ $\quad$ *hyg⁺ her⁺/hyg her* × *hyg⁺ her⁺/hyg her*

F$_2$ $\quad$ 13 percent $\quad$ *hyg her/hyg her*

Because this class represents the combination of two parental chromosomes, it is equal to

$$p(\textit{hyg her}) \times p(\textit{hyg her}) = (^1/_2 \text{ parentals})^2 = 0.13$$

$\quad$ and

$$\text{parentals} = 0.72 \qquad \text{so recombinants} = 1 - 0.72 = 0.28$$

Therefore, a test cross of *hyg⁺ her⁺/hyg her* should give

$\qquad$ 36% $\quad$ *hyg⁺ her⁺/hyg her*
$\qquad$ 36% $\quad$ *hyg her/hyg her*
$\qquad$ 14% $\quad$ *hyg⁺ her/hyg her*
$\qquad$ 14% $\quad$ *hyg her⁺/hyg her*

and 36 percent of the progeny should grow (the *hyg her/hyg her* class).

23. Comparing independent assortment (RF = 0.5) to a recombination frequency of 34 percent (RF = 0.34)

RF	0.5	0.34
P	0.25	0.33
P	0.25	0.33
R	0.25	0.17
R	0.25	0.17

The probability of obtaining these results, assuming independent assortment, will be equal to

0.25 × 0.25 × 0.25 × 0.25 × 0.25 × 0.25 × B = 0.00024B (where B = number of possible birth orders for four parental and two recombinant progeny)

For an RF = 0.34, the probability of this result will be equal to

0.33 × 0.17 × 0.33 × 0.33 × 0.17 × 0.33 × B = 0.00034B

The ratio of the two = 0.00034B/0.00024B = 1.42 and the Lod = 0.15

24. **a.** Meiosis I (crossing-over has occurred between all genes and their centromeres)

b. Impossible

c. Meiosis I (crossing-over has occurred between gene B and its centromere)

d. Meiosis I

e. Meiosis II

f. Meiosis II (crossing-over has occurred between genes A and B and their centromeres)

g. Meiosis II

h. Impossible

i. Mitosis

j. Impossible

25. **a.**

al-2$^+$	*al-2*
al-2$^+$	*al-2*
al-2	*al-2*$^+$
al-2	*al-2*$^+$
al-2$^+$	*al-2*
al-2$^+$	*al-2*
al-2	*al-2*$^+$
al-2	*al-2*$^+$

b. The 8 percent value can be used to calculate the distance between the gene and the centromere. That distance is $1/2$ the percentage of second-division segregation, or 4 percent.

26. **a.** *arg-6 · al-2, arg-6 · al-2*$^+$*, arg-6*$^+$ *· al-2*$^+$*, arg-6*$^+$ *· al-2*$^+$

b. *arg-6*$^+$ *· al-2*$^+$*, arg-6*$^+$ *· al-2, arg-6 · al-2*$^+$*, arg-6 · al-2*

c. *arg-6*$^+$ *· al-2, arg-6*$^+$ *· al-2, arg-6 · al-2*$^+$*, arg-6 · al-2*$^+$

27. The formula for this problem is $f(i) = e^{-m}m^i/i!$ where $m = 2$ and $i = 0, 1,$ or 2.

a. $f(0) = e^{-2}2^0/0! = e^{-2} = 0.135$ or 13.5%

b. $f(1) = e^{-2}2^1/1! = e^{-2}(2) = 0.27$ or 27%

c. $f(2) = e^{-2}2^2/2! = e^{-2}(2) = 0.27$ or 27%

28.   ***Unpacking the Problem***

1.   Fungi are generally haploid.

2.   There are four pairs, or eight ascospores, in each ascus. One member of each pair is presented in the data.

3.   A mating type in fungi is analogous to sex in humans, in that the mating types of two organisms must differ in order to have a mating that produces progeny. Mating type is determined experimentally simply by seeing if progeny result from specific crosses.

4.   The mating types A and a do not indicate dominance and recessiveness. They simply symbolize the mating-type difference.

5.   *arg-1* indicates that the organism requires arginine for growth. Testing for the genotype involves isolating nutritional mutants and then seeing if arginine supplementation will allow for growth.

6.   *arg-1*$^+$ indicates that the organism is wild type and does not require supplemental arginine for growth.

7.   Wild type refers to the common form of an organism in its natural population.

8.   Mutant means that, for the trait being studied, an organism differs from the wild type.

9.   The actual function of the alleles in this problem does not matter in solving the problem.

10.   Linear tetrad analysis refers to the fact that the ascospores in each ascus are in a linear arrangement that reflects the order in which the two meiotic divisions occurred to produce them. By tracking traits and correlating them with position, it is possible to detect crossing-over events that occurred at the tetrad (four-strand, homologous pairing) stage prior to the two meiotic divisions.

11.   Linear tetrad analysis allows for the mapping of centromeres in relation to genes, which cannot be done with unordered tetrad analysis.

12.   A cross is made in *Neurospora* by placing the two organisms in the same test tube or Petri dish and allowing them to grow. Gametes develop and

fertilization, followed by meiosis, mitosis, and ascus formation, occurs. The asci are isolated, and the ascospores are dissected out of them with the aid of a microscope. The ascus has an octad, or eight spores, within it, and the spores are arranged in four (tetrad) pairs.

13. Meiosis occurs immediately following fertilization in *Neurospora*.

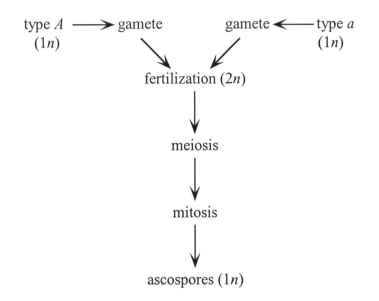

14. Meiosis produced the ascospores that were analyzed.

15. The cross is $A \cdot arg\text{-}1 \times a \cdot arg\text{-}1^+$.

16. Although there are eight ascospores, they occur in pairs. Each pair represents one chromatid of the originally paired chromosomes. By convention, both members of a pair are represented by a single genotype.

17. The seven classes represent the seven types of outcomes. The specific outcomes can be classified as follows:

Class	1	2	3	4	5	6	7
outcome	PD	NPD	T	T	PD	NPD	T
A/a	I	I	I	II	II	II	II
arg-1$^+$/arg-1	I	I	II	I	II	II	II

where PD = parental ditype, NPD = nonparental ditype, T = tetratype, I = first-division segregation, and II = second-division segregation.

Other classes can be detected, but they indicate the same underlying process. For example, the following three asci are equivalent.

1	2	3
A arg	*A arg*$^+$	*A arg*$^+$
A arg$^+$	*A arg*	*A arg*
a arg	*a arg*	*a arg*$^+$
a arg$^+$	*a arg*$^+$	*a arg*

In the first ascus, a crossover occurred between chromatids 2 and 3, while in the second ascus it occurred between chromatids 1 and 3, and in the third ascus the crossover was between chromatids 1 and 4. A fourth equivalent ascus would contain a crossover between chromatids 2 and 3. All four indicate a crossover between the second gene and its centromere and all are tetratypes.

18. This is exemplified in the answer to (17) above.

19. The class is identical with class 1 in the problem, but inverted.

20. Linkage arrangement refers to the relative positions of the two genes and the centromere along the length of the chromosome.

21. A genetic interval refers to the region between two loci, whose size is measured in map units.

22. It is not known whether the two loci are on separate chromosomes or are on the same chromosome. The general formula for calculating the distance of a locus to its centromere is to measure the percentage of tetrads that show second-division segregation patterns for that locus and divide by two.

23. Recall that there are eight ascospores per ascus. By inspection, the frequency of recombinant *A arg-1*$^+$ ascospores is $4(125) + 2(100) + 2(36) + 4(4) + 2(6) = 800$. There is also the reciprocal recombinant genotype *a arg-1*.

24. Class 1 is parental; class 2 is nonparental ditype. Because they occur at equal frequencies, the two genes are not linked.

Solution to the Problem

a. The cross is $A \cdot arg\text{-}1 \times a \cdot arg\text{-}1^+$. Use the classification of asci in part (17) above. First decide if the two genes are linked by using the formula PD>>NPD, when the genes are linked, while PD = NPD when they are not linked. PD = 127 + 2 = 129 and NPD = 125 + 4 = 129, which means that the two genes are not linked. Alternatively,

$$\text{RF} = 100\%(^1/_2\,\text{T} + \text{NPD})/\text{total asci}$$

$$= 100\%[(^1/_2)(100 + 36 + 6) + (125 + 4)]/400 = 50\%.$$

Next calculate the distance between each gene and its centromere using the formula RF = $100\%(^1/_2$ number of tetrads exhibiting M_{II} segregation)/(total number of asci).

$$A\text{–centromere} = 100\%(^1/_2)(36 + 2 + 4 + 6)/400$$

$$= 100\%(24/400) = 6 \text{ m.u.}$$

$$arg^+\text{–centromere} = 100\%(^1/_2)(100 + 2 + 4 + 6)/400$$

$$= 100\%(56/400) = 14 \text{ m.u.}$$

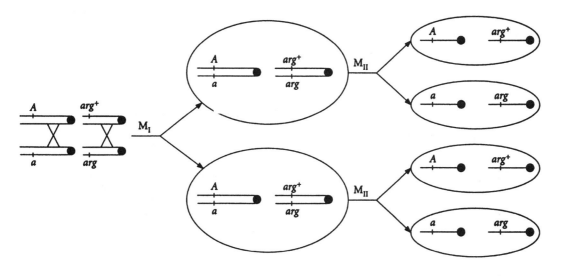

b.   Class 6 can be obtained if a single crossover occurred between chromatids 2 and 3 between each gene and its centromere.

29.   Before beginning this problem, classify all asci as PD, NPD, or T and determine whether there is M_I or M_{II} segregation for each gene:

	Asci type						
	1	2	3	4	5	6	7
type	PD	NPD	T	T	PD	NPD	T
gene a	I	I	I	II	II	II	II
gene b	I	I	II	I	II	II	II

If PD >> NPD, linkage is indicated. The distance between a gene and its centromere = 100% $(^1/_2)(M_{II})$/total. The distance between two genes = 100% $(^1/_2T + NPD)$/total.

Cross 1: PD = NPD and RF = 50%; the genes are not linked.
a–centromere: 100% $(^1/_2)(0)/100 = 0$ m.u. Gene a is close to the centromere.
b–centromere: 100% $(^1/_2)(32)/100 = 16$ m.u.

Cross 2: PD >> NPD; the genes are linked.
a–b: 100% $[^1/_2(15) + 1]/100 = 8.5$ m.u.
a–centromere: 100% $(^1/_2)(0)/100 = 0$ m.u. Gene a is close to the centromere.
b–centromere: 100% $(^1/_2)(15)/100 = 7.5$ m.u.

Cross 3: PD >> NPD; the genes are linked.
a–b: 100% $[^1/_2(40) + 3]/100 = 23$ m.u.
a–centromere: 100% $(^1/_2)(2)/100 = 1$ m.u.
b–centromere: 100% $(^1/_2)(40 + 2)/100 = 21$ m.u.

Cross 4: PD > > NPD; the genes are linked.
a–b: 100% $[^1/_2(20) + 1]/100 = 11$ m.u.
a–centromere: 100% $(^1/_2)(10)/100 = 5$ m.u.
b–centromere: 100% $(^1/_2)(18 + 8 + 1)/100 = 13.5$ m.u.

Cross 5: PD = NPD (and RF = 49%); the genes are not linked.
a–centromere: 100% $(^1/_2)(22 + 8 + 10 + 20)/99 = 30.3$ m.u.
b–centromere: 100% $(^1/_2)(24 + 8 + 10 + 20)/99 = 31.3$ m.u.
These values are approaching the 67 percent theoretical limit of loci exhibiting M_{II} patterns of segregation and should be considered cautiously.

Cross 6: PD >> NPD; the genes are linked.
a–b: 100% $[^1/_2(1 + 3 + 4) + 0]/100 = 4$ m.u.
a–centromere: 100% $(^1/_2)(3 + 61 + 4)/100 = 34$ m.u.
b–centromere: 100% $(^1/_2)(1 + 61 + 4)/100 = 33$ m.u.

These values are at the 67 percent theoretical limit of loci exhibiting M_{II} patterns of segregation and therefore both loci can be considered unlinked to the centromere.

Cross 7: PD >> NPD; the genes are linked.
a–b: 100% $[^1/_2(3 + 2) + 0]/100 = 2.5$ m.u.
a–centromere: 100% $(^1/_2)(2)/100 = 1$ m.u.
b–centromere: 100% $(^1/_2)(3)/100 = 1.5$ m.u.

Cross 8: PD = NPD; the genes are not linked.
a–centromere: 100% $(^1/_2)(22 + 12 + 11 + 22)/100 = 33.5$ m.u.
b–centromere: 100% $(^1/_2)(20 + 12 + 11 + 22)/100 = 32.5$ m.u.
Same as cross 5.

Cross 9: PD >> NPD; the genes are linked.
a–b: 100% $[^1/_2(10 + 18 + 2) + 1]/100 = 16$ m.u.
a–centromere: 100% $(^1/_2)(18 + 1 + 2)/100.= 10.5$ m.u.
b–centromere: 100% $(^1/_2)(10 + 1 + 2)/100 = 6.5$ m.u.

Cross 10: PD = NPD; the genes are not linked.
a–centromere: 100% $(^1/_2)(60 + 1 + 2 + 5)/100 = 34$ m.u.
b–centromere: 100% $(^1/_2)(2 + 1 + 2 + 5)/100 = 5$ m.u.

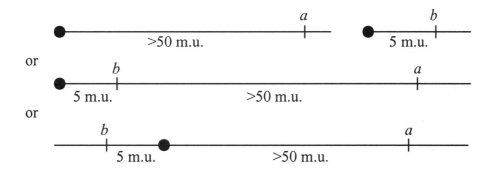

Cross 11: PD = NPD; the genes are not linked.
a–centromere: 100% $(^1/_2)(0)/100 = 0$ m.u.
b–centromere: 100% $(^1/_2)(0)/100 = 0$ m.u.

30. The number of recombinants is equal to NPD + $^1/_2$T. The uncorrected map distance is based on RF = (NPD + $^1/_2$T)/total. The corrected map distance = 50(T + 6NPD)/total.

Cross 1:
recombinant frequency $= 4\% + ^1/_2(45\%) = 26.5\%$
uncorrected map distance $= [4\% + ^1/_2(45\%)]/100\% = 26.5$ m.u.
corrected map distance $= 50[45\% + 6(4\%)]/100\% = 34.5$ m.u.

Cross 2:
recombinant frequency $= 2\% + ^1/_2(34\%) = 19\%$
uncorrected map distance $= [2\% + ^1/_2(34\%)]/100\% = 19$ m.u.
corrected map distance $= 50[34\% + 6(2\%)]/100\% = 29$ m.u.

Cross 3:
recombinant frequency $= 5\% + \frac{1}{2}(50\%) = 30\%$
uncorrected map distance $= [5\% + \frac{1}{2}(50\%)]/100\% = 30$ m.u.
corrected map distance $= 50[50\% + 6(5\%)]/100\% = 40$ m.u.

CHALLENGING PROBLEMS

31. a.   All of these genes are linked. To determine this, each gene pair is examined separately. For example, are A and B linked?

$A\ B = 140 + 305 = 445$
$a\ b = 145 + 310 = 455$
$a\ B = 42 + 6 = 48$
$A\ b = 43 + 9 = 52$

Conclusion: the two genes are linked and 10 m.u. apart.

Are A and D linked?

$A\ D = 0$
$a\ d = 0$
$A\ d = 43 + 140 + 9 + 305 = 497$
$a\ D = 42 + 145 + 6 + 310 = 503$

Conclusion: the two genes show no recombination and at this resolution, are 0 m.u. apart.

Are B and C linked?

$B\ C = 42 + 140 = 182$
$b\ c = 43 + 145 = 188$
$B\ c = 6 + 305 = 311$
$b\ C = 9 + 310 = 319$

Conclusion: the two genes are linked and 37 m.u. apart.

Are C and D linked?

$C\ D = 42 + 310 = 350$
$c\ d = 43 + 305 = 348$
$C\ d = 140 + 9 = 149$
$c\ D = 145 + 6 = 151$

Conclusion: the two genes are linked and 30 m.u. apart. Therefore, all four genes are linked.

b. and c. Because *A* and *D* show no recombination, first rewrite the progeny omitting *D* and *d* (or omitting *A* and *a*).

a B C	42
A b c	43
A B C	140
a b c	145
a B c	6
A b C	9
A B c	305
a b C	310
	1000

Note that the progeny now look like those of a typical three point testcross, with *A B c* and *a b C* the parental types (most frequent) and *a B c* and *A b C* the double recombinants (least frequent). The gene order is *B A C*. This is determined either by the map distances or by comparing double recombinants with the parentals; the gene that switches in reference with the other two is the gene in the center (*B A c* → *B a c*, *b a C* → *b A C*).

Next, rewrite the progeny again, this time putting the genes in the proper order, and classify the progeny.

B a C	42	CO *A–B*
b A c	43	CO *A–B*
B A C	140	CO *A–C*
b a c	145	CO *A–C*
B a c	6	DCO
b A C	9	DCO
B A c	305	parental
b a C	310	parental

To construct the map of these genes, use the following formula:

$$\text{distance between two genes} = \frac{(100\%)\,(\text{number of single CO} + \text{number of DCO})}{\text{total number of progeny}}$$

For the *A* to *B* distance

$$= \frac{(100\%)(42 + 43 + 6 + 9)}{1000} = 10 \text{ m.u.}$$

For the *A* to *C* distance

$$= \frac{(100\%)(140 + 145 + 6 + 9)}{1000} = 30 \text{ m.u.}$$

The map is

The parental chromosomes actually were *B (A,d) c/b (a,D) C*, where the parentheses indicate that the order of the genes within is unknown.

d.   Interference = 1 – (observed DCO/expected DCO)

$$= 1 - (6 + 9)/[(0.10)(0.30)(1000)]$$
$$= 1 - 15/30 = 0.5$$

32.   The verbal description indicates the following cross and result:

P *N/– · A/– × n/n · O/O*

F$_1$ *N/n · A/O × N/n · A/O*

The results indicate linkage, so the cross and results can be rewritten:
P *N A/– – × n O/n O*

F$_1$ *N A/ n O × N A/n O*

F$_2$ 66%   *N A/– – or N –/– A*
            16%   *n O/n O*
             9%   *n A/n –*
             9%   *N O/– O*

Only one genotype is fully known: 16 percent *n O/n O*, a combination of two parental gametes. The frequency of two parental gametes coming together is the frequency of the first times the frequency of the second. Therefore, the frequency of each *n O* gamete is the square root of 0.16, or 0.4. Within an organism the two parental gametes occur in equal frequency. Therefore, the frequency of *N A* is also 0.4. The parental total is 0.8, leaving 0.2 for all recombinants. Therefore, *N O* and *n A* occur at a frequency of 0.1 each. The two genes are 20 m.u. apart.

33.   **a. and b.** The data indicate that the progeny males have a different phenotype than the females. Therefore, all the genes are on the X chromosome. The two most frequent phenotypes in the males indicate the genotypes of the X chromosomes in the female, and the two least frequent phenotypes in the males indicate the gene order. Data from only the males are used to determine map distances. The cross is:

P $x\,z\,y^+/x^+\,z^+\,y \times x^+\,z^+\,y^+/Y$

F_1 males

430	$x\,z\,y^+/Y$	parental
441	$x^+\,z^+\,y/Y$	parental
39	$x\,z\,y/Y\ CO\ z\text{--}y$	
30	$x^+\,z^+\,y^+/Y$	CO $z\text{--}y$
32	$x^+\,z\,y^+/Y$	CO $x\text{--}z$
27	$x\,z^+\,y/Y$	CO $x\text{--}z$
1	$x^+\,z\,y/Y$	DCO
0	$x\,z^+\,y^+/Y$	DCO

c. $z\text{--}y$: $100\%(39 + 30 + 1)/1000 = 7.0$ m.u.
   $x\text{--}z$: $100\%(32 + 27 + 1)/1000 = 6.0$ m.u.

c.c.   = observed DCO/expected DCO
   $= 1/[(0.06)(0.07)(1000)] = 0.238$

34. The data given for each of the three-point testcrosses can be used to determine the gene order by realizing that the rarest recombinant classes are the result of double cross-over events. By comparing these chromosomes to the "parental" types, the alleles that have switched represent the gene in the middle.

For example, in (1), the most common phenotypes (+ + + and $a\ b\ c$) represent the parental allele combinations. Comparing these to the rarest phenotypes of this data set (+ $b\ c$ and a + +) indicates that the a gene is recombinant and must be in the middle. The gene order is $b\ a\ c$.

For (2), + $b\ c$ and a + + (the parentals) should be compared to + + + and $a\ b\ c$ (the rarest recombinants) to indicate that the a gene is in the middle. The gene order is $b\ a\ c$.

For (3), compare + b + and a + c with $a\ b$ + and + + c, which gives the gene order $b\ a\ c$.

For (4), compare + + c and $a\ b$ + with + + + and $a\ b\ c$, which gives the gene order $a\ c\ b$.

For (5), compare + + + and $a\ b\ c$ with + + c and $a\ b$ +, which gives the gene order $a\ c\ b$.

35. The gene order is $a\ c\ b\ d$.

Recombination between a and c occurred at a frequency of

$100\%(139 + 3 + 121 + 2)/(669 + 139 + 3 + 121 + 2 + 2{,}280 + 653 + 2{,}215)$
$= 100\%(265/6{,}082) = 4.36\%$

Recombination between b and c in cross 1 occurred at a frequency of

$100\%(669 + 3 + 2 + 653)/(669 + 139 + 3 + 121 + 2 + 2{,}280 + 653 + 2{,}215)$
$= 100\%(1{,}327/6{,}082) = 21.82\%$

Recombination between b and c in cross 2 occurred at a frequency of
$100\%(8 + 14 + 153 + 141)/(8 + 441 + 90 + 376 + 14 + 153 + 64 + 141)$
$= 100\%(316/1{,}287) = 24.55\%$

The difference between the two calculated distances between b and c is not surprising because each set of data would not be expected to yield exactly identical results. Also, many more offspring were analyzed in cross 1. Combined, the distance would be

$100\%[(316 + 1{,}327)/(1{,}287 + 6{,}082)] = 22.3\%$

Recombination between b and d occurred at a frequency of
$100\%(8 + 90 + 14 + 64)/(8 + 441 + 90 + 376 + 14 + 153 + 64 + 141)$
$= 100\%(176/1{,}287) = 13.68\%$

The general map is:

36. Part (a) of this problem is solved two ways, once in the standard way, once in a way that emphasizes a more mathematical approach.

The cross is

P $P\,A\,R/P\,A\,R \times p\,a\,r/p\,a\,r$

F$_1$ $P\,A\,R/p\,a\,r \times p\,a\,r/p\,a\,r$, a three-point testcross

a.   In order to find what proportion will have the Vulcan phenotype for all three characteristics, we must determine the frequency of parentals. Crossing-over occurs 15 percent of the time between P and A, which means it does not occur 85 percent of the time. Crossing-over occurs 20 percent of the time between A and R, which means that it does not occur 80 percent of the time.

p (no crossover between either gene)

= p(no crossover between P and A) × p(no crossover between A and R)
= (0.85)(0.80) = 0.68

Half the parentals are Vulcan, so the proportion that are completely Vulcan is $^1/_2$(0.68) = 0.34

Mathematical method

Number of parentals = 1 – (single CO individuals – DCO individuals)
= 1 – {[0.15 + 0.20 – 2(0.15)(0.20)] – (0.15) (0.20)} = 0.68

Because half the parentals are Earth alleles and half are Vulcan, the frequency of children with all three Vulcan characteristics is $^1/_2$(0.68) = 0.34

b. Same as above, 0.34

c. To yield Vulcan ears and hearts and Earth adrenals, a crossover must occur in both regions, producing double crossovers. The frequency of Vulcan ears and hearts and Earth adrenals will be half the DCOs, or $^1/_2$(0.15) (0.20) = 0.015

d. To yield Vulcan ears and an Earth heart and adrenals, a single crossover must occur between P and A, and no crossover can occur between A and R. The frequency will be

p(CO P–A) × p(no CO A–R) = (0.15)(0.80) = 0.12

Of these, $^1/_2$ are P a r and $^1/_2$ are p A R. Therefore, the proportion with Vulcan cars and an Earth heart and adrenals is 0.06

37. a. To obtain a plant that is a b c/a b c from selfing of $A.$ b c/a B C, both gametes must be derived from a crossover between A and B. The frequency of the a b c gamete is

$^1/_2$ p(CO A–B) × p(no CO B–C) = $^1/_2$(0.20)(0.70) = 0.07

Therefore, the frequency of the homozygous plant will be $(0.07)^2$ = 0.0049

b. The cross is A b c/a B C × a b c/a b c.

To calculate the progeny frequencies, note that the parentals are equal to all those that did not experience a crossover. Mathematically this can be stated as

parentals   = p(no CO A–B) × p(no CO B–C)

$$= (0.80)(0.70) = 0.56$$

Because each parental should be represented equally

$A\ b\ c = {}^1/_2(0.56) = 0.28$
$a\ B\ C = {}^1/_2(0.56) = 0.28$

As calculated above, the frequency of the $a\ b\ c$ gamete is

$$^1/_2\ p(\text{CO } A–B) \times p(\text{no CO } B–C) = {}^1/_2(0.20)(0.70) = 0.07$$

as is the frequency of $A\ B\ C$.

The frequency of the $A\ b\ C$ gamete is

$$^1/_2\ p(\text{CO } B–C) \times p(\text{no CO } A–B) = {}^1/_2(0.30)(0.80) = 0.12$$

as is the frequency of $a\ B\ c$.

Finally, the frequency of the $A\ B\ c$ gamete is

$$^1/_2\ p(\text{CO } A–B) \times p(\text{CO } B–C) = {}^1/_2(0.20)(0.30) = 0.03$$

as is the frequency of $a\ b\ C$.

So for 1,000 progeny, the expected results are

$A\ b\ c$	280
$a\ B\ C$	280
$A\ B\ C$	70
$a\ b\ c$	70
$A\ b\ C$	120
$a\ B\ c$	120
$A\ B\ c$	30
$a\ b\ C$	30

c. Interference = 1 – observed DCO/expected DCO
0.2 = 1 – observed DCO/(0.20)(0.30)

observed DCO = $(0.20)(0.30) – (0.20)(0.20)(0.30) = 0.048$

The $A–B$ distance = 20% = 100% $[p(\text{CO } A–B) + p(\text{DCO})]$.
Therefore, $p(\text{CO } A–B) = 0.20 – 0.048 = 0.152$

Similarly, the $B–C$ distance = 30% = 100% $[p(\text{CO } B–C) + p(\text{DCO})]$
Therefore, $p(\text{CO } B–C) = 0.30 – 0.048 = 0.252$

The p(parental) $= 1 - p$(CO A–B) $- p$(CO B–C) $- p$(observed DCO)
$$= 1 - 0.152 - 0.252 - 0.048 = 0.548$$

So for 1,000 progeny, the expected results are

$A\ b\ c$	274
$a\ B\ C$	274
$A\ B\ C$	76
$a\ b\ c$	76
$A\ b\ C$	126
$a\ B\ c$	126
$A\ B\ c$	24
$a\ b\ C$	24

38. **a.** Blue sclerotic (B) appears to be an autosomal dominant disorder. Hemophilia (h) appears to be an X-linked recessive disorder.

b. If the individuals in the pedigree are numbered as generations I through IV and the individuals in each generation are numbered clockwise, starting from the top right-hand portion of the pedigree, their genotypes are

I: b/b ; H/h, B/b ; H/Y

II: B/b ; H/Y, B/b ; H/Y, b/b ; H/Y, B/b ; H/h, b/b ; H/Y, B/b ; H/h, B/b ; H/h, B/b ; $H/-$, b/b ; $H/-$

III: b/b ; $H/-$, B/b ; $H/-$, b/b ; h/Y, b/b ; H/Y, B/b ; H/Y, B/b ; $H/-$, B/b ; H/Y, B/b ; h/Y, B/b ; $H/-$, b/b ; H/Y, B/b ; $H/-$, b/b ; H/Y, B/b ; $H/-$, B/b ; H/Y, B/b ; h/Y, b/b ; H/Y, b/b ; H/Y, b/b ; $H/-$, b/b ; H/Y, b/b ; H/Y, B/b ; $H/-$, B/b ; H/Y, B/b ;h/Y

IV: b/b ; $H/-$, B/b ; $H/-$, B/b ; $H/-$, b/b ; H/h, b/b ; H/h, b/b ; H/Y, b/b ; H/H, b/b ; H/Y, b/b ; H/h, b/b ; H/H, b/b ; H/H, b/b ; H/Y, b/b ; H/Y, b/b ; H/H, b/b ; H/Y, b/b ; H/Y, B/b ; H/Y, b/b ; H/Y, b/b ; $H/-$, b/b ; H/Y, b/b ; H/Y, b/b ; $H/-$, b/b ; H/H, b/b ; $H/-$, b/b ; $H/-$, b/b ; H/Y, b/b ; H/Y, b/b ; H/Y, b/b ; H/h, B/b ; H/h, B/b ; H/Y, b/b ; H/Y, B/b ; H/Y, b/b ; H/h

c. There is no evidence of linkage between these two disorders. Because of the modes of inheritance for these two genes, no linkage would be expected.

d. The two genes exhibit independent assortment.

e. No individual could be considered intrachromosomally recombinant. However, a number show interchromosomal recombination, for example, all individuals in generation III that have both disorders.

39. If h = hemophilia and b = colorblindness, the genotypes for individuals in the pedigree can be written as

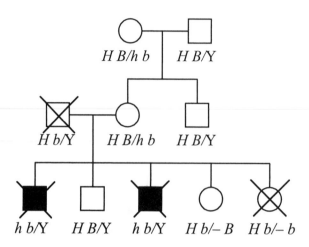

The mother of the two women in question would produce the following gametes:

0.45	$H\,B$
0.45	$h\,b$
0.05	$h\,B$
0.05	$H\,b$

Woman III-4 can be either $H\,b/H\,B$ (0.45 chance) or $H\,b/h\,B$ (0.05 chance), because she received B from her mother. If she is $H\,b/h\,B$ [0.05/(0.45 + 0.05) = 0.10 chance], she will produce the parental and recombinant gametes with the same probabilities as her mother. Thus, her child has a 45 percent chance of receiving $h\,B$, a 5 percent chance of receiving $h\,b$, and a 50 percent chance of receiving a Y from his father. The probability that her child will be a hemophiliac son is (0.1)(0.5)(0.5) = 0.025 = 2.5 percent.

Woman III-5 can be either $H\,b/H\,b$ (0.05 chance) or $H\,b/h\,b$ (0.45 chance), because she received b from her mother. If she is $H\,b/h\,b$ [0.45/(0.45 + 0.05) = 0.90 chance], she has a 50 percent chance of passing h to her child, and there is a 50 percent chance that the child will be male. The probability that she will have a son with hemophilia is (0.9)(0.5)(0.5) = 0.225 = 22.5 percent.

40.   a.   Cross 1 reduces to

P $A/A \cdot B/B \cdot D/D \times a/a \cdot b/b \cdot d/d$

F_1 $A/a \cdot B/b \cdot D/d \times a/a \cdot b/b \cdot d/d$

The test cross progeny indicate these three genes are linked.

Test cross	A B D	316	parental
progeny	a b d	314	parental
	A B d	31	CO B–D
	a b D	39	CO B–D
	A b d	130	CO A–B
	a B D	140	CO A–B
	A b D	17	DCO
	a B d	13	DCO

A–B: 100%(130 + 140 + 17 + 13)/1000 = 30 m.u.
B–D: 100%(31 + 39 + 17 + 13)/1000 = 10 m.u.

Cross 2 reduces to

P   A/A · C/C · E/E × a/a · c/c · e/e

F_1 A/a · C/c · E/e × a/a · c/c · e/e

The test cross progeny indicate these three genes are linked.

Test cross	A C E	243	parental
progeny	a c e	237	parental
	A c e	62	CO A–C
	a C E	58	CO A–C
	A C e	155	CO C–E
	a c E	165	CO C–E
	a C e	46	DCO
	A c E	34	DCO

A–C: 100% (62 + 58 + 46 + 34)/1000 = 20 m.u.
C–E: 100% (155 + 165 + 46 + 34)/1000 = 40 m.u.

The map that accommodates all the data is

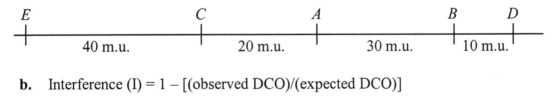

b.   Interference (I) = 1 – [(observed DCO)/(expected DCO)]

For cross 1: I = 1 – {30/[(0.30)(0.10)(1000)]} = 1 – 1 = 0, no interference

For cross 2: I = 1 – {80/[(0.20)(0.40)(1000)]} = 1 – 1 = 0, no interference

41.   **a.**   The first F_1 is *L H/l h* and the second is *l H/L h*. For progeny that are *l h/l h*, they have received a "parental" chromosome from the first F_1 and a "recombinant" chromosome from the second F_1. The genes are 16 percent apart so the chance of a parental chromosome is $^1/_2(100 - 16\%) = 42\%$ and the chance of a recombinant chromosome is $^1/_2(16\%) = 8\%$.

The chance of both events $= 42\% \times 8\% = 3.36\%$

b. To obtain $Lh/l\ h$ progeny, either a parental chromosome from each parent was inherited *or* a recombinant chromosome from each parent was inherited. The total probability will therefore be

$(42\% \times 42\%) + (8\% \times 8\%) = (17.6\% + 0.6\%) = 18.2\%$

42. Crossing-over occurs 8 percent of the time between w and s, which means it does not occur 92 percent of the time. Crossing-over occurs 14 percent of the time between s and e, which means that it does not occur 86 percent of the time.

a. and b. The frequency of parentals $= p$ (no crossover between either gene)

$$= p(\text{no CO } w\text{–}s) \times p(\text{no CO } s\text{–}e) = (0.92)(0.86)$$
$$= 0.791$$

or

$$^1/_2(0.791) = 0.396 \text{ each}$$

c. and d. The frequency that will show recombination between w and s only

$$= p(\text{CO } w\text{–}s) \times p(\text{no CO } s\text{–}e) = (0.08)(0.86) = 0.069$$

or

$$^1/_2(0.069) = 0.035 \text{ each}$$

e. and f. The frequency that will show recombination between s and e only

$$= p(\text{CO } s\text{–}e) \times p(\text{no CO } w\text{–}s) = (0.14)(0.92) = 0.128$$

or

$$^1/_2(0.128) = 0.064 \text{ each}$$

g. and h. The frequency that will show recombination between w and s and s and e

$$= p(\text{CO } w\text{–}s) \times p(\text{CO } s\text{–}e) = (0.08)(0.14) = 0.011$$

or

$$^1/_2(0.011) = 0.006 \text{ each}$$

43. This problem is analogous to meiosis in organisms that form linear tetrads. Let red $= R$ and blue $= r$. This can now be compared to meiosis in an organism that is R/r. The patterns, their frequencies, and the division of segregation are given below. Notice that the probabilities change as each ball/allele is selected. This occurs when there is sampling without replacement.

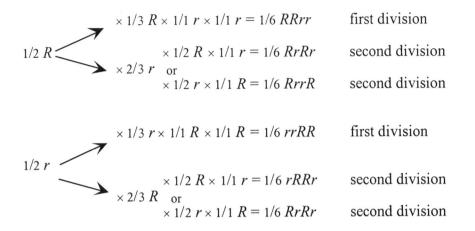

$\times 1/3\ R \times 1/1\ r \times 1/1\ r = 1/6\ RRrr$ first division

$1/2\ R$

$\times 1/2\ R \times 1/1\ r = 1/6\ RrRr$ second division
$\times 2/3\ r$ or
$\times 1/2\ r \times 1/1\ R = 1/6\ RrrR$ second division

$\times 1/3\ r \times 1/1\ R \times 1/1\ R = 1/6\ rrRR$ first division

$1/2\ r$

$\times 1/2\ R \times 1/1\ r = 1/6\ rRRr$ second division
$\times 2/3\ R$ or
$\times 1/2\ r \times 1/1\ R = 1/6\ RrRr$ second division

These results indicate one-third first-division segregation and two-thirds second-division segregation.

44. As the problem suggests, calculate the frequencies of the various possibilities. The percentage of tetrads without crossing over is 88% × 80% = 70.4%. The percentage of tetrads with a single crossover in region (i) and none in region (ii) is 12% × 80% = 9.6%. The percentage of tetrads with a single crossover in region (ii) and none in region (i) is 20% × 88% = 17.6% and the percentage of tetrads with crossovers in both regions is 12% × 20% = 2.4%.

Now work out the patterns of segregation that result in each case

For no crossovers

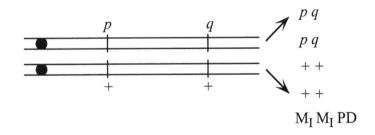

$p\ q$
$p\ q$
$+\ +$
$+\ +$

$M_I\ M_I\ PD$

For a single crossover in region (i)

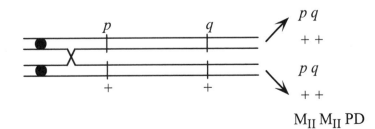

$p\ q$
$+\ +$
$p\ q$
$+\ +$

$M_{II}\ M_{II}\ PD$

For a single crossover in region (ii)

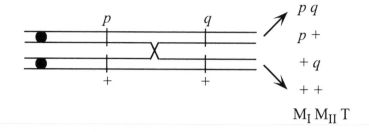

$p\ q$
$p\ +$
$+\ q$
$+\ +$

$M_I\ M_{II}\ T$

For double crossovers, there are four types, all equally likely

two-strand

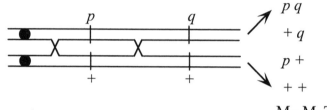

$p\ q$
$+\ q$
$p\ +$
$+\ +$

$M_{II}\ M_I\ T$

four-strand

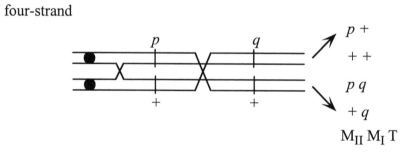

$p\ +$
$+\ +$
$p\ q$
$+\ q$

$M_{II}\ M_I\ T$

and two different three-strand

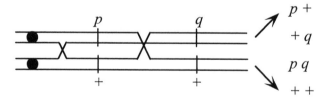

$p\ +$
$+\ q$
$p\ q$
$+\ +$

$M_{II}\ M_{II}\ T$

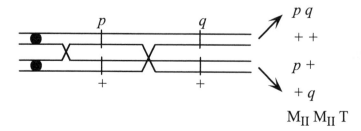

$p\ q$
$+\ +$
$p\ +$
$+\ q$

$M_{II}\ M_{II}\ T$

(a) M_I M_I PD is the result of no crossovers = 70.4%

(b) M_I M_I NPD is not found as a result

(c) M_I M_{II} T is the result of a single crossover in region (ii) = 17.6%

(d) M_{II} M_I T is the result of the two- and four-strand double crossovers = 1.2%

(e) M_{II} M_{II} PD is the result of a single crossover in region (i) = 9.6%

(f) M_{II} M_{II} NPD is not found as result

(g) M_{II} M_{II} T is the result of both three-strand double crossovers = 1.2%

45. a. and b. The data support the independent assortment of two genes (call them *arg1* and *arg2*). The cross becomes *arg1* ; *arg2+* × *arg1+* ; *arg2* and the resulting tetrads are:

4 : 0 (PD)	3 : 1 (T)	2 : 2 (NPD)
arg1 ; *arg2+*	*arg1* ; *arg2+*	*arg1* ; *arg2*
arg1 ; *arg2+*	*arg1+* ; *arg2*	*arg1* ; *arg2*
arg1+ ; *arg2*	*arg1* ; *arg2*	*arg1+* ; *arg2+*
arg1+ ; *arg2*	*arg1+* ; *arg2+*	*arg1+* ; *arg2+*

Because PD = NPD, the genes are unlinked.

5

The Genetics of Bacteria
and Their Viruses

BASIC PROBLEMS

1. An Hfr strain has the fertility factor F integrated into the chromosome. An F^+ strain has the fertility factor free in the cytoplasm. An F^- strain lacks the fertility factor.

2. All cultures of F^+ strains have a small proportion of cells in which the F factor is integrated into the bacterial chromosome and are, by definition, Hfr cells. These Hfr cells transfer markers from the host chromosome to a recipient during conjugation.

3. **a.** Hfr cells involved in conjugation transfer host genes in a linear fashion. The genes transferred depend on both the Hfr strain and the length of time during which the transfer occurred. Therefore, a population containing several different Hfr strains will appear to have an almost random transfer of host genes. This is similar to generalized transduction, in which the viral protein coat forms around a specific amount of DNA rather than specific genes. In generalized transduction, any gene can be transferred.

   b. F′ factors arise from improper excision of an Hfr from the bacterial chromosome. They can have only specific bacterial genes on them because the integration site is fixed for each strain. Specialized transduction resembles this in that the viral particle integrates into a specific region of the bacterial chromosome and then, upon improper excision, can take with it only specific bacterial genes. In both cases, the transferred gene exists as a second copy.

4. Generalized transduction occurs with lytic phages that enter a bacterial cell, fragment the bacterial chromosome, and then, while new viral particles are being assembled, improperly incorporate some bacterial DNA within the viral protein coat. Because the amount of DNA, not the information content of the

DNA, is what governs viral particle formation, any bacterial gene can be included within the newly formed virus. In contrast, specialized transduction occurs with improper excision of viral DNA from the host chromosome in lysogenic phages. Because the integration site is fixed, only those bacterial genes very close to the integration site will be included in a newly formed virus.

5. While the interrupted-mating experiments will yield the gene order, it will be relative only to fairly distant markers. Thus, the precise location cannot be pinpointed with this technique. Generalized transduction will yield information with regard to very close markers, which makes it a poor choice for the initial experiments because of the massive amount of screening that would have to be done. Together, the two techniques allow, first, for a localization of the mutant (interrupted-mating) and, second, for precise determination of the location of the mutant (generalized transduction) within the general region.

6. This problem is analogous to forming long gene maps with a series of three-point testcrosses. Arrange the four sequences so that their regions of overlap are aligned:

$$\overline{M-Z}-X-W-C$$
$$W-C-N-A-L$$
$$A-L-B-R-U$$
$$B-R-U-\underline{M-Z}$$

The regions with the bars above or below are identical in sequence (and "close" the circular chromosome). The correct order of markers on this circular map is

$$-M-Z-X-W-C-N-A-L-B-R-U-$$

7. First, carry out a series of crosses in which you select in a long mating each of the auxotrophic markers. Thus, select for arg^+ T^r. In each case score for penicillin resistance. Although not too informative, these crosses will give the marker that is closest to pen^r by showing which marker has the highest linkage. Then, do a second cross concentrating on the two markers on either side of the pen^r locus. Suppose that the markers are ala and glu. You can first verify the order by taking the cross in which you selected for ala^+, the first entering marker, and scoring the percentage of both pen^r and glu^+. Because of the gradient of transfer, the percentage of pen^r should be higher than the percentage of glu^+ among the selected ala^+ recombinants.

Then, take the mating in which glu^+ was the selected marker. Because this marker enters last, one can use the cross data to determine the map units by determining the percentage of colonies that are ala^+ pen^r, and by the number of ala^- pen^r colonies.

8. **a.** Determine the gene order by comparing *arg⁺ bio⁺ leu⁻* with *arg⁺ bio⁻ leu⁺*. If the order were *arg leu bio*, four crossovers would be required to get *arg⁺ leu⁻ bio⁺*, while only two would be required to get *arg⁺ leu⁺ bio⁻*. If the order is *arg bio leu*, four crossovers would be required to get *arg⁺ bio⁻ leu⁺*, and only two would be required to get *arg⁺ bio⁺ leu⁻*. There are eight recombinants that are *arg⁺ bio⁺ leu⁻* and none that are *arg⁺ bio⁻ leu⁺*. On the basis of the frequencies of these two classes, the gene order is *arg bio leu*.

b. The *arg-bio* distance is determined by calculating the percentage of the exconjugants that are *arg⁺ bio⁻ leu⁻*. These cells would have had a crossing-over event between the *arg* and *bio* genes.
RF = 100%(48)/376 = 12.76 m.u.

Similarly, the *bio-leu* distance is estimated by the *arg⁺ bio⁺ leu⁻* colony type.
RF = 100%(8)/376 = 2.12 m.u.

9. The most straightforward way would be to pick two Hfr strains that are near the genes in question but are oriented in opposite directions. Then, measure the time of transfer between two specific genes, in one case when they are transferred early and in the other when they are transferred late. For example,

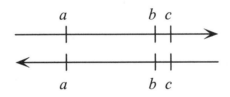

10. The best explanation is that the integrated F factor of the Hfr looped out of the bacterial chromosome abnormally and is now an F′ that contains the *pro⁺* gene. This F′ is rapidly transferred to F⁻ cells, converting them to *pro⁺* (and F⁺).

11. The high rate of integration and the preference for the same site originally occupied by the F factor suggest that the F′ contains some homology with the original site. The source of homology could be a fragment of the F factor, or more likely, it is homology with the chromosomal copy of the bacterial gene that is also present on the F′.

12. First, carry out a cross between the Hfr and F⁻, and then select for colonies that are *ala⁺ str^r*. If the Hfr donates the *ala* region late, then redo the cross but now interrupt the mating early and select for *ala⁺*. This selects for an F′, because this Hfr would not have transferred the *ala* gene early.

If the Hfr instead donates this region early, then use a Rec⁻ strain that cannot incorporate a fragment of the donor chromosome by recombination. Any *ala⁺*

colonies from the cross should then be used in a second mating to another *ala⁻* strain to see whether they can donate the *ala* gene easily, which would indicate that there is F′ *ala*. (This would also require another marker to differentiate the donor and recipient strains. For example, the *ala⁻* strain could be tetracycline^r and selection would be for *ala⁺ tet^r*.)

13.   **a. and b.**

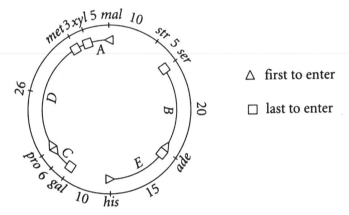

c.   A: Select for *mal⁺*
    B: Select for *ade⁺*
    C: Select for *pro⁺*
    D: Select for *pro⁺*
    E: Select for *his⁺*

14.   **a.**   If the two genes are far enough apart to be located on separate DNA fragments, then the frequency of double transformants should be the product of the frequency of the two single transformants, or (4.3%) × (0.40%) = 0.017%. The observed double transformant frequency is 0.17 percent, a factor of 10 greater than expected. Therefore, the two genes are located close enough together to be cotransformed at a rate of 0.17 percent.

b.   Here, when the two genes must be contained on separate pieces of DNA, the rate of cotransformation is much lower, confirming the conclusion in part (a).

15.   The expected number of double recombinants is (0.01)(0.002)(100,000) = 2. Interference = 1 − (observed DCO/expected DCO) = 1 − 5/2 = −1.5. By definition, the interference is negative.

16.   **a.**   The parental genotypes are + + + and *m r tu*. For determining the *m–r* distance, the recombinant progeny are

m + tu	162
m + +	520
+ r tu	474
+ r +	172
	1328

Therefore, the map distance is 100%(1328)/10,342 = 12.8 m.u.

Using the same approach, the *r–tu* distance is 100%(2152)/10,342 = 20.8 m.u., and the *m–tu* distance is 100%(2812)/10,342 = 27.2 m.u.

b. Because genes *m* and *tu* are the farthest apart, the gene order must be *m r tu*.

c. The coefficient of coincidence (c.o.c.) compares the actual number of double crossovers to the expected number, (where c.o.c. = observed double crossovers/expected double crossovers). For these data, the expected number of double recombinants is (0.128)(0.208)(10,342) = 275. Thus, c.o.c. = (162 + 172)/275 = 1.2. This indicates that there are more double crossover events than predicted and suggests that the occurrence of one crossover makes a second crossover between the same DNA molecules more likely to occur.

17. a. I: minimal plus proline and histidine
II: minimal plus purines and histidine
III: minimal plus purines and proline

b. The order can be deduced from cotransfer rates. It is *pur–his–pro*.

c. The closer the two genes, the higher the rate of cotransfer. *his* and *pro* are closest

d. *pro*⁺ transduction requires a crossover on both sides of the *pro* gene. Because *his* is closer to *pro* than *pur*, you get the following:

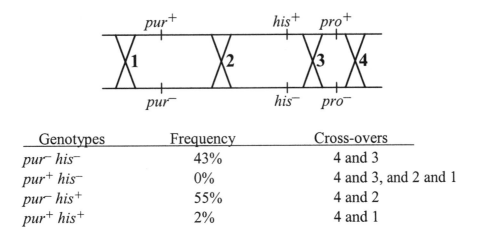

Genotypes	Frequency	Cross-overs
pur⁻ his⁻	43%	4 and 3
pur⁺ his⁻	0%	4 and 3, and 2 and 1
pur⁻ his⁺	55%	4 and 2
pur⁺ his⁺	2%	4 and 1

As can be seen, a *pur⁺ his⁻ pro⁺* genotype requires four crossovers and as expected, would occur less frequently (in this example, 0%).

18.   In a small percent of the cases, *gal⁺* transductants can arise by recombination between the *gal⁺* DNA of the λdgal transducing phage and the *gal⁻* gene on the chromosome. This will generate *gal⁺* transductants without phage integration.

19.   **a.**   This appears to be specialized transduction. It is characterized by the transduction of specific markers based on the position of the integration of the prophage. Only those genes near the integration site are possible candidates for misincorporation into phage particles that then deliver this DNA to recipient bacteria.

   b.   The only media that supported colony growth were those lacking either cysteine or leucine. These selected for *cys⁺* or *leu⁺* transductants and indicate that the prophage is located in the *cys-leu* region.

20.   **a.**   This is simply calculated as the percentage of *pur⁺* colonies that are also *nad⁺*

$$= 100\%(3 + 10)/50 = 26\%$$

   b.   This is calculated as the percentage of *pur⁺* colonies that are also *pdx⁻*

$$= 100\%(10 + 13)/50 = 46\%$$

   c.   *pdx* is closer, as determined by cotransduction rates.

   d.   From the cotransduction frequencies, you know that *pdx* is closer to *pur* than *nad* is, so there are two gene orders possible: *pur pdx nad* or *pdx pur nad*. Now, consider how a bacterial chromosome that is *pur⁺ pdx⁺ nad⁺* might be generated, given the two gene orders: if *pdx* is in the middle, 4 crossovers are required to get *pur⁺ pdx⁺ nad⁺*; if pur is in the middle, only 2 crossovers are required (see next page). The results indicate that there are fewer *pur⁺ pdx⁺ nad⁺* transductants than any other class suggesting that this class, is "harder" to generate than the others. This implies that *pdx* is in middle and the gene order is *pur pdx nad*.

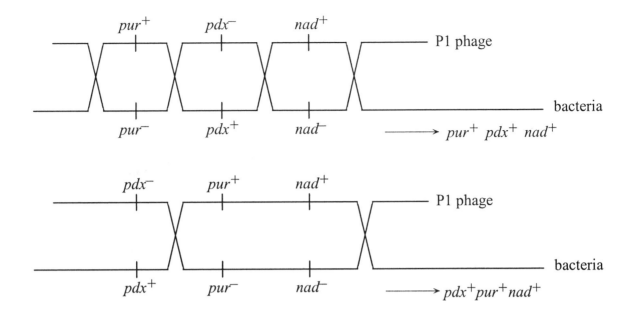

21. **a.** Owing to the medium used, all colonies are cys^+ but either + or – for the other two genes.

b. (1) $cys^+ leu^+ thr^+$ and $cys^+ leu^+ thr^-$ (supplemented with threonine)
(2) $cys^+ leu^+ thr^+$ and $cys^+ leu^- thr^+$ (supplemented with leucine)
(3) $cys^+ leu^+ thr^+$ (no supplements)

c. Because none grew on minimal medium, no colony was $leu^+ thr^+$. Therefore, medium (1) had $cys^+ leu^+ thr^-$, and medium (2) had $cys^+ leu^- thr^+$. The remaining cultures were $cys^+ leu^- thr^-$, and this genotype occurred in 100% – 56% – 5% = 39% of the colonies.

d. *cys* and *leu* are cotransduced 56 percent of the time, while *cys* and *thr* are cotransduced only 5 percent of the time. This indicates that *cys* is closer to *leu* than it is to *thr*. Because no $leu^+ cys^+ thr^+$ cotransductants are found, it indicates that *cys* is in the middle.

22. Prototrophic strains of *E. coli* will grow on minimal media, while auxotrophic strains will only grow on media supplemented with the required molecule(s). Thus, strain 3 is prototrophic (wild-type), strain 4 is met^-, strain 1 is arg^-, and strain 2 is $arg^- met^-$.

23. **Unpacking the Problem**

1. *E. coli* is a bacterium and a prokaryote.

2. *E. coli* can be grown in suspension or on an agar medium. The latter method allows for the identification of individual colonies, each a clone of descendants from a single cell (and visible to the naked eye when it reaches more than 10^7 cells).

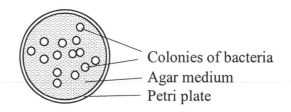

Colonies of bacteria
Agar medium
Petri plate

3. Naturally, *E. coli* is an enteric bacterium living symbiotically within the gut of host organisms (like us).

4. Minimal medium consists of inorganic salts, a carbon source for energy, and water.

5. Prototroph refers to the wild-type phenotype, or in other words, an organism that can grow on minimal media. Auxotroph refers to a mutant that can grow only on a medium supplemented with one or more specific nutrients not required by the wild-type strain.

6. In this experiment, the Hfr and the exconjugants that can grow on minimal medium are prototrophs, whereas the recipient F⁻ and the exconjugants that do not grow on minimal medium are auxotrophs.

7. Unknown strains would be grown as individual colonies on medium enriched with proline and thiamine, and then cells from each colony could be picked (by a sterile toothpick, for example) and placed individually onto medium supplemented with either thiamine or proline or onto minimal medium. Proline and thiamine auxotrophs would be identified on the basis of growth patterns. For example, a *pro⁻* strain will grow only on medium supplemented with proline.
Instead of the labor-intensive method of individually picking cells, replica plating can be used to transfer some cells of each colony from a master plate (supplemented with proline and thiamine) to plates that contain the various media described above. The physical arrangement (and positional patterns) of colonies is used to identify the various colonies as they are transferred from plate to plate.

8. Proline is an amino acid and thiamine is a B_1 vitamin. Their chemical nature does not matter to the experiment other than that they are necessary chemicals for cell growth that prototrophs can synthesize from ingredients in minimal medium and specific auxotrophic mutants cannot.

9.

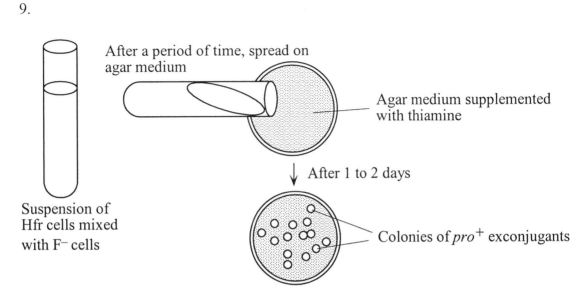

After a period of time, spread on agar medium

Agar medium supplemented with thiamine

After 1 to 2 days

Suspension of Hfr cells mixed with F⁻ cells

Colonies of *pro⁺* exconjugants

10. Interrupted-mating experiments are used to roughly map genes onto the circular bacterial chromosome.

11. The Hfr and F⁻ strains are mixed together in solution, and then at various times, samples are removed and put into a kitchen blender, vortexed (blender is turned on) for a few seconds to disrupt conjugation, and then plated onto a medium containing the appropriate supplements. The amount of time that has passed from the mixing of the strains to mating disruption is used as a measurement for mapping. The time of first appearance of a specific gene from the Hfr in the F⁻ cell gives the gene's relative position in minutes. Typically, the F⁻ cells are streptomycin-resistant and the Hfr cells are streptomycin-sensitive. The antibiotic is used in the various media to kill the Hfr cells (which are otherwise prototrophic) and allow only those F⁻ cells that have received the appropriate gene or genes from the Hfr to grow. In this case, it would be discovered that some of the F⁻ cells would become *thi⁺* in samples taken earlier in the experiment than samples taken when they first become *pro⁺*.

12. In this experiment, there is no attempt to disrupt conjugation. The two strains are mixed and at some later (unspecified) time, plated onto medium containing thiamine. This selects for strains that are *pro⁺*, because proline is not present in this medium.

13. Exconjugants are recipient cells (F⁻) that now contain alleles from the donor (Hfr). Typically, the F⁻ cells are streptomycin-resistant and the Hfr cells are streptomycin-sensitive. The antibiotic is used in the various media to kill the Hfr cells and allow only the appropriate F⁻ exconjugants to grow.

14. The statement *"pro* enters after *thi"* is one of gene position and order relative to the transfer of the bacterial chromosome by a particular Hfr. For the Hfr in this experiment, transfer occurs such that the *pro* gene is transferred after the *thi* gene. Because this Hfr is also *pro*⁺, it is this specific allele that is entering.

15. In this experiment, "fully supplemented" medium contains proline and thiamine.

16. All exconjugants are *pro*⁺, because that is the way they were selected. Thus, those that do not grow on minimal medium must require thiamine.

17. Genetic exchange in prokaryotes does not take place between two whole genomes as it does in eukaryotes. It takes place between one complete circular genome, the F⁻, and an incomplete linear genomic fragment donated by the Hfr. In this way, exchange of genetic information is nonreciprocal (from Hfr to F⁻). Only even numbers of crossovers are allowed between the two DNAs, because the circular chromosome would become linear otherwise. This results in unidirectional exchange, because part of the DNA of the recipient chromosome is replaced by the DNA of the donor, while the other product (the rest of the donor DNA now with some recombined recipient DNA) is nonviable and lost.

18. In this experiment, the map distance is calculated by selecting for the last marker to enter (in this case *pro*⁺) and then determining how often the earlier unselected marker (in this case *thi*⁺) is also present. Look at the following diagram:

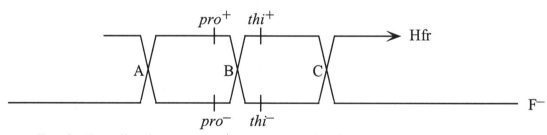

For the F⁻ cell to become *pro*⁺, two recombination events have to occur—one in the region to the left (marked A) and one in either region to the right (marked B or C.) Thus the percentage of *pro*⁺ (second recombination within either B or C) that are *thi*⁻ (second event only within region B) can be used to determine map distance where, 1% = 1 map unit.

Solution to the Problem

a. The two genotypes being cultured are *pro*⁺ *thi*⁻ (grows only on media supplemented with thiamine) and *pro*⁺ *thi*⁺ (grows on minimal media).

b. Two recombination events must occur, one on either side of *pro* (because exconjugants were plated on medium supplemented with thiamine, only *pro*+ cells would have grown). The *pro*+ *thi*- strains would have had recombination in regions labeled A and B, and the *pro*+ *thi*+ strains would have had recombination in regions labeled A and C.

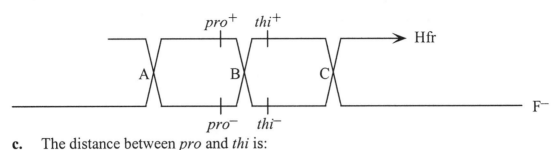

c. The distance between *pro* and *thi* is:

$$= \frac{100\%(\text{the number of colonies that are } pro^+\, thi^-)}{\text{total number of } pro^+ \text{ colonies}}$$

$$= 100\%(40)/360 = 11.1\%$$

24. No. Closely linked loci would be expected to be cotransduced; the greater the contransduction frequency, the closer the loci are. Because only 1 of 858 *metE*+ was also *pyrD*+, the genes are not closely linked. The lone *metE*+ *pyrD*+ could be the result of contransduction, or it could be a spontaneous mutation of *pyrD* to *pyrD*+, or the result of co-infection by two separate transducing phage.

CHALLENGING PROBLEMS

25. To interpret the data, the following results are expected:

Cross	Result
F+ × F-	(L) low number of recombinants
Hfr × F-	(M) many recombinants
Hfr × Hfr	(0) no recombinants
Hfr × F+	(0) no recombinants
F+ × F+	(0) no recombinants
F- × F-	(0) no recombinants

The only strains that show both the (L) and the (M) result when crossed are 2, 3, and 7. These must be F- because that is the only cell type that can participate in a cross and give either recombination result. Hfr strains will result in only (M) or (0), and F+ will result in only (L) or (0) when crossed. Thus, strains 1 and 8 are F+, and strains 4, 5, and 6 are Hfr.

26. a.

Agar type	Selected genes
1	c^+
2	a^+
3	b^+

b. The order of genes is revealed in the sequence of colony appearance. Because colonies first appear on agar type 1, which selects for c^+, c must be first. Colonies next appear on agar type 3, which selects for b^+, indicating that b follows c. Allele a^+ appears last. The gene order is $c\ b\ a$. The three genes are roughly equally spaced.

c. In this problem you are looking at the cotransfer of the selected gene with the d^- allele (both from the Hfr). Cells that are d^- do not grow because the medium is lacking D and selecting for those cells that are d^+. Therefore, the farther a gene is from gene d, the less likely cotransfer of the selected gene will occur with d^- and the more likely that colonies will grow (remain d^+). From the data, d is closest to b (only $^8/_{100}$ did not cotransfer d^- with b^+.) It is also closer to a than it is to c. Thus the gene order is $c\ b\ d\ a$ (or $a\ d\ b\ c$).

d. With no A or B in the agar, the medium selects for $a^+\ b^+$, and the first colonies should appear at about 17.5 minutes.

27. a. To survive on the selective medium, all cultures must be ery^r. Keep in mind that cells from all 300 colonies were each tested under four separate conditions.

263 colonies grew when only erythromycin is added, so these must be arg^+ $aro^+\ ery^r$. The remaining 37 cultures are mutant for one or both genes. One additional colony grew if arginine was also added to the medium ($264 - 263 = 1$). It must be $arg^-\ aro^+\ ery^r$. A total of 290 colonies are arg^+ because they grew when erythromycin and aromatic amino acids were added to the medium. Of these, 27 are aro^- ($290 - 263 = 27$). The genotypes and their frequencies are summarized below:

263	$ery^r\ arg^+\ aro^+$
27	$ery^r\ arg^+\ aro^-$
1	$ery^r\ arg^-\ aro^+$
9	$ery^r\ arg^-\ aro^-$
300	

b. Recombination in the aro–arg region is represented by two genotypes: aro^+ arg^- and $aro^-\ arg+$. The frequency of recombination is

$$100\%(1 + 27)/300 = 9.3 \text{ m.u.}$$

Recombination in the ery–arg region is represented by two genotypes: aro^+ arg^- and aro^- arg^-. The frequency of recombination is

$$100\%(1 + 9)/300 = 3.3 \text{ m.u.}$$

Recombination in the *ery–aro* region is represented by three genotypes: arg^+ aro^-, arg^- aro^-, and arg^- aro^+. Recall that the DCO must be counted twice. The frequency of recombination is

$$100\%(27 + 9 + 2)/300 = 12.6 \text{ m.u.}$$

    c.   The ratio is 28:10, or 2.8:1.0.

28. **a.** To determine which genes are close, compare the frequency of double transformants. Pairwise testing gives low values whenever B is involved but fairly high rates when any drug but B is involved. This suggests that the gene for B resistance is not close to the other three genes.

    b. To determine the relative order of genes for resistance to A, C, and D, compare the frequency of double and triple transformants. The frequency of resistance to AC is approximately the same as resistance to ACD. This strongly suggests that D is in the middle. Also, the frequency of AD co-resistance is higher than AC (suggesting that the gene for A resistance is closer to D than to C), and the frequency of CD is higher than AC (suggesting that C is closer to D than to A).

29. **a. and b.**

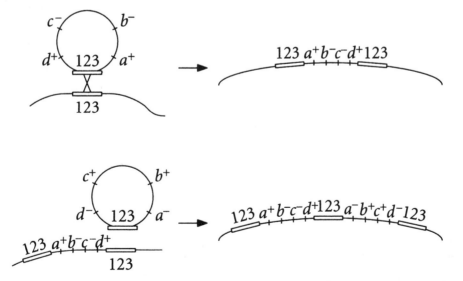

c.

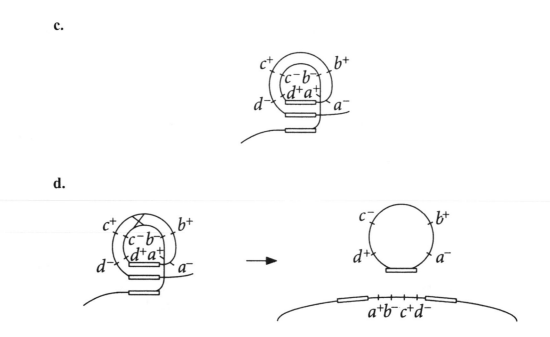

d.

30. If *trp1* and *trp2* are alleles of the same locus, then a cross between strains A and B will not result in Trp⁺ cells; if they are not allelic, strain B cells that have received the F′ from strain A will be Trp+ (complementation).

31. If a compound is not added and growth occurs, the *E. coli* recipient cell must have received the wild-type genes for production of those nutrients by transduction. Thus, the BCE culture selects for cells that are now a^+ and d^+, the BCD culture selects for cells that are a^+ and e^+, and the ABD culture selects for cells that are c^+ and e^+. These genes can be aligned (see below) to give the map order of *d a e c*. (Notice that *b* is never cotransduced and is therefore distant from this group of genes.)

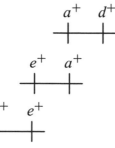

32. To isolate the specialized transducing particles of phage ϕ80 that carried *lac⁺*, the researchers would have had to lysogenize the strain with ϕ80, induce the phage with UV, and then use these lysates to transduce a Lac⁻ strain to Lac⁺. Lac⁺ colonies would then be used to make a new lysate, which should be highly enriched for the *lac⁺* transducing phage.

6

From Gene to Phenotype

1. Lactose is composed of one molecule of galactose and one molecule of glucose. A secondary cure would result if all galactose and lactose were removed from the diet. The disorder would be expected not to be dominant, because one good copy of the gene should allow for at least some, if not all, breakdown of galactose. In fact, the disorder is recessive.

2. Assuming homozygosity for the normal gene, the mating is $A/A \cdot b/b \times a/a \cdot B/B$. The children would be normal, $A/a \cdot B/b$ (see Problem 12).

3.

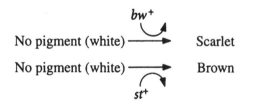

    Scarlet plus brown results in red.

4. Growth will be supported by a particular compound if it is later in the pathway than the enzymatic step blocked in the mutant. Restated, the more mutants a compound supports, the later in the pathway it must be. In this example, compound G supports growth of all mutants and can be considered the end product of the pathway. Alternatively, compound E does not support the growth of any mutant and can be considered the starting substrate for the pathway. The data indicate the following:

a. and b.

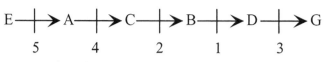

vertical lines indicate the step where each mutant is blocked

c. A heterokaryon of double mutants 1, 3 and 2, 4 would grow as the first would supply functional 2 and 4, and the second would supply functional 1 and 3.

A heterokaryon of the double mutants 1, 3 and 3, 4 would not grow as both are mutant for 3.

A heterokaryon of the double mutants 1, 2 and 2, 4 and 1, 4 would grow as the first would supply functional 4, the second would supply functional 1, and the last would supply functional 2.

5. a. If enzyme A was defective or missing (m_2/m_2), red pigment would still be made and the petals would be red.

b. Purple, because it has a wild-type allele for each gene, and you are told that the mutations are recessive.

c.

9	$M_1/-$; $M_2/-$	purple
3	m_1/m_1 ; $M_2/-$	blue
3	$M_1/-$; m_2/m_2	red
1	m_1/m_1 ; m_2/m_2	white

d. The mutant alleles do not produce functional enzyme. However, enough functional enzyme must be produced by the single wild-type allele of each gene to synthesize normal levels of pigment.

6. a. If enzyme B is missing, a white intermediate will accumulate and the petals will be white.

b. If enzyme D is missing, a blue intermediate will accumulate and the petals will be blue.

c.   P b/b ; D/D × B/B ; d/d

   F_1 B/b ; D/d purple

d.   P b/b ; D/D × B/B ; d/d

   F_1 B/b ; D/d × B/b ; D/d

F$_2$ 9 *B/–* ; *D/–*   purple
        3 *b/b* ; *D/–*   white
        3 *B/–* ; *d/d*   blue
        1*b/b* ; *d/d*   white

The ratio of purple : blue : white would be 9:3:4.

7. The woman must be *A/O*, so the mating is *A/O* × *A/B*. Their children will be

Genotype	Phenotype
$\frac{1}{4}$ *A/A*	A
$\frac{1}{4}$ *A/B*	AB
$\frac{1}{4}$ *A/O*	A
$\frac{1}{4}$ *B/O*	B

8. You are told that the cross of two erminette fowls results in 22 erminette, 14 black, and 12 pure white. Two facts are important: (1) the parents consist of only one phenotype, yet the offspring have three phenotypes, and (2) the progeny appear in an approximate ratio of 1:2:1. These facts should tell you immediately that you are dealing with a heterozygous × heterozygous cross involving one gene and that the erminette phenotype must be the heterozygous phenotype.

When the heterozygote shows a different phenotype from either of the two homozygotes, the heterozygous phenotype results from incomplete dominance or codominance. Because two of the three phenotypes contain black, either fully or in an occasional feather, you might classify erminette as an instance of incomplete dominance because it is intermediate between fully black and fully white. Alternatively, because erminette has both black and white feathers, you might classify the phenotype as codominant. Your decision will rest on whether you look at the whole animal (incomplete dominance) or at individual feathers (codominance). This is yet another instance where what you conclude is determined by how you observe.

To test the hypothesis that the erminette phenotype is a heterozygous phenotype, you could cross an erminette with either, or both, of the homozygotes. You should observe a 1:1 ratio in the progeny of both crosses.

9. **a.** The original cross and results were

    P   long, white × round, red

    F$_1$ oval, purple

	F$_2$	9 long, red	19 oval, red	8 round, white
		15 long, purple	32 oval, purple	16 round, purple
		<u>8 long, white</u>	<u>16 oval, white</u>	<u>9 round, red</u>
		32 long	67 oval	33 round

The data show that, when the results are rearranged by shape, a 1:2:1 ratio is observed for color within each shape category. Likewise, when the data are rearranged by color, a 1:2:1 ratio is observed for shape within each color category.

9 long, red	15 long, purple	8 round, white
19 oval, red	32 oval, purple	16 oval, white
<u>9 round, red</u>	<u>16 round, purple</u>	<u>8 long, white</u>
37 red	63 purple	32 white

A 1:2:1 ratio is observed when there is a heterozygous × heterozygous cross. Therefore, the original cross was a dihybrid cross. Both oval and purple must represent an incomplete dominant phenotype.

Let L = long, L' = round, R = red and R' = white. The cross becomes

P $\quad L/L\;;\;R'/R' \times L'/L'\;;\;R/R$

F$_1$ $\quad L/L'\;;\;R/R' \times L/L'\;;\;R/R'$

F$_2$ $\quad \frac{1}{4}\,L/L \times$

$\frac{1}{4}\,R/R =$ $\quad \frac{1}{16}$ long, red
$\frac{1}{2}\,R/R' =$ $\quad \frac{1}{8}$ long, purple
$\frac{1}{4}\,R'/R' =$ $\quad \frac{1}{16}$ long, white

$\frac{1}{2}\,L/L' \times$

$\frac{1}{4}\,R/R =$ $\quad \frac{1}{8}$ oval, red
$\frac{1}{2}\,R/R' =$ $\quad \frac{1}{4}$ oval, purple
$\frac{1}{4}\,R'/R' =$ $\quad \frac{1}{8}$ oval, white

$\frac{1}{4}\,L'/L' \times$

$\frac{1}{4}\,R/R =$ $\quad \frac{1}{16}$ round, red
$\frac{1}{2}\,R/R' =$ $\quad \frac{1}{8}$ round, purple
$\frac{1}{4}\,R'/R' =$ $\quad \frac{1}{16}$ round, white

b. A long, purple × oval, purple cross is as follows

P $\quad L/L\;;\;R/R' \times L/L'\;;\;R/R'\,'$

F$_1$ $\quad \frac{1}{2}\,L/L \times$

$\frac{1}{4}\,R/R =$ $\quad \frac{1}{8}$ long, red
$\frac{1}{2}\,R/R' =$ $\quad \frac{1}{4}$ long, purple
$\frac{1}{4}\,R'/R' =$ $\quad \frac{1}{8}$ long, white

$$\frac{1}{2} L/L' \times \quad \begin{matrix} \frac{1}{4} R/R = & \frac{1}{8} \text{ oval, red} \\ \frac{1}{2} R/R' = & \frac{1}{4} \text{ oval, purple} \\ \frac{1}{4} R'/R' = & \frac{1}{8} \text{ oval, white} \end{matrix}$$

10. From the cross $c^+/c^{ch} \times c^{ch}/c^h$ the progeny are

$\frac{1}{4}$	c^+/c^{ch}	full color
$\frac{1}{4}$	c^+/c^h	full color
$\frac{1}{4}$	c^{ch}/c^{ch}	chinchilla
$\frac{1}{4}$	c^{ch}/c^h	chinchilla

Thus, 50 percent of the progeny will be chinchilla.

11. **a.** The data indicate that there is a single gene with multiple alleles. All the ratios produced are 3:1 (complete dominance), 1:2:1 (incomplete of codominance), or 1:1 (test cross). The order of dominance is

$$\text{black} > \text{sepia} > \text{cream} > \text{albino}$$

Cross	Parents	Progeny	Conclusion
Cross 1:	$b/a \times b/a$	$3\ b/-$: $1\ a/a$	black is dominant to albino.
Cross 2:	$b/s \times a/a$	$1\ b/a$: $1\ s/a$	black is dominant to sepia; sepia is dominant to albino.
Cross 3:	$c/a \times c/a$	$3\ c/-$: $1\ a/a$	cream is dominant to albino.
Cross 4:	$s/a \times c/a$	$1\ c/a$: $2\ s/-$: $1\ a/a$	sepia is dominant to cream.
Cross 5:	$b/c \times a/a$	$1\ b/a$: $1\ c/a$	black is dominant to cream.
Cross 6:	$b/s \times c/-$	$1\ b/-$: $1\ s/-$	"–" can be c or a.
Cross 7:	$b/s \times s/-$	$1\ b/s$: $1\ s/-$	"–" can be s, c, or a.
Cross 8:	$b/c \times s/c$	$1\ s/c$: $2\ b/-$: $1\ c/c$	
Cross 9:	$s/c \times s/c$	$3\ s/-$: $1\ c/c$	
Cross 10:	$c/a \times a/a$	$1\ c/a$: $1\ a/a$	

   b. The progeny of the cross $b/s \times b/c$ will be $\frac{3}{4}$ black ($\frac{1}{4}\ b/b$, $\frac{1}{4}\ b/c$, $\frac{1}{4}\ b/s$) : $\frac{1}{4}$ sepia (s/c).

12. Both codominance (=) and classical dominance (>) are present in the multiple allelic series for blood type: $A = B$, $A > O$, $B > O$.

Parents' phenotype	Parents' possible genotypes	Parents' possible children
a. AB × O	$A/B \times O/O$	A/O, B/O
b. A × O	A/A or $A/O \times O/O$	A/O, O/O
c. A × AB	A/A or $A/O \times A/B$	A/A, A/B, A/O, B/O
d. O × O	$O/O \times O/O$	O/O

The possible genotypes of the children are

Phenotype	Possible genotypes
O	*O/O*
A	*A/A, A/O*
B	*B/B, B/O*
AB	*A/B*

Using the assumption that each set of parents had one child, the following combinations are the only ones that will work as a solution.

Parents	Child
a. AB × O	B
b. A × O	A
c. A × AB	AB
d. O × O	O

13. *M* and *N* are codominant alleles. The rhesus group is determined by classically dominant alleles. The *ABO* alleles are mixed codominance and classical dominance (see Problem 6).

Person	Blood type			Possible paternal contribution		
husband	O	M	Rh$^+$	*O*	*M*	*R* or *r*
wife's lover	AB	MN	Rh$^-$	*A* or *B*	*M* or *N*	*r*
wife	A	N	Rh$^+$	*A* or *O*	*N*	*R* or *r*
child 1	O	MN	Rh$^+$	*O*	*M*	*R* or *r*
child 2	A	N	Rh$^+$	*A* or *O*	*N*	*R* or *r*
child 3	A	MN	Rh$^-$	*A* or *O*	*M*	*r*

The wife must be *A/O* ; *N/N* ; *R/r*. (She has a child with type O blood and another child that is Rh$^-$ so she must carry both of these recessive alleles.) Only the husband could donate O to child 1. Only the lover could donate A and N to child 2. Both the husband and the lover could have donated the necessary alleles to child 3.

14. The key to solving this problem is in the statement that breeders cannot develop a pure–breeding stock and that a cross of two platinum foxes results in some normal progeny. Platinum must be dominant to normal color and heterozygous (*A/a*). An 82:38 ratio is very close to 2:1. Because a 1:2:1 ratio is expected in a heterozygous cross, one genotype is nonviable. It must be the *A/A*, homozygous platinum, genotype that is nonviable, because the homozygous recessive genotype is normal color (*a/a*). Therefore, the platinum allele is a pleiotropic allele that governs coat color in the heterozygous state and is lethal when homozygous.

15. a. Because Pelger crossed with normal stock results in two phenotypes in a 1:1 ratio, either Pelger or normal is heterozygous (*A/a*) and the other is homozygous (*a/a*) recessive. The problem states that normal is true–breeding, or *a/a*. Pelger must be *A/a*.

b. The cross of two Pelger rabbits results in three phenotypes. This means that the Pelger anomaly is dominant to normal. This cross is *A/a* × *A/a*, with an expected ratio of 1:2:1. Because the normal must be *a/a*, the extremely abnormal progeny must be *A/A*. There were only 39 extremely abnormal progeny because the others died before birth.

c. The Pelger allele is pleiotropic. In the heterozygous state, it is dominant for nuclear segmentation of white blood cells. In the homozygous state, it is a lethal.

You could look for the nonsurviving fetuses in utero. Because the hypothesis of embryonic death when the Pelger allele is homozygous predicts a one-fourth reduction in litter size, you could also do an extensive statistical analysis of litter size, comparing normal × normal with Pelger × Pelger.

d. By analogy with rabbits, the absence of a homozygous Pelger anomaly in humans can be explained as recessive lethality. Also, because one in 1000 people have the Pelger anomaly, a heterozygous × heterozygous mating would be expected in only one of 1 million.

($^1/_{1000}$ × $^1/_{1000}$) random matings, and then only one in four of the progeny would be expected to be homozygous. Thus, the homozygous Pelger anomaly is expected in only 1 of 4 million births. This is extremely rare and might not be recognized.

e. By analogy with rabbits, among the children of a man and a woman with the Pelger anomaly, two-thirds of the surviving progeny would be expected to show the Pelger anomaly and one-third would be expected to be normal. The developing fetus that is homozygous for the Pelger allele would not be expected to survive until birth.

16. a. The sex ratio is expected to be 1:1.

b. The female parent was heterozygous for an X-linked recessive lethal allele. This would result in 50 percent fewer males than females.

c. Half of the female progeny should be heterozygous for the lethal allele and half should be homozygous for the nonlethal allele. Individually mate the F_1 females and determine the sex ratio of their progeny.

17. Note that a cross of the short-bristled female with a normal male results in two phenotypes with regard to bristles and an abnormal sex ratio of two females : one male. Furthermore, all the males are normal, while the females are normal and short in equal numbers. Whenever the sexes differ with respect to phenotype among the progeny, an X-linked gene is implicated. Because only the normal phenotype is observed in males, the short-bristled phenotype must be heterozygous, and the allele must be a recessive lethal. Thus the first cross was $A/a \times a/Y$.

Long-bristled females (a/a) were crossed with long-bristled males (a/Y). All their progeny would be expected to be long-bristled (a/a or a/Y). Short-bristled females (A/a) were crossed with long-bristled males (a/Y). The progeny expected are

$1/4$ A/a	short-bristled females
$1/4$ a/a	long-bristled females
$1/4$ a/Y	long-bristled males
$1/4$ A/Y	nonviable

18. In order to do this problem, you should first restate the information provided. The following two genes are independently assorting

h/h = hairy	s/s = no effect
H/h = hairless	S/s suppresses H/h, giving hairy
H/H = lethal	S/S = lethal

a. The cross is H/h ; $S/s \times H/h$; S/s. Because this is a typical dihybrid cross, the expected ratio is 9:3:3:1. However, the problem cannot be worked in this simple fashion because of the epistatic relationship of these two genes. Therefore, the following approach should be used.

For the H gene, you expect $1/4$ H/H : $1/2$ H/h : $1/4$ h/h. For the S gene, you expect $1/4$ S/S : $1/2$ S/s : $1/4$ s/s. To get the final ratios, multiply the frequency of the first genotype by the frequency of the second genotype.

$1/4$ H/H		all progeny die regardless of the S gene		
		$1/4$ S/S =	$1/8$ H/h ; S/S	die
$1/2$ H/h	$\times$	$1/2$ S/s =	$1/4$ H/h ; S/s	hairy
		$1/4$ s/s =	$1/8$ H/h ; s/s	hairless
		$1/4$ S/S =	$1/16$ h/h ; S/S	die
$1/4$ h/h	$\times$	$1/2$ S/s =	$1/8$ h/h ; S/s	hairy
		$1/4$ s/s =	$1/16$ h/h ; s/s	hairy

Of the $^9/_{16}$ living progeny, the ratio of hairy to hairless is 7:2.

b. This cross is H/h ; $s/s \times H/h$; S/s. A 1:2:1 ratio is expected for the H gene and a 1:1 ratio is expected for the S gene.

$^1/_4$ H/H			all progeny die regardless of the S gene

$^1/_2$ H/h	$\times$	$^1/_2$ $S/s =$	$^1/_4$ H/h ; S/s	hairy
		$^1/_2$ $s/s =$	$^1/_4$ H/h ; s/s	hairless

$^1/_4$ h/h	$\times$	$^1/_2$ $S/s =$	$^1/_8$ h/h ; S/s	hairy
		$^1/_2$ $s/s =$	$^1/_8$ h/h ; s/s	hairy

Of the $^3/_4$ living progeny, the ratio of hairy to hairless is 2:1.

19. a. The mutations are in two different genes as the heterokaryon is prototrophic (the two mutations complemented each other).

b. $leu1^+$; $leu\ 2^-$ and $leu1^-$; $leu2^+$

c. With independent assortment, expect

$^1/_4$	$leu1^+$; $leu\ 2^-$
$^1/_4$	$leu1^-$; $leu2^+$
$^1/_4$	$leu1^-$; $leu\ 2^-$
$^1/_4$	$leu1^+$; $leu\ 2^+$

20. a. The first type of prototroph is due to reversion of the original $ade1$ mutant to wild type. The second type of prototroph is due to a new mutation (call it sup^m) in an unlinked gene that suppresses the $ade1^-$ phenotype. For the results of crossing type 2 prototrophs to wild type

$ade1^-$; $sup^m \times ade1^+$; sup

and the results would be

$^1/_4$	$ade1^+$; sup	prototroph
$^1/_4$	$ade1^+$; sup^m	prototroph
$^1/_4$	$ade1^-$; sup^m	prototroph
$^1/_4$	$ade1^-$; sup	auxotroph

b. type 1 would be $ade1^+$; sup
type 2 would be $ade1^-$; sup^m

c. For the results of crossing type 2 prototrophs to the original mutant

$$ade1^-; sup^m \times ade1^-; sup$$

and the results would be

$1/2$	$ade1^-; sup$	auxotroph
$1/2$	$ade1^-; sup^m$	prototroph

21. **a.** colorless

b. magenta

c. colorless

d. $p/p ; Q/Q$
$P/P ; q/q$
$p/p ; q/q$

e. 9 red : 4 colorless : 3 magenta

22. The suggestion from the data is that the two albino lines had mutations in two different genes. When the extracts from the two lines were placed in the same test tube, they were capable of producing color because the gene product of one line was capable of compensating for the absence of a gene product from the second line.

a. The most obvious control is to cross the two pure-breeding lines. The cross would be $A/A ; b/b \times a/a ; B/B$. The progeny will be $A/a ; B/b$, and all should be reddish purple.

b. The most likely explanation is that the red pigment is produced by the action of at least two different gene products. When petals of the two plants were ground together, the different defective enzyme of each plant was complemented by the normal enzyme of the other.

c. The genotypes of the two lines would be $A/A ; b/b$ and $a/a ; B/B$.

d. The F_1 would be all be pigmented, $A/a ; B/b$. This is an example of complementation. The mutants are defective for different genes. The F_2 would be

9	$A/- ; B/-$	pigmented
3	$a/a ; B/-$	white
3	$A/- ; b/b$	white
1	$a/a ; b/b$	white

23. a. This is an example where one phenotype in the parents gives rise to three phenotypes in the offspring. The "frizzle" is the heterozygous phenotype and shows incomplete dominance.

$$P \qquad A/a \text{ (frizzle)} \times A/a \text{ (frizzle)}$$

$$F_1 \qquad 1\ A/A \text{ (normal)} : 2\ A/a \text{ (frizzle)} : 1\ a/a \text{ (woolly)}$$

b. If *A/A* (normal) is crossed to *a/a* (woolly), all offspring will be *A/a* (frizzle).

24. *Unpacking the Problem*

1. The character being studied is petal color.

2. The wild-type phenotype is blue.

3. A variant is a phenotypic difference from wild type that is observed.

4. There are two variants: pink and white.

5. *In nature* means that the variants did not appear in laboratory stock and, instead, were found growing wild.

6. Possibly, the variants appeared as a small patch or even a single plant within a larger patch of wild type.

7. Seeds would be grown to check the outcome from each cross.

8. Given that no sex linkage appears to exist (sex is not specified in parents or offspring), *blue × white* means the same as *white × blue*. Similar results would be expected because the trait being studied appears to be autosomal.

9. The first two crosses show a 3:1 ratio in the F_2, suggesting the segregation of one gene. The third cross has a 9:4:3 ratio for the F_2, suggesting that two genes are segregating.

10. Blue is dominant to both white and pink.

11. *Complementation* refers to generation of wild-type progeny from the cross of two strains that are mutant in different genes.

12. The ability to make blue pigment requires two enzymes that are individually defective in the pink or white strains. The F_1 progeny of this cross is blue, since each has inherited one nonmutant allele for both genes and can therefore produce both functional enzymes.

13. Blueness from a pink × white cross arises through complementation.

14. The following ratios are observed: 3:1, 9:4:3.

15. There are monohybrid ratios observed in the first two crosses.

16. There is a modified 9:3:3:1 ratio in the third cross.

17. A monohybrid ratio indicates that one gene is segregating, while a dihybrid ratio indicates that two genes are segregating.

18. 15:1, 12:3:1, 9:6:1, 9:4:3, 9:7, 1:2:1, 2:1

19. There is a modified dihybrid ratio in the third cross.

20. A modified dihybrid ratio most frequently indicates the interaction of two or more genes.

21. Recessive epistasis is indicated by the modified dihybrid ratio.

22.

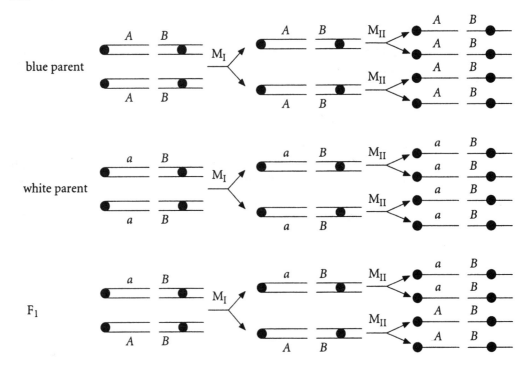

Solution to the Problem

a. Let A = wild-type, a = white, B = wild-type, and b pink.

Cross 1:	P	blue × white	*A/A ; B/B × a/a ; B/B*
	F$_1$	all blue	all *A/a ; B/B*
	F$_2$	3 blue : 1 white	3 *A/– ; B/B* : 1 *a/a ; B/B*

Cross 2:	P	blue × pink	*A/A ; B/B × A/A ; b/b*
	F$_1$	all blue	all *A/A ; B/b*
	F$_2$	3 blue : 1 pink	3 *A/A ; B/–* : 1 *A/A ; b/b*

Cross 3:	P	pink × white	*A/A ; b/b × a/a ; B/B*
	F$_1$	all blue	all *A/a ; B/b*
	F$_2$	9 blue	9 *A/– ; B/–*
		4 white	3 *a/a ; B/–* : 1 *a/a ; b/b*
		3 pink	3 *A/– ; b/b*

When the allele *a* is homozygous, the expression of alleles *B* or *b* is blocked or masked. The white phenotype is epistatic to the pigmented phenotypes. It is likely that the product of the *A* gene produces an intermediate that is then modified by the product of the *B* gene. If the plant is *a/a*, this intermediate is not made and the phenotype of the plant is the same regardless of the ability to produce functional *B* product.

b. The cross is

	F$_2$	blue × white
	F$_3$	$3/8$ blue
		$1/8$ pink
		$4/8$ white

Begin by writing as much of each genotype as can be assumed:

	F$_2$	*A/– ; B/– × a/a ; –/–*
	F$_3$	$3/8$ *A/– ; B/–*
		$1/8$ *A/– ; b/b*
		$4/8$ *a/a ; –/–*

Notice that both *a/a* and *b/b* appear in the F$_3$ progeny. In order for these homozygous recessives to occur, each parent must have at least one *a* and one *b*. Using this information, the cross becomes

	F$_2$	*A/a ; B/b × a/a ; –/b*
	F$_3$	$3/8$ A/a ; *B/b*
		$1/8$ *A/a ; b/b*
		$4/8$ *a/a ; b/–*

The only remaining question is whether the white parent was homozygous recessive, *b/b*, or heterozygous, *B/b*. If the white parent had been homozygous recessive, then the cross would have been a testcross of the blue parent, and the progeny ratio would have been

1 blue : 1 pink : 2 white, or 1 *A/a* ; *B/b* : 1 *A/a* ; *b/b* : 1 *a/a* ; *B/b* : 1 *a/a* ; *b/b*.

This was not observed. Therefore, the white parent had to have been heterozygous, and the F_2 cross was *A/a* ; *B/b* × *a/a* ; *B/b*.

25. It is possible to produce black offspring from two pure-breeding recessive albino parents if albinism results from mutations in two different genes. If the cross is designated

$$A/A \; ; \; b/b \times a/a \; ; \; B/B$$

all offspring would be

$$A/a \; ; \; B/b$$

and they would have a black phenotype because of complementation.

26. The data indicate that white is dominant to solid purple. Note that the F_2 are in an approximate 12:3:1 ratio. In order to achieve such a ratio, epistasis must be involved.

    a. Because a modified 9:3:3:1 ratio was obtained in the F_2, the F_1 had to be a double heterozygote. Solid purple occurred at one-third the rate of white, which means that it will be in the form of either *D/–* ; *e/e* or *d/d* ; *E/–*. In order to achieve a double heterozygote in the F_1, the original white parent also has to be either *D/–* ; *e/e* or *d/d* ; *E/–*.

Arbitrarily assume that the original cross was *D/D* ; *e/e* (white) × *d/d* ; *E/E* (purple). The F_1 would all be *D/d* ; *E/e*. The F_2 would be

9 *D/–* ; *E/–*	white, by definition
3 *d/d* ; *E/–*	purple, by definition
3 *D/–* ; *e/e*	white, by deduction
1 *d/d* ; *e/e*	spotted purple, by deduction

Under these assumptions, *D* blocks the expression of both *E* and *e*. The *d* allele has no effect on the expression of *E* and *e*. *E* results in solid purple, while *e* results in spotted purple. It would also be correct, of course, to assume the opposite set of epistatic relationships (*E* blocks the expression of *D* or *d*, *D* results in solid purple, and *d* results in spotted purple).

b. The cross is white × solid purple. While the solid purple genotype must be d/d ; $E/-$, as defined in part (a), the white genotype can be one of several possibilities. Note that the progeny phenotypes are in a 1:2:1 ratio and that one of the phenotypes, spotted, must be d/d ; e/e. In order to achieve such an outcome, the purple genotype must be d/d ; E/e. The white genotype of the parent must contain both a D and a d allele in order to produce both white ($D/-$) and spotted plants (d/d). At this point, the cross has been deduced to be D/d ; $-/-$ (white) × d/d ; E/e (purple).

If the white plant is E/E, the progeny will be

$^1/_2$ D/d ; $E/-$ white
$^1/_2$ d/d ; $E/-$ solid purple

This was not observed. If the white plant is E/e, the progeny will be

$^3/_8$ D/d ; $E/-$ white
$^1/_8$ D/d ; e/e white
$^3/_8$ d/d ; $E/-$ solid purple
$^1/_8$ d/d ; e/e spotted purple

The phenotypes were observed, but in a different ratio. If the white plant is e/e, the progeny will be

$^1/_4$ D/d ; E/e white
$^1/_4$ D/d ; e/e white
$^1/_4$ d/d ; E/e solid purple
$^1/_4$ d/d ; e/e spotted purple

This was observed in the progeny. Therefore, the parents were D/d ; e/e (white) × d/d ; E/e (purple).

27. **a. and b.** Crosses 1–3 show a 3:1 ratio, indicating that brown, black, and yellow are all alleles of one gene. Crosses 4–6 show a modified 9:3:3:1 ratio, indicating that at least two genes are involved. Those crosses also indicate that the presence of color is dominant to its absence. Furthermore, epistasis must be involved for there to be a modified 9:3:3:1 ratio.

By looking at the F_1 of crosses 1–3, the following allelic dominance relationships can be determined: black > brown > yellow. Arbitrarily assign the following genotypes for homozygotes: B^l/B^l = black, B^r/B^r = brown, B^y/B^y = yellow.

By looking at the F_2 of crosses 4–6, a white phenotype is composed of two categories: the double homozygote and one class of the mixed homozygote/heterozygote. Let lack of color be caused by *c/c*. Color will therefore be *C/–*.

Parents	F_1	F_2
1 B^r/B^r ; C/C × B^y/B^y ; C/C	B^r/B^y ; C/C	3 $B^r/–$; C/C : 1 B^y/B^y ; C/C
2 B^l/B^l ; C/C × B^r/B^r ; C/C	B^l/B^r ; C/C	3 $B^l/–$; C/C : 1 B^r/B^r ; C/C
3 B^l/B^l ; C/C × B^y/B^y ; C/C	B^l/B^y ; C/C	3 $B^l/–$; C/C : 1 B^y/B^y ; C/C
4 B^l/B^l ; c/c × B^y/B^y ; C/C	B^l/B^y ; C/c	9 $B^l/–$; $C/–$: 3 B^y/B^y ; $C/–$: 3 $B^l/–$; c/c : 1 B^y/B^y ; c/c
5 B^l/B^l ; c/c × B^r/B^r ; C/C	B^l/B^r ; C/c	9 $B^l/–$; $C/–$: 3 B^r/B^r ; $C/–$: 3 $B^l/–$; c/c : 1 B^r/B^r ; c/c
6 B^l/B^l ; C/C × B^y/B^y ; c/c	B^l/B^y ; C/c	9 $B^l/–$; $C/–$: 3 B^y/B^y ; $C/–$: 3 $B^l/–$; c/c : 1 B^y/B^y ; c/c

28. It is possible to produce normally pigmented offspring from albino parents if albinism results from mutations in either of two different genes. If the cross is designated

$$A/A \cdot b/b \times a/a \cdot B/B$$

then all the offspring would be

$$A/a \cdot B/b$$

and they would have a pigmented phenotype because of complementation.

29. The first step in each cross is to write as much of the genotype as possible from the phenotype.

Cross 1: $A/–$; $B/–$ × a/a ; b/b → 1 $A/–$; $B/–$: 2 ?/a; ?/b : 1 a/a ; b/b

Because the double recessive appears, the blue parent must be A/a ; B/b. The $^1/_2$ purple then must be A/a ; b/b and a/a ; B/b.

Cross 2: ?/? ; ?/? × ?/? ; ?/? → 1 $A/–$; $B/–$: 2 ?/? ; ?/? : 1 a/a ; b/b

The two parents must be, in either order, A/a ; b/b and a/a ; B/b. The two purple progeny must be the same. The blue progeny are A/a ; B/b.

Cross 3: $A/-$; $B/-$ × $A/-$; $B/-$ →   3 $A/-$; $B/-$: 1 $?/?$; $?/?$

The only conclusions possible here are that one parent is either A/A or B/B and the other parent is B/b if the first is A/A or A/a if the first is B/B.

Cross 4: $A/-$; $B/-$ × $?/?$; $?/?$ →   3 $A/-$; $B/-$: 4 $?/?$; $?/?$: 1 a/a ; b/b

The purple parent can be either A/a ; b/b or a/a ; B/b for this answer. Assume the purple parent is A/a ; b/b. The blue parent must be A/a ; B/b. The progeny are

$$\begin{array}{llll}
^3/_4\ A/- & \times & ^1/_2\ B/b = {}^3/_8\ A/-\ ;\ B/b & \text{blue} \\
 & & ^1/_2\ b/b = {}^3/_8\ A/-\ ;\ b/b & \text{purple} \\[8pt]
^1/_4\ a/a & \times & ^1/_2\ B/b = {}^1/_8\ a/a\ ;\ B/b & \text{purple} \\
 & & ^1/_2\ b/b = {}^1/_8\ a/a\ ;\ b/b & \text{scarlet}
\end{array}$$

Cross 5: $A/-$; b/b × a/a ; b/b →   1 $A/-$; b/b : 1 a/a ; b/b

As written this is a testcross for gene A. The purple parent and progeny are A/a ; b/b. Alternatively, the purple parent and progeny could be a/a ; B/b.

30. The F_1 progeny of cross 1 indicate that sun red is dominant to pink. The F_2 progeny, which are approximately in a 3:1 ratio, support this. The same pattern is seen in crosses 2 and 3, with sun red dominant to orange and orange dominant to pink. Thus, we have a multiple allelic series with sun red > orange > pink. In all three crosses, the parents must be homozygous.

If c^{sr} = sun red, c^o = orange, and c^p = pink, then the crosses and the results are

Cross	Parents	F_1	F_2
1	c^{sr}/c^{sr} × c^p/c^p	c^{sr}/c^p	3 $c^{sr}/-$: 1 c^p/c^p
2	c^o/c^o × c^{sr}/c^{sr}	c^{sr}/c^o	3 $c^{sr}/-$: 1 c^o/c^o
3	c^o/c^o × c^p/c^p	c^o/cc^p	3 $c^o/-$: 1 c^p/c^p

Cross 4 presents a new situation. The color of the F_1 differs from that of either parent, suggesting that two separate genes are involved. An alternative explanation is either codominance or incomplete dominance. If either codominance or incomplete dominance is involved, then the F_2 will appear in a 1:2:1 ratio. If two genes are involved, then a 9:3:3:1 ratio, or some variant of it, will be observed. Because the wild-type phenotype appears in the F_1 and F_2, it appears that complementation is occurring. This requires two genes. The progeny actually are in a 9:4:3 ratio. This means that two genes are involved and that there is epistasis. Furthermore, for three phenotypes to be present in the F_2, the two F_1 parents must have been heterozygous.

Let a stand for the scarlet gene and A for its colorless allele, and assume that there is a dominant allele, C, that blocks the expression of the alleles that we have been studying to this point.

Cross 4: P c^o/c^o ; A/A × C/C ; a/a

F_1 C/c^o ; A/a

F_2 9 $C/-$; $A/-$ yellow
              3 $C/-$; a/a scarlet
              3 c^o/c^o ; $A/-$ orange
              1 c^o/c^o ; a/a orange (epistasis, with c^o blocking the expression of a/a)

31. a.   P A/A (agouti) × a/a (nonagouti)
               gametes A and a

               F_1 A/a (agouti)
               gametes A and a

               F_2 1 A/A (agouti) : 2 A/a (agouti) : 1 a/a (nonagouti)

b.   P B/B (wild type) × b/b (cinnamon)
               gametes B and b

               F_1 B/b (wild type)
               gametes B and b

               F_2 1 B/B (wild type) : 2 B/b (wild type) : 1 b/b (cinnamon)

c.   P A/A ; b/b (cinnamon or brown agouti) × a/a ; B/B (black nonagouti)
               gametes A ; b and a ; B

               F_1 A/a ; B/b (wild type, or black agouti)

d.   9 $A/-$; $B/-$ black agouti
               3 a/a ; $B/-$ black nonagouti
               3 $A/-$; b/b cinnamon
               1 a/a ; b/b chocolate

e. P A/A ; b/b (cinnamon) $\times$ a/a ; B/B (black nonagouti)

gametes A ; b and a ; B

F_1 A/a ; B/b (wild type)

gametes A ; B, A ; b, a ; B, and a ; b

F_2 9 $A/-$; $B/-$ wild type
                    1 A/A ; B/B
                    2 A/a ; B/B
                    2 A/A ; B/b
                    4 A/a ; B/b
            3   a/a ; $B/-$ black nonagouti
                    1 a/a ; B/B
                    2 a/a ; B/b
            3       $A/-$; b/b cinnamon
                    1 A/A ; b/b
                    2 A/a ; b/b
            1   a/a ; b/b chocolate

f. P A/a ; B/b $\times$ A/A ; b/b A/a ; B/b $\times$ a/a ; B/B
                 (wild type) (cinnamon)   (wild type) (black nonagouti)

F_1 $^1/_4$ A/A ; B/b wild type $^1/_4$ A/a ; B/B wild type
         $^1/_4$ A/a ; B/b wild type $^1/_4$ A/a ; B/b wild type
         $^1/_4$ A/A ; b/b cinnamon $^1/_4$ a/a ; B/B black nonagouti
         $^1/_4$ A/a ; b/b cinnamon $^1/_4$ a/a ; B/b black nonagouti

g. P A/a ; B/b $\times$ a/a ; b/b
                 (wild type) (chocolate)

F_1 $^1/_4$ A/a ; B/b wild type
         $^1/_4$ A/a ; b/b cinnamon
         $^1/_4$ a/a ; B/b black nonagouti
         $^1/_4$ a/a ; b/b chocolate

h. To be albino, the mice must be c/c, but the genotype with regard to the A and B genes can be determined only by looking at the F_2 progeny.

Cross 1:   P c/c ; $?/?$; $?/?$ $\times$ C/C ; A/A ; B/B

F_1 C/c ; $A/-$; $B/-$

F_2 87 wild type $C/-$; $A/-$; $B/-$
          32 cinnamon $C/-$; $A/-$; b/b
          39 albino c/c ; $?/?$; $?/?$

For cinnamon to appear in the F$_2$, the F$_1$ parents must be *B/b*. Because the wild type is *B/B*, the albino parent must have been *b/b*. Now the F$_1$ parent can be written *C/c* ; *A/–* ; *B/b*. With such a cross, one-fourth of the progeny would be expected to be albino (*c/c*), which is what is observed. Three-fourths of the remaining progeny would be black, either agouti or nonagouti, and one-fourth would be either cinnamon, if agouti, or chocolate, if nonagouti. Because chocolate is not observed, the F$_1$ parent must not carry the allele for nonagouti. Therefore, the F$_1$ parent is *A/A* and the original albino must have been *c/c* ; *A/A* ; *b/b*.

Cross 2:   P   *c/c* ; *?/?* ; *?/?* × *C/C* ; *A/A* ; *B/B*

            F$_1$ *C/c* ; *A/–* ; *B/–*

            F$_2$ 62 wild type   *C/–* ; *A/–* ; *B/–*
                      18 albino   *c/c* ; *?/?* ; *?/?*

This is a 3:1 ratio, indicating that only one gene is heterozygous in the F$_1$. That gene must be *C/c*. Therefore, the albino parent must be *c/c* ; *A/A* ; *B/B*.

Cross 3:   P   *c/c* ; *?/?* ; *?/?* × *C/C* ; *A/A* ; *B/B*

            F$_1$ *C/c* ; *A/–* ; *B/–*

            F$_2$ 96 wild type   *C/–* ; *A/–* ; *B/–*
                      30 black   *C/–* ; *a/a* ; *B/–*
                      41 albino   *c/c* ; *?/?* ; *?/?*

For a black nonagouti phenotype to appear in the F$_2$, the F$_1$ must have been heterozygous for the *A* gene. Therefore, its genotype can be written *C/c* ; *A/a* ; *B/–* and the albino parent must be *c/c* ; *a/a* ; *?/?*. Among the colored F$_2$ a 3:1 ratio is observed, indicating that only one of the two genes is heterozygous in the F$_1$. Therefore, the F$_1$ must be *C/c* ; *A/a* ; *B/B* and the albino parent must be *c/c* ; *a/a* ; *B/B*.

Cross 4:   P   *c/c* ; *?/?* ; *?/?* × *C/C* ; *A/A* ; *B/B*

            F$_1$ *C/c* ; *A/–* ; *B/–*

            F$_2$ 287 wild type   *C/–* ; *A/–* ; *B/–*
                      86 black   *C/–* ; *a/a* ; *B/–*
                      92 cinnamon   *C/–* ; *A/–* ; *b/b*
                      29 chocolate   *C/–* ; *a/a* ; *b/b*
                      164 albino   *c/c* ; *?/?* ; *?/?*

To get chocolate F_2 progeny the F_1 parent must be heterozygous for all genes and the albino parent must be *c/c* ; *a/a* ; *b/b*.

32. To solve this problem, first restate the information.

A/–	yellow	*A/–* ; *R/–*	gray
R/–	black	*a/a* ; *r/r*	white

The cross is gray × yellow, or *A/–* ; *R/–* × *A/–* ; *r/r*. The F_1 progeny are

$^3/_8$ yellow	$^1/_8$ black
$^3/_8$ gray	$^1/_8$ white

For white progeny, both parents must carry an *r* and an *a* allele. Now the cross can be rewritten as: *A/a* ; *R/r* × *A/a* ; *r/r*

33. **a.** The stated cross is

P   single-combed (*r/r* ; *p/p*) × walnut-combed (*R/R* ; *P/P*)

F_1 *R/r* ; *P/p*   walnut

F_2 9 *R/–* ; *P/–*   walnut
           3 *r/r* ; *P/–*   pea
           3 *R/–* ; *p/p*   rose
           1 *r/r* ; *p/p*   single

b. The stated cross is

P   Walnut-combed × rose-combed

and the F_1 progeny are

Phenotypes		Possible genotypes
$^3/_8$	rose	*R/–* ; *p/p*
$^3/_8$	walnut	*R/–* ; *P/–*
$^1/_8$	pea	*r/r* ; *P/–*
$^1/_8$	single	*r/r* ; *p/p*

The 3 *R/–* : 1 *r/r* ratio indicates that the parents were heterozygous for the *R* gene. The 1 *P/–* : 1 *p/p* ratio indicates a testcross for this gene. Therefore, the parents were *R/r* ; *P/p* and *R/r* ; *p/p*.

c. The stated cross is

P   walnut-combed × rose-combed

F$_1$ walnut ($R/-$; P/p)

To get this result, one of the parents must be homozygous R, but both need not be, and the walnut parent must be homozygous P/P.

d.   The following genotypes produce the walnut phenotype

R/R ; P/P, R/r ; P/P, R/R ; P/p, R/r ; P/p

34.   Notice that the F$_1$ shows a difference in phenotype correlated with sex. At least one of the two genes is X-linked. The F$_2$ ratio suggests independent assortment between the two genes. Because purple is present in the F$_1$, the parental white-eyed male must have at least one P allele. The presence of white eyes in the F$_2$ suggests that the F$_1$ was heterozygous for pigment production, which means that the male also must carry the a allele. A start on the parental genotypes can now be made

P A/A ; $P/P \times a/-$; $P/-$, where – could be either a Y chromosome or a second allele.

The question now is, which gene is X-linked? If the A gene is X-linked, the cross is

P A/A ; $p/p \times a/$Y ; P/P

F$_1$ A/a ; $P/p \times A/$Y ; P/p

All F$_2$ females will inherit the A allele from their father. Under this circumstance, no white-eyed females would be observed. Therefore, the A gene cannot be X-linked. The cross is

P A/A ; $p/p \times a/a$; $P/$Y

F$_1$ A/a ; P/p purple-eyed females
        A/a ; $p/$Y   red-eyed males

F$_2$ Females   Males
        $^3/_8 A/-$; P/p purple $^3/_8 A/-$; $P/$Y purple
        $^3/_8 A/-$; p/p red $^3/_8 A/-$; $p/$Y red
        $^1/_8 a/a$; P/p white $^1/_8 a/a$; $P/$Y white
        $^1/_8 a/a$; p/p white $^1/_8 a/a$; $p/$Y white

35.   The results indicate that two genes are involved (modified 9:3:3:1 ratio), with white blocking the expression of color by the other gene. The ratio of white : color is 3:1, indicating that the F$_1$ is heterozygous (W/w). Among colored dogs, the ratio is 3 black : 1 brown, indicating that black is dominant to brown and the F$_1$ is heterozygous (B/b). The original brown dog is w/w ; b/b and the original white dog is W/W ; B/B. The F$_1$ progeny are W/w ; B/b and the F$_2$ progeny are

9	$W/–$; $B/–$	white
3	w/w ; $B/–$	black
3	$W/–$; b/b	white
1	w/w ; b/b	brown

36. a. The cross is

P $\quad$ td ; su (wild type) $\times$ td^+ ; su^+ (wild type)

F$_1$	1 td ; su	wild type
	1 td ; su^+	requires tryptophan
	1 td^+ ; su^+	wild type
	1 td^+ ; su	wild type

b. 1 tryptophan–dependent : 3 tryptophan–independent

37. P $\quad$ A/A ; B/B ; C/C ; D/D ; S/S $\times$ a/a ; b/b ; c/c ; d/d ; s/s

F$_1$ $\quad$ A/a ; B/b ; C/c ; D/d ; S/s

F$_2$ $\quad$ A/a ; B/b ; C/c ; D/d ; S/s $\times$ A/a ; B/b ; C/c ; D/d ; S/s

$(^3/_4\ A/–)(^3/_4\ B/–)(^3/_4\ C/–)(^3/_4\ D/–)(^3/_4\ S/–) = {}^{243}/_{1024}$	agouti
$(^3/_4\ A/–)(^3/_4\ B/–)(^3/_4\ C/–)(^3/_4\ D/–)(^1/_4\ s/s) = {}^{81}/_{1024}$	spotted agouti
$(^3/_4\ A/–)(^3/_4\ B/–)(^3/_4\ C/–)(^1/_4\ d/d)(^3/_4\ S/–) = {}^{81}/_{1024}$	dilute agouti
$(^3/_4\ A/–)(^3/_4\ B/–)(^3/_4\ C/–)(^1/_4\ d/d)(^1/_4\ s/s) = {}^{27}/_{1024}$	dilute spotted agouti
$(^3/_4\ A/–)(^3/_4\ B/–)(^1/_4\ c/c)(^3/_4\ D/–)(^3/_4\ S/–) = {}^{81}/_{1024}$	albino
$(^3/_4\ A/–)(^3/_4\ B/–)(^1/_4\ c/c)(^3/_4\ D/–)(^1/_4\ s/s) = {}^{27}/_{1024}$	albino
$(^3/_4\ A/–)(^3/_4\ B/–)(^1/_4\ c/c)(^1/_4\ d/d)(^3/_4\ S/–) = {}^{27}/_{1024}$	albino
$(^3/_4\ A/–)(^3/_4\ B/–)(^1/_4\ c/c)(^1/_4\ d/d)(^1/_4\ s/s) = {}^9/_{1024}$	albino
$(^3/_4\ A/–)(^1/_4\ b/b)(^3/_4\ C/–)(^3/_4\ D/–)(^3/_4\ S/–) = {}^{81}/_{1024}$	cinnamon
$(^3/_4\ A/–)(^1/_4\ b/b)(^3/_4\ C/–)(^3/_4\ D/–)(^1/_4\ s/s) = {}^{27}/_{1024}$	cinnamon spotted
$(^3/_4\ A/–)(^1/_4\ b/b)(^3/_4\ C/–)(^1/_4\ d/d)(^3/_4\ S/–) = {}^{27}/_{1024}$	dilute cinnamon
$(^3/_4\ A/–)(^1/_4\ b/b)(^3/_4\ C/–)(^1/_4\ d/d)(^1/_4\ s/s) = {}^9/_{1024}$	dilute spotted cinnamon
$(^3/_4\ A/–)(^1/_4\ b/b)(^1/_4\ c/c)(^3/_4\ D/–)(^3/_4\ S/–) = {}^{27}/_{1024}$	albino
$(^3/_4\ A/–)(^1/_4\ b/b)(^1/_4\ c/c)(^3/_4\ D/–)(^1/_4\ s/s) = {}^9/_{1024}$	albino
$(^3/_4\ A/–)(^1/_4\ b/b)(^1/_4\ c/c)(^1/_4\ d/d)(^3/_4\ S/–) = {}^9/_{1024}$	albino
$(^3/_4\ A/–)(^1/_4\ b/b)(^1/_4\ c/c)(^1/_4\ d/d)(^1/_4\ s/s) = {}^3/_{1024}$	albino
$(^1/_4\ a/a)(^3/_4\ B/–)(^3/_4\ C/–)(^3/_4\ D/–)(^3/_4\ S/–) = {}^{81}/_{1024}$	black

$(^1/_4 \, a/a)(^3/_4 \, B/-)(^3/_4 \, C/-)(^3/_4 \, D/-)(^1/_4 \, s/s) = {}^{27}/_{1024}$ black spotted

$(^1/_4 \, a/a)(^3/_4 \, B/-)(^3/_4 \, C/-)(^1/_4 \, d/d)(^3/_4 \, S/-) = {}^{27}/_{1024}$ dilute black

$(^1/_4 \, a/a)(^3/_4 \, B/-)(^3/_4 \, C/-)(^1/_4 \, d/d)(^1/_4 \, s/s) = {}^{9}/_{1024}$ dilute spotted black

$(^1/_4 \, a/a)(^3/_4 \, B/-)(^1/_4 \, c/c)(^3/_4 \, D/-)(^3/_4 \, S/-) = {}^{27}/_{1024}$ albino

$(^1/_4 \, a/a)(^3/_4 \, B/-)(^1/_4 \, c/c)(^3/_4 \, D/-)(^1/_4 \, s/s) = {}^{9}/_{1024}$ albino

$(^1/_4 \, a/a)(^3/_4 \, B/-)(^1/_4 \, c/c)(^1/_4 \, d/d)(^3/_4 \, S/-) = {}^{9}/_{1024}$ albino

$(^1/_4 \, a/a)(^3/_4 \, B/-)(^1/_4 \, c/c)(^1/_4 \, d/d)(^1/_4 \, s/s) = {}^{3}/_{1024}$ albino

$(^1/_4 \, a/a)(^1/_4 \, b/b)(^3/_4 \, C/-)(^3/_4 \, D/-)(^3/_4 \, S/-) = {}^{27}/_{1024}$ brown

$(^1/_4 \, a/a)(^1/_4 \, b/b)(^3/_4 \, C/-)(^3/_4 \, D/-)(^1/_4 \, s/s) = {}^{9}/_{1024}$ brown spotted

$(^1/_4 \, a/a)(^1/_4 \, b/b)(^3/_4 \, C/-)(^1/_4 \, d/d)(^3/_4 \, S/-) = {}^{9}/_{1024}$ dilute brown

$(^1/_4 \, a/a)(^1/_4 \, b/b)(^3/_4 \, C/-)(^1/_4 \, d/d)(^1/_4 \, s/s) = {}^{3}/_{1024}$ dilute spotted brown

$(^1/_4 \, a/a)(^1/_4 \, b/b)(^1/_4 \, c/c)(^3/_4 \, D/-)(^3/_4 \, S/-) = {}^{9}/_{1024}$ albino

$(^1/_4 \, a/a)(^1/_4 \, b/b)(^1/_4 \, c/c)(^3/_4 \, D/-)(^1/_4 \, s/s) = {}^{3}/_{1024}$ albino

$(^1/_4 \, a/a)(^1/_4 \, b/b)(^1/_4 \, c/c)(^1/_4 \, d/d)(^3/_4 \, S/-) = {}^{3}/_{1024}$ albino

$(^1/_4 \, a/a)(^1/_4 \, b/b)(^1/_4 \, c/c)(^1/_4 \, d/d)(^1/_4 \, s/s) = {}^{1}/_{1024}$ albino

38.   **a. and b.** The two starting lines are i/i ; D/D ; M/M ; W/W and I/I ; d/d ; m/m ; w/w, and you are seeking i/i ; d/d ; m/m ; w/w. There are many ways to proceed, one of which follows below.

   I. i/i ; D/D ; M/M ; W/W × I/I ; d/d ; m/m ; w/w

   II. I/i ; D/d ; M/m ; W/w × I/i ; D/d ; M/m ; W/w

   III.   Select i/i ; d/d ; m/m ; w/w, which has a probability of $(^1/_4)^4 = {}^1/_{256}$.

   c. and d. In the first cross, all progeny chickens will be of the desired genotype to proceed to the second cross. Therefore, the only problems are to be sure that progeny of both sexes are obtained for the second cross, which is relatively easy, and that enough females are obtained to make the time required for the desired genotype to appear feasible. Because chickens lay one to two eggs a day, the more females who are egg–laying, the faster the desired genotype will be obtained. In addition, it will be necessary to obtain a male and a female of the desired genotype in order to establish a pure breeding line.

   Assume that each female will lay two eggs a day, that money is no problem, and that excess males cause no problems. By hatching 200 eggs from the first cross, approximately 100 females will be available for the second cross. These 100 females will produce 200 eggs each day. Thus, in one week a total of 1,400 eggs will be produced. Of these 1,400 eggs, there will be approximately 700 of each sex. At a probability of $^1/_{256}$, the desired genotype should be achieved 2.7 times for each sex within that first week.

39. Pedigrees like this are quite common. They indicate lack of penetrance due to epistasis or environmental effects. Individual A must have the dominant autosomal gene, even though she does not express the trait, as both her children are affected.

40. In cross 1, the following can be written immediately

P $M/–$; $D/–$; w/w (dark purple) $\times$ m/m ; $?/?$; $?/?$ (white with yellowish spots)

F$_1$ $^1/_2$ $M/–$; $D/–$; w/w dark purple
         $^1/_2$ $M/–$; d/d ; w/w light purple

All progeny are colored, indicating that no W allele is present in the parents. Because the progeny are in a 1:1 ratio, only one of the genes in the parents is heterozygous. Also, the light purple progeny, d/d, indicates which gene that is. Therefore, the genotypes must be

P M/M ; D/d ; w/w $\times$ m/m ; d/d ; w/w

F$_1$ $^1/_2$ M/m ; D/d ; w/w
         $^1/_2$ M/m ; d/d ; w/w

In cross 2, the following can be written immediately

P m/m ; $?/?$; $?/?$ (white with yellowish spots) $\times$ $M/–$; d/d ; w/w (light purple)

F$_1$ $^1/_2$ $M/–$; $?/?$; $W/–$ white with purple spots
         $^1/_4$ $M/–$; $D/–$; w/w dark purple
         $^1/_4$ $M/–$; d/d ; w/w light purple

For light and dark purple to appear in a 1:1 ratio among the colored plants, one of the parents must be heterozygous D/d. The ratio of white to colored is 1:1, a testcross, so one of the parents is heterozygous W/w. All plants are purple, indicating that one parent is homozygous M/M. Therefore, the genotypes are

P m/m ; D/d ; W/w $\times$ M/M ; d/d ; w/w

F$_1$ $^1/_2$ M/m ; $–/d$; W/w
         $^1/_4$ M/m ; D/d ; w/w
         $^1/_4$ M/m ; d/d ; w/w

41. a. Note that the first two crosses are reciprocal and that the male offspring differ in phenotype between the two crosses. This indicates that the gene is on the X chromosome.

Also note that the F$_1$ females in the first two crosses are sickle. This indicates that sickle is dominant to round. The third cross also indicates that oval is dominant to sickle. Therefore, this is a multiple allelic series with oval > sickle > round.

Let W^o = oval, W^s = sickle, and W^r = round. The three crosses are

Cross 1: $W^s/W^s \times W^r/Y \rightarrow W^s/W^r$ and W^s/Y

Cross 2: $W^r/W^x \times W^s/Y \rightarrow W^s/W^r$ and W^r/Y

Cross 3: $W^s/W^s \times W^o/Y \rightarrow W^o/W^s$ and W^s/Y

b. $W^o/W^s \times W^r/Y$

$1/4$ W^o/W^r	female oval
$1/4$ W^s/W^r	female sickle
$1/4$ W^o/Y	male oval
$1/4$ W^s/Y	male sickle

42. a. First note that there is a phenotypic difference between males and females, indicating X linkage. This means that the male progeny express both alleles of the female parent. A beginning statement of the genotypes could be as follows, where H indicates the gene and the numbers 1, 2, and 3 indicate the variants.

P $H^1/H^2 \times H^3/Y$

F$_1$

$1/4$ H^1/H^3	one-banded female
$1/4$ H^3/H^2	three-banded female
$1/4$ H^1/Y	one-banded male
$1/4$ H^2/Y	two-banded male

Because both female F$_1$ progeny obtain allele H^3 from their father, yet only one has a three-banded pattern, there is obviously a dominance relationship among the alleles. The mother indicates that H^1 is dominant to H^2. The female progeny must be H^1/H^3 and H^3/H^2, and they have a one-banded and three-banded pattern, respectively. H^3 must be dominant to H^2 because the daughter with that combination of alleles has to be the three-banded daughter. In other words, there is a multiple–allelic series with $H^1 > H^3 > H^2$.

b. The cross is

P $\quad H^3/H^2 \times H^1/Y$

F_1

$\quad\quad\quad$ $1/4\ H^1/H^3$ $\quad\quad$ one-banded female

$\quad\quad\quad$ $1/4\ H^1/H^2$ $\quad\quad$ one-banded female

$\quad\quad\quad$ $1/4\ H^2/Y$ $\quad\quad$ two-banded male

$\quad\quad\quad$ $1/4\ H^3/Y$ $\quad\quad$ three-banded male

43. **a.** The first two crosses indicate that wild-type is dominant to both platinum and aleutian. The third cross indicates that two genes are involved rather than one gene with multiple alleles, because, a 9:3:3:1 ratio is observed.

Let platinum be a, aleutian be b, and wild type be $A/-$; $B/-$.

Cross 1:	P	A/A ; $B/B \times a/a$; B/B	wild-type × platinum
	F_1	A/a ; B/B	all wild-type
	F_2	3 $A/-$; B/B : 1 a/a ; B/B	3 wild-type : 1 platinum
Cross 2:	P	A/A ; $B/B \times A/A$; b/b	wild-type × aleutian
	F_1	A/A ; B/b	all wild-type
	F_2	3 A/A ; $B/-$: 1 A/A ; b/b	3 wild-type : 1 aleutian
Cross 3:	P	a/a ; $B/B \times A/A$; b/b	platinum × aleutian
	F_1	A/a ; B/b	all wild-type
	F_2	9 $A/-$; $B/-$	wild-type
		3 $A/-$; b/b	aleutian
		3 a/a ; $B/-$	platinum
		1 a/a ; b/b	sapphire

b. sapphire × platinum $\quad\quad\quad\quad\quad\quad\quad\quad$ sapphire × aleutian

$\quad$ P $\quad$ a/a ; $b/b \times a/a$; B/B $\quad\quad\quad\quad$ a/a ; b/b × A/A ; b/b

$\quad$ F_1 $\quad$ a/a ; B/b $\quad\quad$ platinum $\quad\quad\quad\quad$ A/a ; b/b $\quad\quad\quad$ aleutian

$\quad$ F_2 $\quad$ 3 a/a ; $B/-$ $\quad\quad$ platinum $\quad\quad\quad$ 3 $A/-$; b/b $\quad\quad$ aleutian

$\quad\quad\quad\quad$ 1 a/a ; b/b $\quad\quad$ sapphire $\quad\quad\quad$ 1 a/a ; b/b $\quad\quad$ sapphire

44. **a.** The genotypes are

P $\quad\quad$ B/B ; $i/i \times b/b$; I/I

F_1 $\quad\quad$ B/b ; I/i $\quad\quad\quad$ hairless

F_2 9 $B/-$; $I/-$ hairless
       3 $B/-$; i/i straight
       3 b/b ; $I/-$ hairless
       1 b/b ; i/i bent

b. In order to solve this problem, first write as much as you can of the progeny genotypes.

     4 hairless $-/-$; $I/-$
     3 straight $B/-$; i/i
     1 bent b/b ; i/i

Each parent must have a b allele and an i allele. The partial genotypes of the parents are

$$-/b \; ; -/i \; \times \; -/b \; ; -/i$$

At least one parent carries the B allele, and at least one parent carries the I allele. Assume for a moment that the first parent carries both. The partial genotypes become

$$B/b \; ; I/i \; \times \; -/b \; ; -/i$$

Note that $1/2$ the progeny are hairless. This must come from a $I/i \times i/i$ cross. Of those progeny with hair, the ratio is 3:1, which must come from a $B/b \times B/b$ cross. The final genotypes are therefore

$$B/b \; ; I/i \; \times \; B/b \; ; i/i$$

45. a. The first question is whether there are two genes or one gene with three alleles. Note that a black × eyeless cross produces black and brown progeny in one instance but black and eyeless progeny in the second instance. Further note that a black × black cross produces brown, a brown × eyeless cross produces brown, and a brown × black cross produces eyeless. The results are confusing enough that the best way to proceed is by trial and error.

There are two possibilities: one gene or two genes.

One gene. Assume one gene for a moment. Let the gene be E and assume the following designations
      E^1 = black
      E^2 = brown
      E^3 = eyeless

If you next assume, based on the various crosses, that black > brown > eyeless, genotypes in the pedigree become

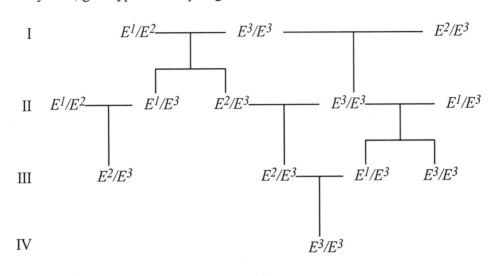

b. The genotype of individual II-3 is E^2/E^3.

Two genes. If two genes are assumed, then questions arise regarding whether both are autosomal or if one might be X-linked. Eye color appears to be autosomal. The presence or absence of eyes appears could be X-linked. Let

B = black X^E = normal eyes
b = brown X^e = eyeless

The pedigree then is

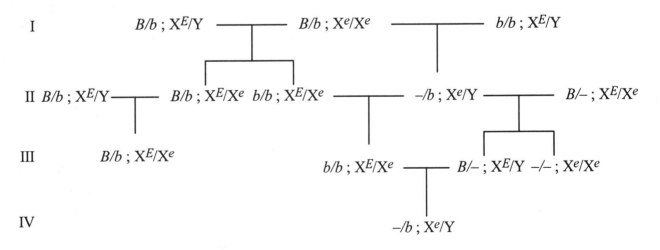

With this interpretation of the pedigree, individual II–3 is b/b ; X^E/X^e.

Without further data, it is impossible to choose scientifically between the two possible explanations. A basic rule in science is that the more simple answer should be used until such time as data exist that would lead to a rejection of it. In this case, the one-gene explanation is essentially as complex as the two-gene explanation, with one exception. The exception is that eye color is fundamentally different from a lack of eyes. Therefore, the better guess is that two genes are involved.

Questions 46-48 all ask you to compare expected results to actual data to determine if a certain hypothesis is correct. In Chapter 2, you learned how to use a specific analytical test (χ^2) to determine if actual observations deviate from expectations purely on the basis of chance. This would be a good time to review that material to confirm your answers.

46.   For the cross $B/b \times B/b$, you expect $^3/_4$ $B/–$ and $^1/_4$ b/b or for 400 progeny, 300 and 100 respectively. This should be compared to the actual data of 280 and 120. Alternatively, if the B/B genotype was actually lethal, then you would expect $^2/_3$ B/b and $^1/_3$ b/b or for 400 progeny, 267 and 133 respectively. This compares more favorably to the actual data.

47.   There are a total of 159 progeny that should be distributed in 9:3:3:1 ratio if the two genes are assorting independently. You can see that

Observed	Expected
88 $P/–$; $Q/–$	90
32 $P/–$; q/q	30
25 p/p ; $Q/–$	30
14 p/p ; q/q	10

48.   For 160 progeny, there should be 90:70, 130:30, or 120:40 of phenotype 1 : phenotype 2 progeny for the various expected ratios. Only the 9:7 ratio seems feasible and that would indicate that two genes are assorting independently where

$A/–$; $B/–$	phenotype 1
$A/–$; b/b	phenotype 2
a/a ; $B/–$	phenotype 2
a/a ; b/b	phenotype 2

49.   **a.**   Intercrossing mutant strains that all share a common recessive phenotype is the basis of the complementation test. This test is designed to identify the number of different genes that can mutate to a particular phenotype. If the progeny of a given cross still express the mutant phenotype, the mutations fail to complement and are considered alleles of the same gene; if the progeny are wild type, the mutations complement and the two strains carry mutant alleles of separate genes.

b. There are 3 genes represented in these crosses. All crosses except 2 × 3 (or 3 × 2) complement and indicate that the strains are mutant for separate genes. Strains 2 and 3 fail to complement and are mutant for the same gene.

c. Let A and a represent alleles of gene 1; B and b represent alleles of gene 4; and c^2, c^3, and C represent alleles of gene 3.

Line 1: $a/a \cdot B/B \cdot C/C$
Line 2: $A/A \cdot B/B \cdot c^2/c^2$
Line 3: $A/A \cdot B/B \cdot c^3/c^3$
Line 4: $A/A \cdot b/b \cdot C/C$

	Cross	Genotype	Phenotype
F_1s	1 × 2	$A/a \cdot B/B \cdot C/c^2$	wild type
	1 × 3	$A/a \cdot B/B \cdot C/c^3$	wild type
	1 × 4	$A/a \cdot B/b \cdot C/C$	wild type
	2 × 3	$A/A \cdot B/B \cdot c^2/c^3$	mutant
	2 × 4	$A/A \cdot B/b \cdot C/c^2$	wild type
	3 × 4	$A/A \cdot B/b \cdot C/c^3$	wild type

d. With the exception that strain 2 and 3 fail to complement and therefore have mutations in the same gene, this test does not give evidence of linkage. To test linkage, the F_1s should be crossed to tester strains (homozygous recessive strains) and segregation of the mutant phenotype followed. If the genes are unlinked, for example A/a ; $B/b \times a/a$; b/b, then 25 percent of the progeny will be wild type (A/a ; B/b) and 75 percent will be mutant (25 percent A/a ; b/b, 25 percent a/a ; B/b, and 25 percent a/a ; b/b). If the genes are linked ($a\ B/a\ B \times A\ b/A\ b$) then only one half of the recombinants (i.e., less than 25 percent of the total progeny) will be wild type ($A\ B/a\ b$).

e. No. All it tells you is that among these strains, there are three genes represented. If genetic dissection of leg coordination was desired, large screens for the mutant phenotype would be executed with the attempt to "saturate" the genome with mutations in all genes involved in the process.

50. **a.** Complementation refers to gene products within a cell, which is not what is happening here. Most likely, what is known as cross-feeding is occurring, whereby a product made by one strain diffuses to another strain and allows growth of the second strain. This is equivalent to supplementing the medium. Because cross-feeding seems to be taking place, the suggestion is that the strains are blocked at different points in the metabolic pathway.

b. For cross-feeding to occur, the growing strain must have a block that occurs earlier in the metabolic pathway than does the block in the strain from which it is obtaining the product for growth.

c. The *trpE* strain grows when cross-fed by either *trpD* or *trpB,* but the converse is not true (placing *trpE* earlier in the pathway than either *trpD* or *trpB*), and *trpD* grows when cross-fed by *trpB* (placing *trpD* prior to *trpB*). This suggests that the metabolic pathway is

$$trpE \rightarrow trpD \rightarrow trpB$$

d. Without some tryptophan, no growth at all would occur, and the cells would not have lived long enough to produce a product that could diffuse.

CHALLENGING PROBLEMS

51. Note that the F_2 are in an approximate 9:6:1 ratio. This suggests a dihybrid cross in which $A/-$; b/b has the same appearance as a/a ; $B/-$. Let the disc phenotype be the result of $A/-$; $B/-$ and the long phenotype be the result of a/a ; b/b. The crosses are

P A/A ; B/B (disc) × a/a ; b/b (long)

F_1 A/a ; B/b (disc)

F_2 9 $A/-$; $B/-$ (disc)
         3   a/a ; $B/-$ (sphere)
         3     $A/-$; b/b (sphere)
         1   a/a ; b/b (long)

52. a. The best explanation is that Marfan's syndrome is inherited as a dominant autosomal trait, because roughly half of the children of all affected individuals also express the trait. If it were recessive, then all individuals marrying affected spouses would have to be heterozygous for an allele that, when homozygous, causes Marfan's.

b. The pedigree shows both pleiotropy (multiple affected traits) and variable expressivity (variable degree of expressed phenotype). Penetrance is the percentage of individuals with a specific genotype who express the associated phenotype. There is no evidence of decreased penetrance in this pedigree.

c. Pleiotropy indicates that the gene product is required in a number of different tissues, organs, or processes. When the gene is mutant, all tissues needing the gene product will be affected. Variable expressivity of a phenotype for a given genotype indicates modification by one or more other genes, random noise, or environmental effects.

53.

Cross	Results	Conclusion
A/– ; *C/–* ; *R/–* × *a/a* ; *c/c* ; *R/R*	50% colored	Colored or white will depend on the *A* and *C* genes. Because half the seeds are colored, one of the two genes is heterozygous.
A/– ; *C/–* ; *R/–* × *a/a* ; *C/C* ; *r/r*	25% colored	Color depends on *A* and *R* in this cross. If only one gene were heterozygous, 50% would be colored. Therefore, both *A* and *R* are heterozygous. The seed is *A/a* ; *C/C*; *R/r*.
A/– ; *C/–* ; *R/–* × *A/A* ; *c/c* ; *r/r*	50% colored	This supports the above conclusion.

54. a. The *A/A* ; *C/C* ; *R/R* ; *pr/pr* parent produces pigment that is not converted to purple. The phenotype is red. The *a/a* ; *c/c* ; *r/r* ; *Pr/Pr* does not produce pigment. The phenotype is yellow.

b. The F$_1$ will be *A/a* ; *C/c* ; *R/r* ; *Pr/pr*, which will produce pigment. The pigment will be converted to purple.

c. The difficult way to determine the phenotypic ratios is to do a branch diagram, yielding the following results.

$81/256$ *A/–* ; *C/–* ; *R/–* ; *Pr/–* purple $9/256$ *a/a* ; *C/–* ; *R/–* ; *Pr/–* yellow
$27/256$ *A/–* ; *C/–* ; *R/–* ; *pr/pr* red $9/256$ *a/a* ; *C/–* ; *R/–* ; *pr/pr* yellow
$27/256$ *A/–* ; *C/–* ; *r/r* ; *Pr/–* yellow $9/256$ *a/a* ; *C/–* ; *r/r* ; *Pr/–* yellow
$9/256$ *A/–* ; *C/–* ; *r/r* ; *pr/pr* yellow $3/256$ *a/a* ; *C/–* ; *r/r* ; *pr/pr* yellow
$27/256$ *A/–* ; *c/c* ; *R/–* ; *Pr/–* yellow $27/256$ *a/a* ; *c/c* ; *R/–* ; *Pr/–* yellow
$9/256$ *A/–* ; *c/c* ; *R/–* ; *pr/pr* yellow $3/256$ *a/a* ; *c/c* ; *R/–* ; *pr/pr* yellow
$9/256$ *A/–* ; *c/c* ; *r/r* ; *Pr/–* yellow $3/256$ *a/a* ; *c/c* ; *r/r* ; *Pr/–* yellow
$3/256$ *A/–* ; *c/c* ; *r/r* ; *pr/pr* yellow $1/256$ *a/a* ; *c/c* ; *r/r* ; *pr/pr* yellow

The final phenotypic ratio is 81 purple : 27 red : 148 yellow.

The easier method of determining phenotypic ratios is to recognize that four genes are involved in a heterozygous × heterozygous cross. Purple requires a dominant allele for each gene. The probability of all dominant alleles is $(3/4)^4 = 81/256$. Red results from all dominant alleles except for the *Pr* gene. The probability of that outcome is $(3/4)^3(1/4) = 27/256$. The remainder of the outcomes will produce no pigment, resulting in yellow. That probability is $1 - {}^{81}/_{256} - {}^{27}/_{256} = {}^{148}/_{256}$.

d. The cross is A/a ; C/c ; R/r ; $Pr/pr \times a/a$; c/c ; r/r ; pr/pr. Again, either a branch diagram or the easier method can be used. The final probabilities are

$$\text{purple} = (^1/_2)^4 = {}^1/_{16}$$
$$\text{red} = (^1/_2)^3(^1/_2) = {}^1/_{16}$$
$$\text{yellow} = 1 - {}^1/_{16} - {}^1/_{16} = {}^7/_8$$

55. a. This type of gene interaction is called *epistasis*. The phenotype of e/e is epistatic to the phenotypes of $B/-$ or b/b.

b. The progeny of generation I have all possible phenotypes. Progeny II-3 is beige (e/e) so both parents must be heterozygous E/e. Progeny II-4 is brown (b/b) so both parents must also be heterozygous B/b. Progeny III-3 and III-5 are brown so II-2 and II-5 must be B/b. Progeny III-2 and III-7 are beige (e/e) so all their parents must be E/e.

The following are the inferred genotypes

I   1 (B/b E/e)   2 (B/b E/e)

II   1 (b/b E/e)   2 (B/b E/e)   3 ($-/-$ e/e)   4 (b/b $E/-$)   5 (B/b E/e)
      6 (b/b E/e)

III 1 (B/b $E/-$)   2 ($-/b$ e/e)   3 (b/b, $E/-$)   4 (B/b $E/-$)   5 (b/b $E/-$)
      6 (B/b $E/-$)   7 ($-/b$ e/e)

56. a. This is a dihybrid cross resulting in a 13 white : 3 red ratio of progeny in the F_2. This ratio of white to red indicates that the double recessive is not the red phenotype. Instead, the general formula for color is represented by a/a ; $B/-$.

Let line 1 be A/A ; B/B and line 2 be a/a ; b/b. The F_1 is A/a ; B/b. Assume that A blocks color in line 1 and b/b blocks color in line 2. The F_1 will be white because of the presence of A. The F_2 would be

9 $A/-$; $B/-$ white
3 $A/-$; b/b white
3 a/a ; $B/-$ red
1 a/a ; b/b white

b. Cross 1: A/A ; $B/B \times A/a$; B/b $\rightarrow$ all $A/-$; $B/-$ white

Cross 2: a/a ; $b/b \times A/a$; B/b

        ${}^1/_4$ A/a ; B/b white

$^1/_4$ A/a ; b/b white

$^1/_4$ a/a ; b/b white

$^1/_4$ a/a ; B/b red

57. **a.** Note that blue is always present, indicating E/E (blue) in both parents. Because of the ratios that are observed, neither C nor D is varying. In this case, the gene pairs that are involved are A/a and B/b. The parents are A/A ; b/b × a/a ; B/B or A/A ; B/B × a/a ; b/b.

The F_1 are A/a ; B/b and the F_2 are

9	$A/-$; $B/-$	blue + red, or purple
3	$A/-$; b/b	blue + yellow, or green
3	a/a ; $B/-$	blue + white$_2$, or blue
1	a/a ; b/b	blue + white$_2$, or blue

b. Blue is not always present, indicating E/e in the F_1. Because green never appears, the F_1 must be B/B ; C/C ; D/D. The parents are A/A ; e/e × a/a ; E/E or A/A ; E/E × a/a ; e/e.

The F_1 are A/a ; E/e, and the F_2 are

9	$A/-$; $E/-$	red + blue, or purple
3	$A/-$; e/e	red + white$_1$, or red
3	a/a ; $E/-$	white$_2$ + blue, or blue
1	a/a ; e/e	white$_2$ + white$_1$, or white

c. Blue is always present, indicating that the F_1 is E/E. No green appears, indicating that the F_1 is also B/B. The two genes involved are A and D. The parents are A/A ; d/d × a/a ; D/D or A/A ; D/D × a/a ; d/d.

The F_1 are A/a ; D/d and the F_2 are

9	$A/-$; $D/-$	blue + red + white$_4$, or purple
3	$A/-$; d/d	blue + red, or purple
3	a/a ; $D/-$	blue + white$_2$ + white$_4$, or blue
1	a/a ; d/d	white$_2$ + blue + red, or purple

d. The presence of yellow indicates b/b ; e/e in the F_2. Therefore, the parents are B/B ; e/e × b/b ; E/E or B/B ; E/E × b/b ; e/e.

The F_1 are B/b ; E/e and the F_2 are

9	$B/-$; $E/-$	red + blue, or purple

3	$B/-$; e/e	red + white$_1$, or red
3	b/b ; $E/-$	yellow + blue, or green
1	b/b ; e/e	yellow + white$_1$, or yellow

   e. Mutations in D suppress mutations in A.

   f. Recessive alleles of A will be epistatic to mutations in B.

58. a. Note that cross 1 suggests that one gene is involved and that single is dominant to double. Cross 2 supports this conclusion. Now, note that in crosses 3 and 4, a 1:1 ratio is seen in the progeny, suggesting that superdouble is an allele of both single and double. Superdouble must be heterozygous, however, and it must be dominant to both single and double. Because the heterozygous superdouble yields both single and double when crossed with the appropriate plant, it cannot be heterozygous for the single allele. Therefore, it must be heterozygous for the double allele. A multiple allelic series has been detected: superdouble > single > double.

For now, assume that only one gene is involved and attempt to rationalize the crosses with the assumptions made above.

Cross	Parents	Progeny	Conclusion
1	$A^S/A^S \times A^D/A^D$	A^S/A^D	A^S is dominant to A^D
2	$A^S/A^D \times A^S A^D$	$3\ A^S/- : 1\ A^D/A^D$	supports above conclusion
3	$A^D/A^D \times A^{Sd}/A^D$	$1\ A^{Sd}/A^D : 1\ A^D/A^D$	A^{Sd} is dominant to A^D
4	$A^S/A^S \times A^{Sd}/A^D$	$1\ A^{Sd}/A^S : 1\ A^S/A^D$	A^{Sd} is dominant to A^S
5	$A^D/A^D \times A^{Sd}/A^S$	$1\ A^{Sd}/A^D : 1\ A^D/A^S$	supports conclusion of heterozygous superdouble
6	$A^D/A^D \times A^S/A^D$	$1\ A^D/A^D : 1\ A^D/A^S$	supports conclusion of heterozygous superdouble

   b. While this explanation does rationalize all the crosses, it does not take into account either the female sterility or the origin of the superdouble plant from a double-flowered variety.

A number of genetic mechanisms could be proposed to explain the origin of superdouble from the double-flowered variety. Most of the mechanisms will be discussed in later chapters and so will not be mentioned here. However, it can be safely assumed at this point that, whatever the mechanism, it was aberrant enough to block the proper formation of the complex structure of the female flower. Because of female sterility, no homozygote for superdouble can be observed.

59. **a.** All the crosses suggest two independently assorting genes. However, that does not mean that there are a total of only two genes governing eye color. In fact, there are four genes controlling eye color that are being studied here. Let

a/a = defect in the yellow line 1
b/b = defect in the yellow line 2
d/d = defect in the brown line
e/e = defect in the orange line

The genotypes of each line are as follows

yellow 1:	a/a ; B/B ; D/D ; E/E
yellow 2:	A/A ; b/b ; D/D ; E/E
brown:	A/A ; B/B ; d/d ; E/E
orange:	A/A ; B/B ; D/D ; e/e

b. P a/a ; B/B ; D/D ; E/E × A/A ; b/b ; D/D ; E/E yellow 1 × yellow 2

F_1 A/a ; B/b ; D/D ; E/E red

F_2 9 $A/-$; $B/-$; D/D ; E/E red
   3 a/a ; $B/-$; D/D ; E/E yellow
   3 $A/-$; b/b ; D/D ; E/E yellow
   1 a/a ; b/b ; D/D ; E/E yellow

P a/a ; B/B ; D/D ; E/E × A/A ; B/B ; d/d ; E/E yellow 1 × brown

F_1 A/a ; B/B ; D/d ; E/E red

F_2 9 $A/-$; B/B ; $D/-$; E/E red
   3 a/a ; B/B ; $D/-$; E/E yellow
   3 $A/-$; B/B ; d/d ; E/E brown
   1 a/a ; B/B ; d/d ; E/E yellow

P a/a ; B/B ; D/D ; E/E × A/A ; B/B ; D/D ; e/e yellow 1 × orange

F_1 A/a ; B/B ; D/D ; E/e red

F_2 9 $A/-$; B/B ; D/D ; $E/-$ red
   3 a/a ; B/B ; D/D ; $E/-$ yellow
   3 $A/-$; B/B ; D/D ; e/e orange
   1 a/a ; B/B ; D/D ; e/e orange

P A/A ; b/b ; D/D ; E/E × A/A ; B/B ; d/d ; E/E yellow 2 × brown

F$_1$ A/A ; B/b ; D/d ; E/E red

F$_2$ 9 A/A ; $B/-$; $D/-$; E/E red
    3 A/A ; b/b ; $D/-$; E/E yellow
    3 A/A ; $B/-$; d/d ; E/E brown
    1 A/A ; b/b ; d/d ; E/E yellow

P A/A ; b/b ; D/D ; E/E × A/A ; B/B ; D/D ; e/e yellow 2 × orange

F$_1$ A/A ; B/b ; D/D ; E/e red

F$_2$ 9 A/A ; B$/-$; D/D ; $E/-$ red
    3 A/A ; b/b ; D/D ; $E/-$ yellow
    3 A/A ; $B/-$; D/D ; e/e orange
    1 A/A ; b/b ; D/D ; e/e yellow

P A/A ; B/B ; d/d ; E/E × A/A ; B/B ; D/D ; e/e brown × orange

F$_1$ A/A ; B/B ; D/d ; E/e red

F$_2$ 9 A/A ; B/B ; $D/-$; $E/-$ red
    3 A/A ; B/B ; d/d ; $E/-$ brown
    3 A/A ; B/B ; $D/-$; e/e orange
    1 A/A ; B/B ; d/d ; e/e orange

c. When constructing a biochemical pathway, remember that the earliest gene that is defective in a pathway will determine the phenotype of a doubly defective genotype. Look at the following doubly recessive homozygotes from the crosses. Notice that the double defect d/d ; e/e has the same phenotype as the defect a/a ; e/e. This suggests that the E gene functions earlier than do the A and D genes. Using this logic, the following table can be constructed

Genotype	Phenotype	Conclusion
a/a ; B/B ; D/D ; e/e	orange	E functions before A
A/A ; B/B ; d/d ; e/e	orange	E functions before D
a/a ; b/b ; D/D ; E/E	yellow	B functions before A
$A/$A ; b/b ; D/D ; e/e	yellow	B functions before E
a/a ; B/B ; d/d ; E/E	yellow	A functions before D
A/A ; b/b ; d/d ; E/E	yellow	B functions before D

The genes function in the following sequence: *B*, *E*, *A*, *D*. The metabolic path is

$$\text{yellow 2} \underset{B}{\rightarrow} \text{orange} \underset{E}{\rightarrow} \text{yellow} \underset{A}{\rightarrow} \text{brown} \underset{D}{\rightarrow} \text{red}$$

60. a. A trihybrid cross would give a 63:1 ratio. Therefore, there are three *R* loci segregating in this cross.

b.

P $R_1/R_1 ; R_2/R_2 ; R_3/R_3 \times r_1/r_1 ; r_2/r_2 ; r_3/r_3$

F_1 $R_1/r_1 ; R_2/r_2 ; R_3/r_3$

F_2

27	$R_1/- ; R_2/- ; R_3/-$	red
9	$R_1/- ; R_2/- ; r_3/r_3$	red
9	$R_1/- ; r_2/r_2 ; R_3/-$	red
9	$r_1/r_1 ; R_2/- ; R_3/-$	red
3	$R_1/- ; r_2/r_2 ; r_3/r_3$	red
3	$r_1/r_1 ; R_2/- ; r_3/r_3$	red
3	$r_1/r_1 ; r_2/r_2 ; R_3/-$	red
1	$r_1/r_1 ; r_2/r_2 ; r_3/r_3$	white

c. (1) In order to obtain a 1:1 ratio, only one of the genes can be heterozygous. A representative cross would be $R_1/r_1 ; r_2/r_2 ; r_3/r_3 \times r_1/r_1 ; r_2/r_2 ; r_3/r_3$.

(2) In order to obtain a 3 red : 1 white ratio, two alleles must be segregating and they cannot be within the same gene. A representative cross would be $R_1/r_1 ; R_2/r_2 ; r_3/r_3 \times r_1/r_1 ; r_2/r_2 ; r_3/r_3$.

(3) In order to obtain a 7 red : 1 white ratio, three alleles must be segregating, and they cannot be within the same gene. The cross would be $R_1/r_1 ; R_2/r_2 ; R_3/r_3 \times r_1/r_1 ; r_2/r_2 ; r_3/r_3$.

d. The formula is $1 - (1/2)^n$, where n = the number of loci that are segregating in the representative crosses above.

61. a. The trait is recessive (parents without the trait have children with the trait) and autosomal (daughters can inherit the trait from unaffected fathers). Looking to generation III, there is also evidence that there are two different genes that, when defective, result in deaf-mutism.

Assuming that one gene has alleles *A* and *a*, and the other has *B* and *b*. The following genotypes can be inferred:

I-1 and I-2   *A/a ; B/B*   I-3 and I-4   *A/A ; B/b*

II-(1, 3, 4, 5, 6)	A/– ; B/B	II- (9, 10, 12, 13, 14, 15)	A/A ; B/–
II-2 and II-7	a/a ; B/B	I-8 and II-11	A/A ; b/b

b. Generation III shows complementation. All are A/a ; B/b.

62. a. The first impression from the pedigree is that the gene causing blue sclera and brittle bones is pleiotropic with variable expressivity. If two genes were involved, it would be highly unlikely that all people with brittle bones also had blue sclera.

b. Sons and daughters inherit from affected fathers so the allele appears to be autosomal.

c. The trait appears to be inherited as a dominant but with incomplete penetrance. For the trait to be recessive, many of the non-related individuals marrying into the pedigree would have to be heterozygous (e.g., I-1, I-3, II-8, II-11). Individuals II-4, II-14, III-2, and III-14 have descendants with the disorder although they do not themselves express the disorder. Therefore, $4/20$ people that can be inferred to carry the gene, do not express the trait. That is 80 percent penetrance. (Penetrance could be significantly less than that since many possible carriers have no shown progeny.) The pedigree also exhibits variable expressivity. Of the 16 individuals who have blue sclera, 10 do not have brittle bones. Usually, expressivity is put in terms of none, variable, and highly variable, rather than expressed as percentages.

63. a. and b. Assuming that both the *Brown* and the *Van Scoy* lines were homozygous, the parental cross suggests that nonhygienic behavior is dominant to hygienic. A consideration of the specific behavior of the $F_1 \times$ *Brown* progeny, however, suggests that the behavior is separable into two processes; one involves uncapping and one involves removal of dead pupae. If this is true, uncapping and removal of dead pupae, two behaviors that normally go together in the *Brown* line, have been separated in the F_2 progeny, suggesting that they are controlled by different and unlinked genes. Those bees that lack uncapping behavior are still able to express removal of dead pupae if environmental conditions are such that they do not need to uncap a compartment first. Also, the uncapping behavior is epistatic to removal of dead pupae.

Let U = no uncapping, u = uncapping, R = no removal, r = removal

P u/u ; r/r × U/U ; R/R
       (*Brown*) (*Van Scoy*)

       U/u ; R/r × u/u ; r/r (F_1 cross *Brown*)

Progeny $^1/_4$ *U/u* ; *R/r*   nonhygienic

               $^1/_4$ *U/u* ; *r/r*   nonhygienic (no uncapping but removal if
                                         uncapped)

               $^1/_4$ *u/u* ; *R/r*   uncapping but no removal

               $^1/_4$ *u/u* ; *r/r*   hygienic

64. **a. and b.** Cross 1 indicates that orange is dominant to yellow. Crosses 2–4 indicate that red is dominant to orange, yellow, and white. Crosses 5–7 indicate that there are two genes involved in the production of color. Cross 5 indicates that yellow and white are different genes. Cross 6 indicates that orange and white are different genes.

In other words, epistasis is involved and the homozygous recessive white genotype seems to block production of color by a second gene.

Begin by explaining the crosses in the simplest manner possible until such time as it becomes necessary to add complexity. Therefore, assume that orange and yellow are alleles of gene *A* with orange dominant.

Cross 1:   P   *A/A* × *a/a*

               F_1 *A/a*

               F_2 3 *A/–* : 1 *a/a*

Immediately there is trouble in trying to write the genotypes for cross 2, unless red is a third allele of the same gene. Assume the following dominance relationships: red > orange > yellow. Let the alleles be designated as follows

      red A^R

      orange A^O

      yellow A^Y

Crosses 1-3 now become

   P A^O/A^O × A^Y/A^Y A^R/A^R × A^O/A^O A^R/A^R × A^Y/A^Y

   F1 A^O/A^Y A^R/A^O A^R/A^Y

   F2   3 $A^O/–$: 1 A^Y/A^Y 3 $A^R/–$: 1 A^O/A^O 3 $A^R/–$: 1 A^Y/A^Y

Cross 4: To do this cross you must add a second gene. You must also rewrite the above crosses to include the second gene. Let *B* allow color and *b* block color expression, producing white. The first three crosses become

P A^O/A^O ; *B/B* × A^Y/A^Y ; *B/B* A^R/A^R ; *B/B* × A^O/A^O ; *B/B* A^R/A^R ; *B/B* × A^Y/A^Y ; *B/B*

F$_1$ A^O/A^Y ; B/B A^R/A^O ; B/B A^R/A^Y ; B/B

F$_2$ 3 $A^O/-$; B/B : 1 A^Y/A^Y ; B/B 3 $A^R/-$; B/B : 1 A^O/A^O ; B/B 3 $A^R/-$: 1 A^Y/A^Y ; B/B

The fourth cross is

      P A^R/A^R ; B/B $\times$ A^R/A^R ; b/b

      F$_1$ A^R/A^R ; B/b

      F$_2$ 3 A^R/A^R ; $B/-$: 1 A^R/A^R ; b/b

Cross 5: To do this cross, note that there is no orange appearing. Therefore, the two parents must carry the alleles for red and yellow, and the expression of red must be blocked.

      P A^Y/A^Y ; B/B $\times$ A^R/A^R ; b/b

      F$_1$ A^R/A^Y ; B/b

      F$_2$ 9 $A^R/-$; $B/-$ red
                        3 $A^R/-$; b/b white
                        3 A^Y/A^Y; $B/-$ yellow
                        1 A^Y/A^Y ; b/b white

Cross 6: This cross is identical with cross 5 except that orange replaces yellow.

      P A^O/A^O ; B/B $\times$ A^R/A^R ; b/b

      F$_1$ A^R/A^O ; B/b

      F$_2$ 9 $A^R/-$; $B/-$ red
                        3 $A^R/-$; b/b white
                        3 A^O/A^O ; $B/-$ orange
                        1 A^O/A^O ; b/b white

Cross 7: In this cross, yellow is suppressed by b/b.

      P A^R/A^R ; B/B $\times$ A^Y/A^Y ; b/b

      F$_1$ A^R/A^Y ; B/b

F_2 $9\ A^R/- \ ; B/-$ red

                     $3\ A^R/- \ ; b/b$ white

                     $3\ A^Y/A^Y \ ; B/-$ yellow

                     $1\ A^Y/A^Y \ ; b/b$ white

65. The ratios observed in the F_2 of these crosses are all variations of 9:3:3:1, indicating that two genes are assorting and epistasis is occurring.

(1) F_1 $A/a \ ; B/b \ \times \ A/a \ ; B/b$

           F_2 $9\ A/- \ ; B/-$ cream

                    $3\ A/- \ ; b/b$ cream

                    $3\ a/a \ ; B/-$ black

                    $1\ a/a \ ; b/b$ gray

A test cross of the F_1 would be $A/a \ ; B/b \ \times \ a/a \ ; b/b$ and the offspring:

      $^1/_4 \ A/a \ ; B/b$ cream

      $^1/_4 \ a/a \ ; B/b$ black

      $^1/_4 \ A/a \ ; b/b$ cream

      $^1/_4 \ a/a \ ; b/b$ gray

(2) F_1 $A/a \ ; B/b \ \times \ A/a \ ; B/b$

           F_2 $9\ A/- \ ; B/-$ orange

                    $3\ A/- \ ; b/b$ yellow

                    $3\ a/a \ ; B/-$ yellow

                    $1\ a/a \ ; b/b$ yellow

A test cross of the F_1 would be $A/a \ ; B/b \ \times \ a/a \ ; b/b$ and the offspring:

      $^1/_4 \ A/a \ ; B/b$ orange

      $^1/_4 \ a/a \ ; B/b$ yellow

      $^1/_4 \ A/a \ ; b/b$ yellow

      $^1/_4 \ a/a \ ; b/b$ yellow

(3) F_1 $A/a \ ; B/b \ \times \ A/a \ ; B/b$

           F_2 $9\ A/- \ ; B/-$ black

                    $3\ A/- \ ; b/b$ white

                    $3\ a/a \ ; B/-$ black

                    $1\ a/a \ ; b/b$ black

A test cross of the F_1 would be $A/a \ ; B/b \ \times \ a/a \ ; b/b$ and the offspring:

$\frac{1}{4}$ *A/a* ; *B/b*   black

$\frac{1}{4}$ *a/a* ; *B/b*   black

$\frac{1}{4}$ *A/a* ; *b/b*   white

$\frac{1}{4}$ *a/a* ; *b/b*   black

(4)   F$_1$ *A/a* ; *B/b* × *A/a* ; *B/b*

F$_2$ 9 *A/–* ; *B/–*   solid red

3 *A/–* ; *b/b*   mottled red

3 *a/a* ; *B/–*   small red dots

1 *a/a* ; *b/b*   small red dots

A test cross of the F$_1$ would be *A/a* ; *B/b* × *a/a* ; *b/b* and the offspring:

$\frac{1}{4}$ *A/a* ; *B/b*   solid red

$\frac{1}{4}$ *a/a* ; *B/b*   small red dots

$\frac{1}{4}$ *A/a* ; *b/b*   mottled red

$\frac{1}{4}$ *a/a* ; *b/b*   small red dots

66.   a. Intercrossing mutant strains that all share a common recessive phenotype is the basis of the complementation test. This test is designed to identify the number of different genes that can mutate to a particular phenotype. In this problem, if the progeny of a given cross still express the wiggle phenotype, the mutations fail to complement and are considered alleles of the same gene; if the progeny are wild type, the mutations complement and the two strains carry mutant alleles of separate genes.

b. From the data,   1 and 5   fail to complement   gene *A*
2, 6, 8, and 10   fail to complement   gene *B*
3 and 4   fail to complement   gene *C*
7, 11, and 12   fail to complement   gene *D*
9   complements all others   gene *E*

There are five complementation groups (genes) identified by this data.

c.   mutant 1: $a^1/a^1 \cdot b^+/b^+ \cdot c^+/c^+ \cdot d^+/d^+ \cdot e^+/e^+$ (although, only the mutant
alleles are usually listed)
mutant 2: $a^+/a^+ \cdot b^2/b^2 \cdot c^+/c^+ \cdot d^+/d^+ \cdot e^+/e^+$
mutant 5: $a^5/a^5 \cdot b/^+b^+ \cdot c^+/c^+ \cdot d^+/d^+ \cdot e^+/e^+$

1/5 hybrid: $a^1/a^5 \cdot b^+/b^+ \cdot c^+/c^+ \cdot d^+/d^+ \cdot e^+/e^+$ phenotype: wiggles

Conclusion: 1 and 5 are both mutant for gene *A*

2/5 hybrid: $a^+/a^5 \cdot b^+/b^2 \cdot c^+/c^+ \cdot d^+/d^+ \cdot e^+/e^+$ phenotype: wild type

Conclusion: 2 and 5 are mutant for different genes

67.   Assume that the two genes are unlinked. Mossy would be genotypically m^- ; s^+ and spider would be m^+ ; s^-. After mating and sporulation, independent assortment would give three types of tetrads. The doubly mutant spores do not germinate. Because either single mutant is slow-growing, the additive effect of both mutant genes is lethal.

A		B		C	
m^+ ; s^+	wild type	m^+ ; s^+	wild type	m^+ ; s^-	spider
m^+ ; s^+	wild type	m^+ ; s^-	spider	m^+ ; s^-	spider
m^- ; s^-	dead	m^- ; s^+	mossy	m^- ; s^+	mossy
m^- ; s^-	dead	m^- ; s^-	dead	m^- ; s^+	mossy

7

DNA: Structure and Replication

BASIC PROBLEMS

1.   The DNA double helix is held together by two types of bonds, covalent and hydrogen. Covalent bonds occur within each linear strand and strongly bond the bases, sugars, and phosphate groups (both within each component and between components). Hydrogen bonds occur between the two strands and involve a base from one strand with a base from the second in complementary pairing. These hydrogen bonds are individually weak but collectively quite strong.

2.   Conservative replication is a form of DNA synthesis in which the two template strands remain together but dictate the synthesis of two new DNA strands, which then form a second DNA helix. The end point is two double helices, one containing only old DNA and one containing only new DNA. This hypothesis was found to be not correct. Semiconservative replication is a form of DNA synthesis in which the two template strands separate and each dictates the synthesis of a new strand. The end point is two double helices, both containing one new and one old strand of DNA. This hypothesis was found to be correct.

3.   A primer is a short segment of RNA that is synthesized by primase using DNA as a template during DNA replication. Once the primer is synthesized, DNA polymerase then adds DNA to the 3′ end of the RNA. Primers are required because the major DNA polymerase involved with DNA replication is unable to initiate DNA synthesis and, rather, requires a 3′ end. (It is the 3′ -OH group that is required to create the next phosphodiester bond.) The RNA is subsequently removed and replaced with DNA so that no gaps exist in the final product.

4.   Helicases are enzymes that disrupt the hydrogen bonds that hold the two DNA strands together in a double helix. This breakage is required for both RNA and DNA synthesis. Topoisomerases are enzymes that create and relax supercoiling in the DNA double helix. The supercoiling itself is a result of the twisting of the DNA helix that occurs when the two strands separate.

5. Because the DNA polymerase is capable of adding new nucleotides only at the 3′ end of a DNA strand, and because the two strands are antiparallel, at least two molecules of DNA polymerase must be involved in the replication of any specific region of DNA. When a region becomes single-stranded, the two strands have an opposite orientation. Imagine a single-stranded region that runs from right to left. The 5′ end is at the right, with the 3′ end pointing to the left; synthesis can initiate and continue uninterrupted toward the right end of this strand. Remember: new nucleotides are added in a 5′→ 3′ direction, so the template must be copied from its 3′ end. The other strand has a 5′ end at the left with the 3′ end pointing right. Thus, the two strands are oriented in opposite directions (antiparallel), and synthesis (which is 5′→ 3′) must proceed in opposite directions. For the leading strand (say, the top strand) replication is to the right, following the replication fork. It is continuous and may be thought of as moving "downstream." Replication on the bottom strand cannot move in the direction of the fork (to the right) because, for this strand, that would mean adding nucleotides to its 5′ end. Therefore, this strand must replicate discontinuously: as the fork creates a new single-stranded stretch of DNA, this is replicated *to the left* (away from the direction of fork movement). For this lagging strand, the replication fork is always opening new single-stranded DNA for replication *upstream* of the previously replicated stretch, and a new fragment of DNA is replicated back to the previously created fragment. Thus, one (Okazaki) fragment follows the other in the direction of the replication fork, but each fragment is created in the opposite direction.

6. No. The information of DNA is dependent on a faithful copying mechanism. The strict rules of complementarity insure that replication and transcription is reproducible.

7. Helicases are enzymes that disrupt the hydrogen bonds that hold the two DNA strands together in a double helix. This breakage exposes lengths of single-stranded DNA that will act as the template and is required for DNA replication. Therefore, the absence of helicases would prevent the replication process.

8. Theoretically, DNA could be replicated this way but not with the replisome, which is organized to replicate both stands simultaneously. Further, this would leave one strand of the DNA single-stranded where mutagenic events would be more likely, and it would certainly take longer.

9. The chromosome would become hopelessly fragmented.

10. c. DNA synthesis might take longer.

11. d. It does not have to adapt to each of the body's tissues.

12. b. The RNA would be more likely to contain errors.

13. Part of the replisome is the sliding clamp, which encircles the DNA and keeps pol III attached to the DNA molecule. Thus pol III is transformed into a processive enzyme capable of adding tens of thousands of nucleotides.

14. d. Replication would take twice as long.

15. a. Prior to the S phase, each chromosome has two telomeres, so in the case of $2n = 14$, there are 14 chromosomes and 28 telomeres.

    b. After S, each chromosome consists of two chromatids, each with two telomeres, for a total of four telomeres per chromosome. So, for 14 chromosomes, there would be $14 \times 4 = 56$ telomeres.

    c. At prophase, the chromosomes still consist of two chromatids each, so there would be $14 \times 4 = 56$ telomeres.

    d. At telophase, there would be 28 telomeres in each of the soon-to-be daughter cells.

16. If the DNA is double stranded, A = T and G = C and A + T + C + G = 100%. If T = 15%, then C = [100 – 15(2)]/2 = 35%.

17. If the DNA is double stranded, G = C = 24% and A = T = 26%.

18. Six. The first replication start would have two replication forks proceeding to completion, and the now replicated origins would each start replication again. Each would have two more replication forks, for a total of six.

19. Only the DNA molecule that used the poly-T strand as a template would be radioactive. The other daughter molecule would not be radioactive, because it would not have required any dATP for its replication.

    Because each strand of the second molecule contains T, both daughter molecules would require dATP for replication, so each would be radioactive.

20. Yes. DNA replication is also semi-conservative in diploid eukaryotes.

21. a. The bottom strand.

b.

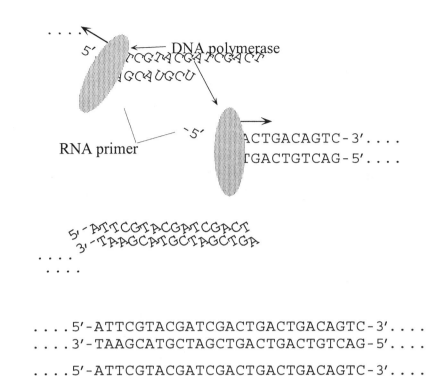

DNA polymerase

RNA primer

5′-ATTCGTACGATCGACT
3′-TAAGCATGCTAGCTGA

. . . .
. . . .

c.

....5′-ATTCGTACGATCGACTGACTGACAGTC-3′....
....3′-TAAGCATGCTAGCTGACTGACTGTCAG-5′....

....5′-ATTCGTACGATCGACTGACTGACAGTC-3′....
....3′-TAAGCATGCTAGCTGACTGACTGTCAG-5′....

d. Yes, but the other replication fork would be moving in the opposite direction and the top strand, as drawn, would now be the leading strand and the bottom strand would now be the lagging strand.

22. The bottom strand will serve as the template for the Okazaki fragment so its sequence will be:

5′ ....CCTTAAGACTAACTACTTACTGGGATC....3′

23. a.

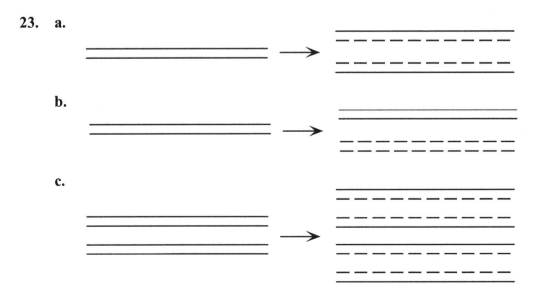

d.

e. Model (b) is ruled out by the experiment. The results are compatible with semiconservative replication, but the exact structure could not be predicted. The other models would all give one band of intermediate density.

24. Sample A must be live S cells because it kills mice when injected and S cells are recovered from the mice.

   Sample C must be live R cells because it has no effect on mice when injected but live R cells are recovered.

   Sample B must be DNA from S cells as it transforms sample C when co-injected.

25. If all samples remain the same, only the line will change when samples B + C are co-injected. Because B contains DNA, it will not transform the R cells, so injected mice will have no response and live R cells will be recovered.

CHALLENGING PROBLEMS

26. Without functional telomerase, the telomeres would shorten at each replication cycle, leading to eventual loss of essential coding information and death. In fact, there are some current observations that decline or loss of telomerase activity plays a role in the mechanism of aging in humans.

27. **a.** A very plausible model is of a triple helix, which would look like a braid, with each strand interacting by hydrogen bonding to the other two.

   b. Replication would have to be terti-conservative. The three strands would separate, and each strand would dictate the synthesis of the other two strands.

   c. The reductional division would have to result in three daughter cells, and the equational would have to result in two daughter cells, in either order. Thus, meiosis would yield six gametes.

28. Chargaff's rules are that A = T and G = C. Because this is not observed, the most likely interpretation is that the DNA is single-stranded. The phage would

first have to synthesize a complementary strand before it could begin to make multiple copies of itself.

8

RNA: Transcription and Processing

BASIC PROBLEMS

1. Because RNA can hybridize to both strands, the RNA must be transcribed from both strands. This does not mean, however, that both strands are used as a template *within each gene*. The expectation is that only one strand is used within a gene but that different genes are transcribed in different directions along the DNA. The most direct test would be to purify a specific RNA coding for a specific protein and then hybridize it to the λ genome. Only one strand should hybridize to the purified RNA.

2. In prokaryotes, translation is beginning at the 5′ end while the 3′ end is still being transcribed. In eukaryotes, processing (capping, splicing) is occurring at the 5′ end while the 3′ end is still being transcribed.

3. There are many examples of proteins that act on nucleic acids, but some mentioned in this chapter are RNA polymerase, GTFs (general transcription factors), σ (sigma factor), rho, TBP (TATA binding protein), and snRNPs (a combination of proteins and snRNAs).

4. Sigma factor, as part of the RNA polyermase holoenzyme, recognizes and binds to the −35 and −10 regions of bacterial promoters. It positions the holoenzyme to correctly initiate transcription at the start site. In eukaryotes, TBP (TATA binding protein) and other GTFs (general transcription factors) have an analagous function.

5. The CTD (carboxy tail domain) of the ß subunit of RNA polymerase II contains binding sites for enzymes and other proteins that are required for RNA processing and is located near the site where nascent RNA emerges. If mutations in this subunit prevent the correct binding and/or localization of the proteins necessary for capping, then this modification will not occur even though all the required enzymes are normal.

6. For a given gene, only one strand of DNA is transcribed. This strand (called the template) will be complementary to the RNA and also to the other strand (called the nontemplate or coding strand). Consequently, the nucleotide sequence of the RNA must be the same as that of the nontemplate strand of the DNA (except that the Ts are instead Us). Ultimately, it is the nucleotide sequence that gives the RNA its function. Transcription of both strands would give two complementary RNAs that would code for completely different polypeptides. However, as you will read about in Chapter 16 of the companion text, double-stranded RNA initiates cellular processes that lead to its degradation.

7. **a., b., c., and d.**

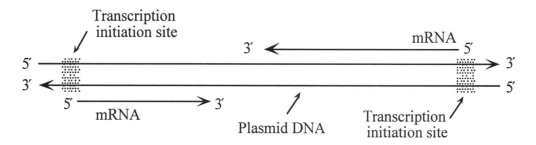

8. Yes. Both replication and transcription is performed by large, multi-subunit molecular machines (the replisome and RNA polymerase II, respectively) and both require helicase activity at the fork of the bubble. However, transcription proceeds in only one direction and only one DNA strand is copied.

9. **a.** False. Sigma factor is required in prokaryotes, not eukaryotes.

    b. True. Processing begins at the 5′ end, while the 3′ end is still being synthesized.

    c. False. Processing occurs in the nucleus and only mature RNA is transported out to the cytoplasm.

    d. False. Hairpin loops or rho factor (in conjunction with the *rut* site) is used to terminate transcription in prokaryotes. In eukaryotes the conserved sequences AAUAAA or AUUAAA, near the 3′ end of the transcript is recognized by an enzyme that cuts off the end of the RNA approximately 20 bases downstream.

    e. True. Multiple RNA polymerases may transcribe the same template simultaneously.

10. **a.** The original sequence represents the −35 and −10 consensus sequences (with the correct number of intervening spaces) of a bacterial promoter. Sigma factor, as part of the RNA polymerase holoenzyme, recognizes and binds to these sequences.

b. The mutated (transposed) sequences would not be a binding site for sigma factor. The two regions are not in the correct orientation to each other and therefore would not be recognized as a promoter.

11. a. The promoters of eukaryotes and prokaryotes do not have the same conserved sequences. In yeast, the promoter would have the required TATA box located about –30 whereas bacteria would have conserved sequences at –35 and –10 that would interact with sigma factor as part of the RNA polymerase holoenzyme.

b. There are two possible reasons that the mRNA is longer than expected. First, many eukaryotic genes contain introns, and bacteria would not have the splicing machinery necessary for their removal. Second, termination of transcription is not the same in bacteria and yeast; the sequences necessary for correct terminaton in *E. coli* would not be expected in the yeast gene.

12.

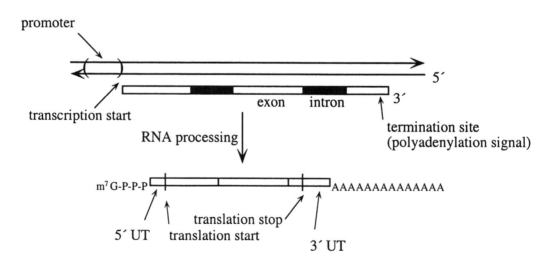

13.

promoter

transcription start

exon   intron

termination site
(polyadenylation signal)

RNA processing

m⁷G-P-P-P

AAAAAAAAAAAAA

5´ UT   translation start   translation stop

3´ UT

14. a. Yes. The exons encode the protein, so null mutations would be expected to map within exons.

b. Possibly. There are sequences near the boundaries of and within introns that are necessary for correct splicing. If these are altered by mutation, correct splicing will be disrupted. Although transcribed, it is likely that translation will not occur.

c. Yes. If the promoter is deleted or altered such that GTFs cannot bind, transcription will be disrupted.

d. Yes. There are sequences near the boundaries of and within introns that are necessary for correct splicing.

CHALLENGING PROBLEMS

15. a. The data cannot indicate whether one or both strands are used for transcription. You do not know how much of the DNA is transcribed nor which regions of DNA are transcribed. Only when the purine/pyrimidine ratio is not unity can you deduce that only one strand is used as template.

b. If the RNA is double-stranded, the percentage of purines (A + G) would equal the percentage of pyrimidines (U + C) and the (A +G)/(U + C) ratio would be 1.0. This is clearly not the case for *E. coli*, which has a ratio of 0.80. The ratio for *B. subtilis* is 1.02. This is consistent with the RNA being double-stranded but does not rule out single-stranded if there are an equal number of purines and pyrimidines in the strand.

16. a.

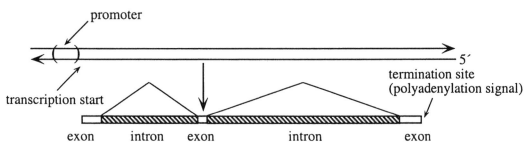

b. Alternative splicing of the primary transcript would result in mRNAs that were only partially identical. In this case, the two transcripts share 224 nucleotides in common. As this is the exact length of the second exon, one possible solution to this problem is that this exon is shared by the two alternatively-spliced mRNAs. The second transcript also contains 2.3 kb of sequence not found in the first. Perhaps what was considered the first intron is actually also part of the second transcript as that would result in the 2524 nucleotides stated as this transcript's length. Of course, other combinations of alternative splicing would also fit the data.

9

Proteins and Their Translation

BASIC PROBLEMS

1. 3´ CGT ACC ACT GCA 5´   DNA double helix (transcribed strand)
    5´ GCA TGG TGA CGT 3´   DNA double helix
    5´ GCA UGG UGA CGU 3´   mRNA transcribed
    3´ CGU ACC ACU GCA 5´   appropriate tRNA anticodon
    NH_3 - Ala - Trp - (stop) - COOH   amino acids incorporated

2. **a. and b.**   5´ UUG GGA AGC 3´
    c. and d.   Assuming the reading frame starts at the first base:

    NH_3 - Leu - Gly - Ser - COOH

    For the bottom strand, the mRNA is 5´ GCU UCC CAA 3´ and assuming the reading frame starts at the first base, the corresponding amino acid chain is NH_3 - Ala - Ser - Gln - COOH.

3. (5) With an insertion, the reading frame is disrupted. This will result in a drastically altered protein from the insertion to the end of the protein (which may be much shorter or longer than wild type because of altered stop signals).

4. A single nucleotide change should result in three adjacent amino acid changes in a protein. One and two adjacent amino acid changes would be expected to be much rarer than the three changes. This is directly the opposite of what is observed in proteins. Also, given any triplet coding for an amino acid, the next triplet could only be one of four. For example, if the first is GGG, then the next must be GGN (N = any base). This puts severe limits on which amino acids could be adjacent to each other. You could check amino acid sequences of various proteins to show that this is not the case.

5.   It suggests very little evolutionary change between *E. coli* and humans with regard to the translational apparatus. The code is universal, the ribosomes are interchangeable, the tRNAs are interchangeable, and the enzymes involved are interchangeable.

6.   There are three codons for isoleucine: 5′ AUU 3′, 5′ AUC 3′, and 5′ AUA 3′. Possible anticodons are 3′ UAA 5′ (complementary), 3′ UAG 5′ (complementary), and 3′ UAI 5′ (wobble). 5′ UAU 3′, although complementary, would also base-pair with 5′ AUG 3′ (methionine) due to wobble and therefore would not be an acceptable alternative.

7.   **a.**   By studying the genetic code table provided in the textbook, you will discover that there are eight cases in which knowing the first two nucleotides does not tell you the specific amino acid.

     b.   If you knew the amino acid, you would not know the first two nucleotides in the cases of Arg, Ser, and Leu.

8.   The codon for amber is UAG. Listed below are the amino acids that would have been needed to be inserted to continue the wild-type chain and their codons:

glutamine	CAA, CAG*
lysine	AAA, AAG*
glutamic acid	GAA, GAG*
tyrosine	UAU*, UAC*
tryptophan	UGG*
serine	AGU, AGC, UCU, UCC, UCA, UCG*

In each case, the codon marked by an asterisk would require a single base change to become UAG.

9.   **a.**   The codons for phenylalanine are UUU and UUC. Only the UUU codon can exist with randomly positioned A and U. Therefore, the chance of UUU is $(1/2)(1/2)(1/2) = 1/8$.

     b.   The codons for isoleucine are AUU, AUC, and AUA. AUC cannot exist. The probability of AUU is $(1/2)(1/2)(1/2) = 1/8$, and the probability of AUA is $(1/2)(1/2)(1/2) = 1/8$. The total probability is thus $1/4$.

     c.   The codons for leucine are UUA, UUG, CUU, CUC, CUA, and CUG, of which only UUA can exist. It has a probability of $(1/2)(1/2)(1/2) = 1/8$.

     d.   The codons for tyrosine are UAU and UAC, of which only UAU can exist. It has a probability of $(1/2)(1/2)(1/2) = 1/8$.

10.   **a.**   1 U : 5 C — The probability of a U is $1/6$, and the probability of a C is $5/6$.

Codon	Amino acid	Probability	Sum
UUU	Phe	$(^1/_6)(^1/_6)(^1/_6) = 0.005$	Phe = 0.028
UUC	Phe	$(^1/_6)(^1/_6)(^5/_6) = 0.023$	
CCC	Pro	$(^5/_6)(^5/_6)(^5/_6) = 0.578$	Pro = 0.694
CCU	Pro	$(^5/_6)(^5/_6)(^1/_6) = 0.116$	
UCC	Ser	$(^1/_6)(^5/_6)(^5/_6) = 0.116$	Ser = 0.139
UCU	Ser	$(^1/_6)(^5/_6)(^1/_6) = 0.023$	
CUC	Leu	$(^5/_6)(^1/_6)(^5/_6) = 0.116$	Leu = 0.139
CUU	Leu	$(^5/_6)(^1/_6)(^1/_6) = 0.023$	

1 Phe : 25 Pro : 5 Ser : 5 Leu

b. Using the same method as above, the final answer is 4 stop : 80 Phe : 40 Leu : 24 Ile : 24 Ser : 20 Tyr : 6 Pro : 6 Thr : 5 Asn : 5 His : 1 Lys : 1 Gln.

c. All amino acids are found in the proportions seen in the code table.

11. Quaternary structure is due to the interactions of subunits of a protein. In this example, the enzyme activity being studied may be from a protein consisting of two different subunits. The polypeptides of the subunits are encoded by separate and unlinked genes.

12. There are a number of mutational changes that can lead to the absence of enzymatic function in the product of a gene. Some of these changes would result in the complete absence of protein product and therefore also the absence of a detectable band on a Western blot. Mutations such as deletions of the gene, for example, would result in the lack of detectable protein. Other mutations that destroy function (missense, for example) may not alter the production of the protein and would be detected on a Western blot. Still other mutations (nonsense, frameshift) could alter the size of the protein yet would still lead to detectable protein.

13. a. Typtophan synthetase is a heterotetramer of two copies each of two different polypeptides, each encoded by a separate gene. Mutations that prevent the synthesis of one subunit would lead to the loss of one of the bands on a Western blot. Mutants that still make both subunits (those with exactly the same bands as wild type) might have mutations that prevent the subunits from interacting or disrupt the active site of the enzyme.

    b. Because the two subunits are encoded by separate genes, the absence of both bands simultaneously would require two independent and rare mutagenic events.

14. Yes. It was not known at the time what number of bases the "plus" and "minus" mutations actually were. If each mutation was two bases, then a codon would

have been six bases. Since the mutations were actually adding or subtracting single bases, the codon is indeed three bases.

15. No. The enzyme may require post-translational modification to be active. Mutations in the enzymes required for these modifications would not map to the isocitrate lyase gene.

16. A nonsense suppressor is a mutation in a tRNA such that its anticodon can base-pair with a stop codon. In this way, a mutant stop codon (nonsense mutation) can be read through and the polypeptide can be fully synthesized. However, the mutant tRNA may be for an amino acid that was not encoded in that position in the original gene. For example, the codon UCG (serine) is instead UAG in the nonsense mutant. The suppressor mutation could be in tRNA for tryptophan such that its anticodon now recognizes UAG instead of UGG. During translation in the double mutant, the machinery puts tryptophan into the location of the mutant stop codon. This allows translation to continue but does place tryptophan into a position that was serine in the wild-type gene. This may create a protein that is not as active and a cell that is "not exactly wild type." Another explanation is that translation of the mutant gene is not as efficient and that premature termination still occurs some of the time. This would lead to less product and again, a state that is "not exactly wild type."

17.

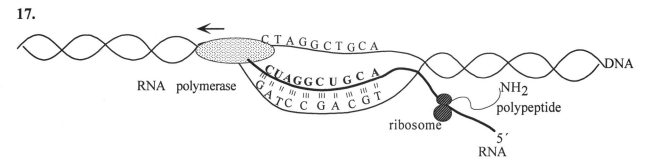

In eukaryotes, transcription occurs within the nucleus while translation occurs in the cytoplasm. Thus the two processes cannot occur together.

18. Assuming that the three mutations of gene P are all nonsense mutations, there are three different possible stop codons that might be the cause (amber, ochre, or opal). A suppressor mutation would be specific to one type of nonsense codon. For example, amber suppressors would suppress amber mutants but not opal or ochre.

19. Initiation of translation in prokaryotes requires specific base-pairing between the 3′ end of the 16s rRNA and a Shine–Delgano sequence found in the 5′ untranslated region of the mRNA. A Shine–Delgano sequence would not be expected (unless by chance) in a eukaryotic mRNA and therefore initiation of translation would not occur.

20. Initiaton of translation in eukaryotes requires initiation factors (eIF4a, b, and G) that associate with the 5′ cap of the mRNA. Because prokaryotic mRNAs are not capped, translation would not initiate.

21. Not likely. Although the steps of translation and the components of ribosomes are similar in both eukaryotes and prokaryotes, the ribosomes are not identical. The sizes of both subunits are larger in eukaryotes and the many specific and intricate interactions that must take place between the small and large subunits would not be possible in a chimeric system.

22. Single amino acid changes can result in changes in protein folding, protein targeting, or post-translational modifications. Any of these changes could give the results indicated.

23. The first indication of rRNAs importance was the discovery of ribozymes. Recently, structural studies have shown that both the decoding center in the 30S subunit and the peptidyl transferase center in the 50S subunit are composed entirely of rRNA and that the important contacts in these centers are all tRNA/rRNA contacts.

CHALLENGING PROBLEMS

24. **a. and b.** The goal of this type of problem is to align the two sequences. You are told that there is a single nucleotide addition and single nucleotide deletion, so look for single base differences that effect this alignment. These should be located where the protein sequence changes (i.e., between Lys-Ser and Asn-Ala). Remember also that the genetic code is redundant. (N = any base)

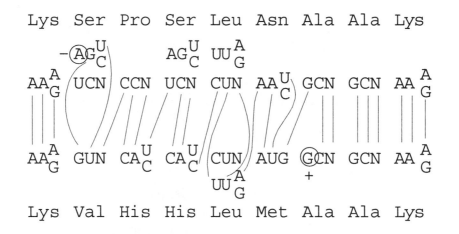

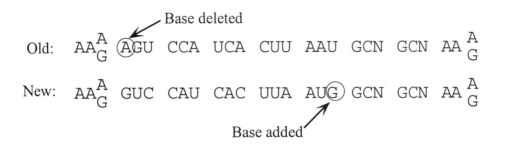

25. Mutant 1: A simple substitution of Arg for Ser exists, suggesting a nucleotide change. Two codons for Arg are AGA and AGG, and one codon for Ser is AGU. The U of the Ser codon could have been replaced by either an A or a G.

Mutant 2: The Trp codon (UGG) changed to a stop codon (UGA or UAG).

Mutant 3: Two frameshift mutations occurred:

```
5´—GCN CCN (-U)GGA GUG AAA AA(+U or C) UGU(or C) CAU(or C)-3´.
```

Mutant 4: An inversion occurred after Trp and before Cys. The DNA original sequence (with both strands shown for the area of inversion) was

```
3´—CGN GGN ACC TCA CTT TTT ACA(or G) GTA(or G)-5´
5´—         —AGT GAA AAA—                    -3´
```

Therefore, the complementary RNA sequence was

```
5´—GCN CCN UGG AGU GAA AAA UGU/C CAU/C-3´
```

The DNA inverted sequence became

```
3´—CGN GGN ACC AAA AAG TGA ACA/G GTA/G-5´
           ^        ^
```

Therefore the complementary RNA sequence was

```
5´—GCN CCN UGG UUU UUC ACU UGU/C CAU/C-3´
           ^        ^
```

26. If the anticodon on a tRNA molecule also was altered by mutation to be four bases long, with the fourth base on the 5´ side of the anticodon, it would suppress the insertion. Alterations in the ribosome can also induce frameshifting.

27. f, d, j, e, c, i, b, h, a, g

28. Cells in long-established culture lines usually are not fully diploid. For reasons that are currently unknown, adaptation to culture frequently results in both karyotypic and gene dosage changes. This can result in hemizygosity for some genes, which allows for the expression of previously hidden recessive alleles.

29. **a. and b.** The sequence of double-stranded DNA is as follows:

```
5´—TAC ATG ATC ATT TCA CGG AAT TTC TAG CAT GTA—3´
3´—ATG TAC TAG TAA AGT GCC TTA AAG ATC GTA CAT—5´
```

First look for stop codons. Next look for the initiating codon, AUG (3´— TAC—5´ in DNA). Only the upper strand contains the necessary codons.

```
DNA          3´ TAC GAT CTT TAA GGC ACT 5´
RNA          5´ AUG CUA GAA AUU CCG UGA 3´
protein         Met Leu Glu Ile Pro stop
```

The DNA strand is read from right to left as written in your text and is written above in reverse order from your text.

c. Remember that polarity must be taken into account. The inversion is

```
DNA 5´ TAC ATG CTA GAA ATT CCG TGA AAT GAT CAT GTA 3´
RNA 3´         —GAU CUU UAA GGC ACU UUA CUA GUA—      5´
amino acids   HOOC 7   6   5   4   3   2   1 —NH₃
```

d. DNA   3´ATG TAC TAG TAA AGT GCC TTA AAG ATC GTA CAT 5´
mRNA 5´UAC AUG AUC AUU UCA CGG AAU UUC UAG 3´
                    1   2   3   4   5   6   7 stop

Codon 4 is 5´—UCA—3´, which codes for Ser. Anticodon 4 would be 3´—AGU—5´ (or 3´—AGI—5´ given wobble).

30. **a.** $(GAU)_n$ codes for Asp_n $(GAU)_n$, Met_n $(AUG)_n$, and $stop_n$ $(UGA)_n$. $(GUA)_n$ codes for Val_n $(GUA)_n$, Ser_n $(AGU)_n$, and $stop_n$ $(UAG)_n$.
One reading frame in each contains a stop codon.

b. Each of the three reading frames contains a stop codon.

c. The way to approach this problem is to focus initially on one amino acid at a time. For instance, line 4 indicates that the codon for Arg might be AGA or GAG. Line 7 indicates it might be AAG, AGA, or GAA. Therefore, Arg is at least AGA. That also means that Glu is GAG (line 4). Lys and Glu can be AAG or GAA (line 7). Because no other combinations except the ones already mentioned result in either Lys or Glu, no further decision can be made with respect to them. However, taking wobble into consideration, Glu may also be GAA, which leaves Lys as AAG.

Next, focus on lines 1 and 5. Ser and Leu can be UCU and CUC. Ser, Leu, and Phe can be UUC, UCU, and CUU. Phe is not UCU, which is seen in both lines. From line 14, CUU is Leu. Therefore, UUC is Phe, and UCU is Ser.

The footnote states that line 13 and 14 are in the correct order. In line 13, if UCU is Ser (see above), then Ile is AUC, Tyr is UAU, and Leu is CUA.

Continued application of this approach will allow the assignment of an amino acid to each codon.

10

Regulation of Gene Transcription

BASIC PROBLEMS

1. The I gene determines the synthesis of a repressor molecule, which blocks expression of the *lac* operon and which is inactivated by the inducer. The presence of the repressor I^+ will be dominant to the absence of a repressor I^-. I^s mutants are unresponsive to an inducer. For this reason, the gene product cannot be stopped from interacting with the operator and blocking the *lac* operon. Therefore, I^s is dominant to I^+.

2. O^c mutants are changes in the DNA sequence of the operator that impair the binding of the *lac* repressor. Therefore, the *lac* operon associated with the O^c operator cannot be turned off. Because an operator controls only the genes on the same DNA strand, it is cis (on the same strand) and dominant (cannot be turned off).

3. **a.** You are told that a, b, and c represent *lacI*, *lacO*, and *lacZ*, but you do not know which is which. Both a^- and c^- have constitutive phenotypes (lines 1 and 2) and therefore must represent mutations in either the operator (*lacO*) or the repressor (*lac I*). b^- (line 3) shows no ß-gal activity and by elimination must represent the *lacZ* gene.

    Mutations in the operator will be cis-dominant and will cause constitutive expression of the *lacZ* gene only if it's on the same chromosome. Line 6 has c^- on the same chromosome as b^+ but the phenotype is still inducible (owing to c^+ in trans). Line 7 has a^- on the same chromosome as b^+ and is constitutive even though the other chromosome is a^+. Therefore a is *lacO*, c is *lacI*, and b is *lacZ*.

    b. Another way of labeling mutants of the operator is to denote that they lead to a constitutive phenotype; $lacO^-$ (or a^-) can also be written as $lacO^c$. There are also mutations of the repressor that fail to bind inducer (allolactose) as opposed to fail to bind DNA. These two classes have quite different phenotypes and are distinguished by *lacI^s* (fails to bind allolactose and leads to a dominant uninducible phenotype in the presence of a wild-type operator) and *lacI^-* (fails to bind DNA and is recessive). It is possible that line 3, line 4, and line 7 have *lacI^s* mutations (because dominance cannot be ascertained in a cell that is also *lacO^c*) but the other *c^-* alleles must be *lacI^-*.

4.

	ß-Galactosidase		**Permease**	
Part	No lactose	Lactose	No lactose	Lactose
a	+	+	–	+
b	+	+	–	–
c	–	–	–	–
d	–	–	–	–
e	+	+	+	+
f	+	+	–	–
g	–	+	–	+

a. The *O^c* mutation leads to the constitutive synthesis of ß-galactosidase because it is cis to a *lacZ^+* gene, but the permease is inducible because the *lacY^+* gene is cis to a wild-type operator.

b. The *lacP^-* mutation prevents transcription so only the genes cis to *lacP^+* will be transcribed. These genes are also cis to *O^c* so the *lacZ^+* gene is transcribed constitutively.

c. The *lacI^s* is a trans-dominant mutation and prevents transcription from either operon.

d. Same as part c.

e. There is no functional repressor made (and one operator is mutant as well).

f. Same as part b.

g. Both operators are wild type and the one functional copy of *lacI* will direct the synthesis of enough repressor to control both operons.

5. A gene is turned off or inactivated by the "modulator" (usually called a *repressor*) in negative control, and the repressor must be removed for transcription to occur. A gene is turned on by the "modulator" (usually called an *activator*) in positive control, and the activator must be added or converted to an active form for transcription to occur.

6. The *lacY* gene produces a permease that transports lactose into the cell. A cell containing a *lacY⁻* mutation cannot transport lactose into the cell, so ß-galactosidase will not be induced.

7. Activation of gene expression by trans-acting factors occurs in both prokaryotes and eukaryotes. In both cases, the trans-acting factors interact with specific DNA sequences that control expression of cis genes.

In prokaryotes, proteins bind to specific DNA sequences, which in turn regulate one or more downstream genes.

In eukaryotes, highly conserved sequences such as CCAAT and various enhancers in conjunction with trans-acting binding proteins increase transcription controlled by the downstream TATA box promoter. Several proteins have been found that bind to the CCAAT sequence, upstream GC boxes, and the TATA sequence in *Drosophila*, yeast, and other organisms. Specifically, the Sp1 protein recognizes the upstream GC boxes of the SV40 promoter and many other genes; GCN4 and GAL4 proteins recognize upstream sequences in yeast; and many hormone receptors bind to specific sites on the DNA (e.g., estrogen complexed to its receptor binding to a sequence upstream of the ovalbumin gene in chicken oviduct cells). Additionally, the structure of some of these trans-acting DNA-binding proteins is quite similar to the structure of binding proteins seen in prokaryotes. Further, protein-protein interactions are important in both prokaryotes and eukaryotes. For the above reasons, eukaryotic regulation is now thought to be very close to the model for regulation of the bacterial *ara* operon.

8. Bacterial operons contain a promoter region that extends approximately 35 bases upstream of the site where transcription is initiated. Within this region is the promoter. Activators and repressors, both of which are trans-acting proteins that bind to the promoter region, regulate transcription of associated genes in cis only.

The eukaryotic gene has the same basic organization. However, the promoter region is somewhat larger. Also, enhancers up to several thousand nucleotides upstream or downstream can influence the rate of transcription. A major difference is that eukaryotes have not been demonstrated to have polycistronic messages.

9. The term epigenetic inheritance is used to describe heritable alterations in which the DNA sequence itself is not changed. Paramutation and parental imprinting are two such examples.

10. An enhancersome is a large protein complex that acts synergistically to activate transcription. Because this is a case where the whole is greater than the sum of its parts, loss of any one of the required proteins may severely impact the required synergy.

11. Imprinted genes are functionally hemizygous. Maternally imprinted genes are inactive when inherited from the mother, and paternally imprinted genes are inactive when inherited from the father. A mutation in one of these genes is dominant when an offspring inherits a mutant allele from one parent and a "normal" but inactivated allele from the other parent.

12. A gene not expressed due to alteration of its DNA sequence will never be expressed and will be inherited generation to generation. An epigenetically inactivated gene may still be regulated. Chromatic structure can change in the course of the cell cycle, for example, when transcription factors modify the histone code. Also, inactivation may change generation to generation as it does for imprinted genes.

13. The inheritance of chromatin structure is thought to be responsible for the inheritance of epigenetic information. This is due to the inheritance of the histone code and may also include inheritance of DNA methylation patterns.

14. Many DNA-protein interactions are shared by prokaryotes and eukaryotes, but the mechanisms by which proteins bound to DNA at great distances from the start of transcription affect that transcription is unique to eukaryotes. Also, mechanisms of gene regulation based on chromatin structure are distinctly eukaryotic.

15. (b) The interaction of three factors to recruit a coactivator is synergistic.

16. Transcription factors can be broadly divided into two classes: recruitment and activation of the transcriptional apparatus, and chromatin remodeling or modifying enzymes. Importantly, many of the transcriptional complexes that form share many subunits and appear to be assembled in a modular fashion. The many different combinatorial interactions that are possible can result in distinct patterns of gene expression.

CHALLENGING PROBLEMS

17. There are numerous possibilities. As one example:

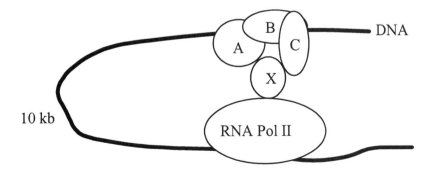

18. The following represents a cell that is mutant for TFC. In this case, CRX does not bind and the activation of transcription does not occur.

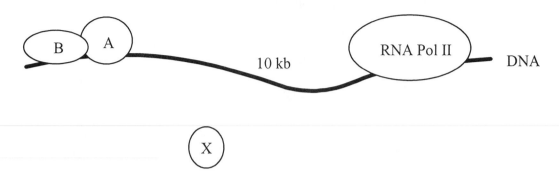

19. The same as 18 but now the binding site for TFC is mutant.

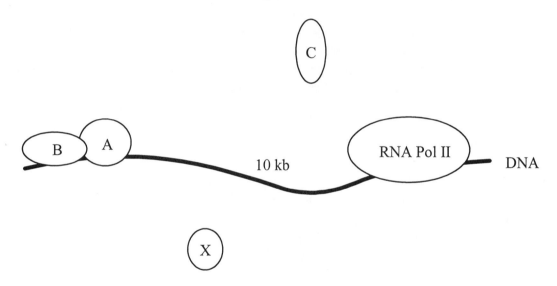

20. While reduced levels of H4 resulted in the presence of fewer nucleosomes that were less tightly packed and the activation of certain inducible genes, it's not obvious that overproduction of H4 would have the opposite effect. Nucleosomes are an octamer of two subunits each of H2A, H2B, H3, and H4. It would not be expected that overproduction of just one histone would result in more nucleosomes and changes in chromatin structure.

21. Normally, the repressor searches for the operator by rapidly binding and dissociating from nonoperator sequences. Even for sequences that mimic the true operator, the dissociation time is only a few seconds or less. Therefore, it is easy for the repressor to find new operators as new strands of DNA are synthesized. However, when the affinity of the repressor for DNA and operator is increased, it takes too long for the repressor to dissociate from sequences on the chromosome that mimic the true operator, and as the cell divides and new operators are synthesized, the repressor never quite finds all of them in time, leading to a partial synthesis of ß-galactosidase. This explains why, in the absence of IPTG, there is some elevated ß-galactosidase synthesis. When IPTG binds to the repressors with increased affinity, it lowers the affinity back to that of the normal repressor (without IPTG bound). Then, the repressor can rapidly dissociate from sequences in the chromosome that mimic the operator and find the true operator. Thus, ß-galactosidase is repressed in the presence of IPTG in strains with repressors that have greatly increased affinity for operator. In summary, because of a kinetic phenomenon, we see a reverse induction curve.

22. Construct a set of reporter genes with the promoter region, the introns, and the region 3′ to the transcription unit of the gene in question, containing different alterations that do not disrupt transcription or processing. Use these reporter genes to make transgenic animals by germ-line transformation. Assay for expression of the reporter gene in various tissues and the kidney of both sexes.

23. If there is an operon governing both genes, then a frameshift mutation could cause the stop codon separating the two genes to be read as a sense codon. Therefore, the second gene product will be incorrect for almost all amino acids. However, there are no known polycistronic messages in eukaryotes. The alternative, and better, explanation is that both enzymatic functions are performed by the same gene product. Here, a frameshift mutation beyond the first function, carbamyl phosphate synthetase, will result in the second half of the protein molecule being nonfunctional.

24. Because very small amounts of the repressor are made, the system as a whole is quite responsive to changes in repressor concentration. In the heterodiploids, repressor heterotetramers may form by association of polypeptides encoded by both I^- and I^+. Only homotetramers of I^+ work properly and their concentration will be reduced by the various heterotetramer combinations of I^- and I^+. This will result in some expression of the *lac* genes in the absence of lactose.

25. The key to this question is to remember that *lacI* mutations will be trans-acting as they produce a protein product, and that *lacP* mutations (like *lacO* mutations) will be cis-acting as they are affecting a binding site for RNA polymerase. For the purposes of this problem, designate the uninducible *lacI* mutations as i^u mutations, and the uninducible *lacP* mutations as p^u mutations. There are quite a number of satisfactory genotypes which can serve to distinguish between the i^u and p^u mutations. Here are a few examples

	Enzyme activity	
Genotype	ß-galactosidase	Permease
1 $i^u\, p^+\, o^+\, z^+\, y^+$	absent	absent
2 $i^+\, p^u\, o^+\, z^+\, y^+$	absent	absent
3 $i^u\, p^+\, o^+\, z^+\, y^- / i^+\, p^+\, o^+\, z^-\, y^+$	absent	absent

$4\ i^+\ p^u\ o^+\ z^+\ y^-\ /\ i^+\ p^+\ o^+\ z^-\ y^+$ absent   inducible

$5\ i^u\ p^+\ o^c\ z^+\ y^-\ /\ i^+\ p^+\ o^+\ z^-\ y^+$ constitutive   absent

$6\ i^+\ p^u\ o^c\ z^+\ y^-\ /\ i^+\ p^+\ o^+\ z^-\ y^+$ absent   inducible

Genotypes 1 and 2 are simply symbolic restatements of the phenotypes of the uninducible mutations. Genotypes 3 and 4 are straightforward tests to distinguish the cis-acting p^u mutations from the trans-acting i^u lesions. The results of genotype 3 reflect the expectation that i^u mutations would be trans-acting and dominant to i^+. This is expected because the i^u-encoded repressor protein molecules would be incapable of being inactivated by binding to inducer; the presence or absence of normal repressor protein is irrelevant. The results of genotype 4, on the other hand, reflect the expectation that p^u mutations would only be cis-acting. Hence, any genes in cis to the p^u allele would be inactive, while any genes in cis to the normal p^+ allele would potentially be transcribed normally (if all other regulatory functions were normal). In a similar fashion, genotypes 5 and 6 distinguish the cis vs. trans action of i^u and p^u mutations. In genotype 5, i^u remains trans-dominant to i^+, but this dominance is overcome by the cis-acting o^c mutation (compare genotypes 3 and 5). In genotype 6, the presence of o^c is irrelevant, as it is in a *lac* operon that contains the p^u mutation preventing RNA polymerase binding (compare genotypes 4 and 6).

26. The S mutation is an alteration in *lacI* such that the repressor protein binds to the operator, regardless of whether inducer is present or not. In other words, it is a mutation that inactivates the allosteric site which binds to inducer, while not affecting the ability of the repressor to bind to the operator site. The dominance of the S mutation is due to the binding of the mutant repressor, even under circumstances when normal repressor does not bind to DNA (that is, in the presence of inducer). The constitutive reverse mutations that map to *lacI* are mutational events which inactivate the ability of this repressor to bind to the operator. The constitutive reverse mutations that map to the operator alter the operator DNA sequence such that it will not permit binding to any repressor molecules (wild-type or mutant repressor).

27. a.

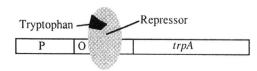

 b. **i.** *trpA* is not synthesized in the presence of tryptophan; it is synthesized in the absence of tryptophan.

   ii. *trpA* is not synthesized in the presence of tryptophan; it is synthesized in the absence of tryptophan. The second operon contains a *trpA⁻* mutation and is unable to make any active tryptophan synthetase enzyme. However, this operon contains a wild-type *trpR* gene, which encodes a functional repressor molecule. Because this gene product is a diffusible molecule (trans-acting), it will act on the other DNA molecule containing the mutant *trpR⁻* gene and bind to the operator on that molecule. Thus, *trpA* gene expression will be repressed when tryptophan binds the repressor, causing the repressor to bind the operator. In the absence of tryptophan, the repressor will not bind to either operator, and transcription will proceed.

   iii. *trpA* is synthesized both in the presence and in the absence of tryptophan. Because the second operon contains a mutant *trpA⁻* gene, no functional tryptophan synthetase enzyme will be made from this DNA molecule. The first operon contains a wild-type *trpA* gene and will be responsible for the intracellular supply of tryptophan synthetase and ultimately, tryptophan. However, this operon contains a mutant (*trpO⁻*) operator region that cannot be bound by the repressor molecules. Because the operator region is cis-acting and does not encode a diffusible gene product, the wild-type *trpO* gene on the other DNA molecule cannot substitute for the mutant operator. Therefore, the *trpA* gene will always be transcribed (constitutive) regardless of the levels of tryptophan in the cell.

8.

	Glucose	Lactose	Lactose + glucose
wild-type	0	100	1
lacI⁻	1	100	1
lacIˢ	0	0	0
lacO⁻	1	100	1
crp⁻	0	1	1

lacI⁻ — leads to absence of negative control by repressor binding but still under the positive control of CAP-cAMP binding

lacIˢ — because the repressor does not bind lactose but still binds DNA, transcription will be blocked under all conditions

lacO⁻ — leads to absence of negative control by repressor binding but still under the positive control of CAP-cAMP binding

crp⁻ — leads to absence of positive control by CAP-cAMP binding but still under the negative control of repressor binding

9. a. D through J — the primary transcript will include all exons and introns.

   b. E, G, I — all introns will be removed.

c. A, C, L — the promoter and enhancer regions will bind various transcription factors that may interact with RNA polymerase.

30. a. Clone A is activated by FGF, but B, C, and D are not. This indicates that the DNA binding site for this activation is located somewhere between the 3′ end of exon 1 [E1(f)] and the 5′ end of exon 3 [E3(f)]. This is the region of DNA found only in clone A.

b. Cortisol represses transcription of clone A and B, but not C. (You do not expect any effect on D, the intact globin gene.) Comparing these clones indicates that the DNA site involved in this repression must be located in the 3′ region of E3(f) or the 3′ flanking sequences of this gene.

c. Activation by EP is seen in clones C and D, but not B. (Again, you do not expect A, the intact c-fos gene to respond to EP.) This indicates that the DNA site involved in this activation must be localized to the 3′ side of E3(g) or the 3′ flanking regions of the globin gene.

11

Gene Isolation and Manipulation

BASIC PROBLEMS

1. **a.** The following are examples of hybridization of single-stranded DNAs discussed in this chapter: use of primers (DNA sequencing, PCR), hybridization of sticky ends in cloning, probe hybridization (Southern blotting, Northern blotting, diagnosis of mutations, identification of clones, and *in situ* hybridization).

    b. The following are examples of proteins that bind to and act on DNA discussed in this chapter: restriction enzymes, DNA polymerase (*Taq* polymerase), reverse transcriptase, and DNA ligase.

2. Reverse transcriptase polymerizes DNA using RNA as a template. This RNA:DNA hybrid is then treated with sodium hydroxide to degrade the RNA. (NaOH hydrolyzes RNA by catalyzing the formation of a 2′-3′cyclic phosphate.)

3. **a.** Recombinant DNA is the term used to describe techniques used to generate hybrid molecules consisting of DNA from two different sources. In this use, recombinant refers to the artificial joining of DNA from different sources.

    b. Recombinant frequency is used to describe the statistical likelihood that two loci in the same organism will assort together or separately during meiosis. Here, recombinant refers to the combinations of loci that are different from the parent.

4. Ligase is an essential enzyme within all cells that seals breaks in the phosphate-sugar backbone of DNA. During DNA replication it joins Okazaki fragments to create a continuous strand, and in cloning, it is used to join the various DNA fragments with the vector. If it was not added, the vector and cloned DNA would simply fall apart.

5. Strictly speaking, a seven-base-pair sequence cannot be a palindrome. This is because the middle base pair must be complementary on the two strands and cannot be identical. However, the restriction site for *Mst*II is CCTNAGG and in essence is a six-base-pair palindrome. For sequences like this, a random seven-base-pair sequence would have a $1/64$ chance of being palindromic. To calculate this, start with any base in the first position. The seventh base must now be complementary to the first so this has a $1/4$ chance. The second position may also be any base but now the sixth base must be complementary to it so again, this has a $1/4$ chance. Finally, the third and fifth bases must be complementary and this again has a $1/4$ chance. Taken together, a palindrome like that recognized by *Mst*II would randomly exist in $1/64$ seven base sequences.

6. Each cycle takes five minutes and doubles the DNA. In one hour there would be 12 cycles so the DNA would be amplified $2^{12} = 4096$-fold amplification.

7. The great advantage of PCR is that fewer procedures are necessary compared with cloning. However, it requires that the sequence be known, and that the primers are each present once in the genome and are sufficiently close (less than 2 kb). If these conditions are met, the primers determine the specificity of the DNA segment amplified and would be most efficient.

8. plasmid (15–20 kb) < cosmid (35–45 kb) < BAC (150–300 kb) < YAC (300 kb and larger)

9. The gene coding for the HGO enzyme was first cloned and characterized from the fungus *Aspergillus nidulans*. Using the inferred amino acid sequence of the protein, computer searches of human cDNA sequences looking for amino acid matches to the fungal gene allowed researchers to isolate the human cDNA clone that encoded the HGO enzyme. It is this enzyme that is defective in individuals affected by alktaponurea.

10. No. It is possible that one of the bands observed represents more than one DNA fragment. If, for example, the restriction fragments from the plasmid were 3, 5, 10, and also another 5 kb, the plasmid would actually have a total size of 23 kb.

11. Resequencing the relevant gene should be done as that will tell you if the original sequence was correct. You can then use the mutant sequence in a gene replacement experiment (depending on the organism) to see if the mutant phenotype is actually the result of the sequence variation you detected.

12. The fetus is heterozygous for sickle-cell anemia. The sickle-cell mutation affects an *Mst*II restriction site such that when digested with this enzyme, the restriction fragment is larger (1.3 kb) than normal (1.1 kb). Because both bands are present, the individual is heterozygous.

13. You could isolate DNA from the suspected transgenic plant and probe for the presence of the transgene by Southern hybridization.

14. Follow the protocol discussed in Figure 11-38 in the companion text. Basically, inject the rat growth hormone gene (RGH) into fertilized eggs of *lit/lit* mice. These eggs are then implanted into a surrogate mother and any resulting offspring mated to test their offspring to see if any are transgenic for RGH. Because RGH will be inherited in a dominant fashion, any large and transgenic offspring will be heterozygous at this point. The transgenic siblings will need to be mated to each other in order to generate mice that are homozygous for the RGH transgene.

15. The commercial cloning of insulin was into bacteria. Bacteria are not capable of processing introns. Genomic DNA would include the introns, while cDNA is a copy of processed (and thus intron-free) mRNA.

16. This problem assumes a random and equal distribution of nucleotides.
    *Alu*I $(1/4)^4$ = on average, once in every 256 nucleotide pairs
    *Eco*RI $(1/4)^6$ = on average, once in every 4096 nucleotide pairs
    *Acy*I $(1/4)^4(1/2)^2$ = on average, once in every 1024 nucleotide pairs

17. **Unpacking the Problem**

    1. Of the two discussed in the text, pBR322 is the closer.

    2. The single *Hin*dIII site in pBP1 allows for a simple opening up of the plasmid so that a DNA fragment made with *Hin*dIII can be inserted.

    3. It is important because it allows for screening for insertions of DNA into the plasmid. If the plasmid simply recircularizes, the transformed bacteria will be tetracycline-resistant. If the plasmid contains "foreign" DNA, the *tet* gene will be disrupted and the strain will be tetracycline-sensitive.

    4. Insertion of donor DNA into the plasmid disrupts the *tet* gene. It is not relevant to the problem but was important in the construction of the library.

    5. A library is a large collection of cloned DNA maintained within easily cultured vectors. For this question, the source of donor DNA was *Hin*dIII-digested fruit fly genomic DNA, and the vector was pBP1. Although it is not relevant to this question, the source of donor DNA is often key to the type of research being conducted.

    6. The gene of interest would have been "found" by using a probe composed of the gene's sequence (typically just a small region is required). This could have been synthesized by "guessing" the DNA sequence on the basis of the gene product's amino acid sequence or by homology to a similar gene from

another organism, etc. For this particular question, how this clone was identified does not matter.

7. An electrophoretic gel is an apparatus to separate fragments of DNA by their size. Generally, the mix of DNA fragments is forced to migrate through an agarose gel by an electric field that is negative at the end where the DNA is placed and positive at the far end. Because DNA is negatively charged, it will move to the far end but at rates that are inversely proportional to its size: small fragments will move more rapidly than large.

8. Ethidium bromide binds to DNA and fluoresces when exposed to UV light. It is used to visualize the location of the various DNA fragments within the gel.

9. The DNA from this gel is not "blotted" onto filter paper in this problem. If it had been, it would have been a Southern blot (because DNA was in the gel).

10. In this gel, DNA molecules of different sizes bound to ethidium bromide are visible under UV illumination.

11. There is only one linear fragment generated when a circle is cut once.

12. If cut twice, two linear fragments are generated.

13. There is a one-to-one relationship between the number of sites cut in a circular plasmid and the number of fragments generated.

14. Because the two enzymes will cut the DNA independently, the total number of fragments will be $n + m$.

15. They were loaded into the wells located at the top of the diagram.

16. Smaller fragments move more rapidly and travel farther per unit time than larger fragments.

17. All the control lanes contain 5 kb of DNA, the size of the plasmid. Both *Hind*III and *Eco*RV cut the plasmid once but at separate locations, as seen in the lane of the double digest (both single digests generate a single band, while the double digest generates two). The lanes with the clone 15-containing plasmid always add up to 7.5 kb, indicating that the donor DNA is 2.5 kb.

18. No. The 5-kb plasmid is cut twice, and the resulting fragments must add up to the total length.

19. No. They represent the cloned DNA cut out from the plasmid by *Hin*dIII and then cut once again by *Eco*RV. The sum of these fragments must equal the whole.

20. It tells you that the fragment that disappeared also contains a restriction site for the second enzyme.

21. A probe will hybridize to any fragments to which it is complementary in sequence.

22. If the two vectors are nonhomologous, the only hybridization observed will be because the gene of interest from the one species is complementary to the gene of interest in the other.

Solution to the problem

a.

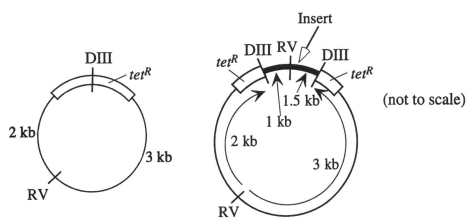

b. The *tet^R* gene used as a probe will detect only those bands that contain *tet^R* DNA. Thus, all bands in control lanes will have sequences complementary to the probe. For the clone 15 digests, the *Hin*dIII 5-kb band will be radioactive and both *Eco*RV bands will be radioactive. For the *Hin*dIII + *Eco*RV double digest, the 3-kb and 2-kb bands will be radioactive.

c. The homologous gene used as a probe will detect only those fragments containing the gene of interest. Thus, no bands will be radioactive in the control lanes, and the clone 15 lanes will all have at least one radioactive band. For *Hin*dIII, the 2.5-kb band (the insert) will be radioactive. For *Eco*RV, the 4.5-kb and 3.0-kb bands will be radioactive. For the *Hin*dIII + *Eco*RV double digest, the 1.5-kb and 1-kb bands will be radioactive.

18. The data indicate that *Eco*RI fragments 1 and 4 contain no *Hin*dII sites, fragment 3 contains one *Hin*dII site, and fragment 2 contains two sites. Conversely, *Hin*dII fragments A, B, and D all contain one *Eco*RI site, and fragment C

contains none. Fragment D contains fragments 1 and 3_1; fragment A contains fragments 3_2 and 2_1; fragment C is the same as fragment 2_2; and fragment B contains fragments 2_3 and 4. The only map consistent with these data is

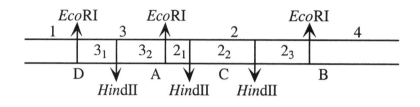

19. **a.** Because the actin protein sequence is known, a probe could be synthesized by "guessing" the DNA sequence based on the amino acid sequence. (This works best if there is a region of amino acids that can be coded with minimal redundancy.) Alternatively, the gene for actin cloned in another species can be used as a probe to find the homologous gene in *Drosophila*. If an expression vector was used, it might also be possible to detect a clone coding for actin by screening with actin antibodies.

    b. Hybridization using the specific tRNA as a probe could identify a clone coding for itself.

20. To answer this problem, you must realize what is being visualized. The 8.5 *Eco*RI fragment is radioactive only at one 5′ end, and only fragments containing that end will be seen by autoradiography. When this fragment is cleaved by other restriction enzymes, the longest fragments will have been cut at sites farthest from the radioactive end. In the following figure, if cut at position labeled 2, the fragment will be longer than any fragment cut at 3, 4, or 5 and shorter than any cut at position 1.

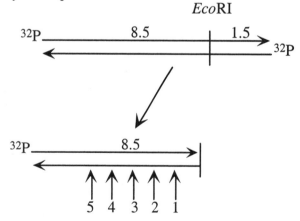

Using this logic, a relative map of the restriction sites of *Hind*II and *Hae*III for this fragment can be generated.

Reading the gels from the top (and from the farthest to the nearest to the labeled end) the order is

(*Eco*RI) - *Hin*dII - *Hae*III - *Hae*III - *Hin*dII - *Hae*III - *Hae*III - *Hae*III - *Hin*dII - *Hin*dII - *Hae*III - *Hin*dII - *Hae*III - labeled end

21. a. The double digest indicates that the 5.0-kb *Hin*dIII fragment also contains a *Sma*I site and the 5.5 *Sma*I fragment also contains a *Hin*dIII site. This suggests the following map:

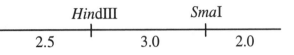

b. Because the only band to disappear is the 3.0-kb fragment, it is the only one that also contains an *Eco*RI site. The appearance of a new 1.5-kb fragment suggests the following:

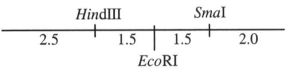

22. Create a library of *Podospora* DNA. This would be accomplished by isolating DNA from *Podospora*, cutting the DNA with *Hae*II, mixing with the vector pBR also cut with *Hae*II, ligating the mixture, and transforming *E. coli*. Only those bacteria that contain the plasmid will be *tet*^{R}. Of these, those that are *kan*^{S} contain plasmids with inserts.

Assuming that the same genes from different species have approximately the same base sequence, the ß-tubulin gene cloned from *Neurospora* can be used as a probe to isolate the ß-tubulin gene from *Podospora*. Identify which clone or clones in the library contain the desired sequence by colony hybridization using the cloned *Neurospora* actin gene as a probe.

23. a. The transformed phenotype would map to the same locus. If gene replacement occurred by a double crossing-over event, the transformed cells would not contain vector DNA. If a single crossing-over took place, the entire vector would now be part of the linear *Neurospora* chromosome.

b. The transformed phenotype would map to a different locus than that of the auxotroph if the transforming gene was inserted ectopically (i.e., at another location).

24. Size, translocations between known chromosomes, and hybridization to probes of known location can all be useful in identifying which band on a PFGE gel corresponds to a particular chromosome.

25.   Conservatively, the amount of DNA necessary to encode this protein of 445 amino acids is 445 × 3 = 1335 base pairs. When compared with the actual amount of DNA used, 60 kb, the gene appears to be roughly 45 times larger than necessary. This "extra" DNA mostly represents the introns that must be correctly spliced out of the primary transcript during RNA processing for correct translation. (There are also comparatively very small amounts of both 5′ and 3′ untranslated regions of the final mRNA that are necessary for correct translation encoded by this 60-kb of DNA.)

26.   The typical procedure is to "knock out" the gene in question and then see if there is any observable phenotype. One methodology to do this is described in in the companion text. A recombinant vector carrying a selectable gene within the gene of interest is used to transform yeast cells. Grown under appropriate conditions, yeast that have incorporated the marker gene will be selected. Many of these will have the gene of interest disrupted by the selectable gene. The phenotype of these cells would then be assessed to determine gene function.

CHALLENGING PROBLEMS

27.   **a.**   There is one *Bgl*II site, and the plasmid is 14 kb.

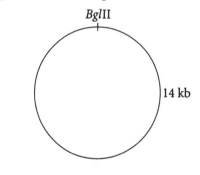

b.   There are two *Eco*RV sites.

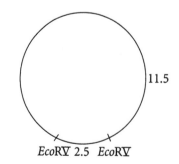

c.   The 11.5 kb *Eco*RV fragment is cut by *Bgl*II. The arrangement of the sites must be as indicated below.

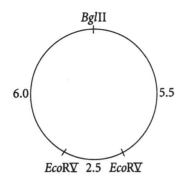

d. The *Bgl*II site must be within the *tet* gene.

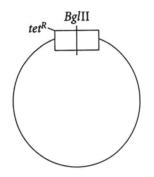

e. There was an insert of 4 kb.

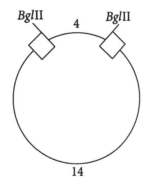

f. There was an *Eco*RV site within the insert.

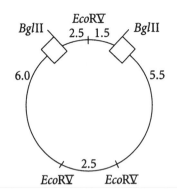

28. **a.** The restriction map of pBR322 with the mouse fragment inserted is shown below. The 2.5-kb and 3.5-kb fragments would hybridize to the pBR322 probe.

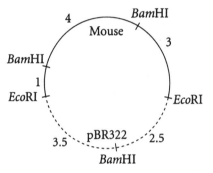

b. A protein 400 amino acids long requires a minimum of 1200 nucleotide bases. Only fragment 3 is long enough (3000 bp) to contain two or more copies of the gene. However, nothing can be said about their orientation.

29. **a.** To ensure that a colony is not, in fact, a prototrophic contaminant, the prototrophic line should be sensitive to a drug to which the recipient is resistant. A simple additional marker would also achieve the same end.

b. Use a nonrevertible auxotroph as the recipient (such as one containing a deletion.)

30. a.

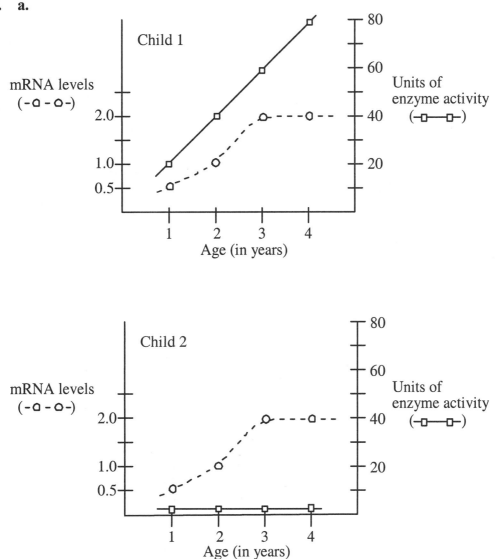

b. and c. Very low levels of active enzyme are caused by the introduction of an *Xho* site within the D gene. Because the size of the nonfunctional enzyme has not changed, the most likely event was a point mutation within the coding region of the gene that also created a new *Xho* site. It is likely that this point mutation also altered the active site destroying enzyme activity.

d. Individual 1 would be defined as homozygous normal, while individual 2 is homozygous mutant. If either were heterozygous, there would be three bands hybridizing to the probe on the Southern blot.

31. a. The gel can be read from the bottom to the top in a 5′-to-3′ direction. The sequence is

5′ T T C G A A A G G T G A C C C C T G G A C C T T T A G A 3′

b. By complementarity, the template was

3′ AAGCTTTCCACTGGGGACCTGGAAATCT 5′

c. The double helix is

5′ TTCGAAAGGTGACCCCTGGACCTTTAGA 3′
3′ AAGCTTTCCACTGGGGACCTGGAAATCT 5′

d. Open reading frames have no stop codons. There are three frames for each strand, for a total of six possible reading frames. For the strand read from the gel, the transcript would be

5′ UC**UAA**AGGUCCAGGGGUCACCUUUCGAA 3′

And for the template strand

5′ UUCGAAAGG**UGA**CCCCUGGACCUU**UAG**A 3′

Stop codons are in bold and underlined. There are a total of four open reading frames of the six possible.

32. The region of DNA that encodes tyrosinase in "normal" mouse genomic DNA contains two *Eco*RI sites. Thus, after *Eco*RI digestion, three different-sized fragments hybridize to the cDNA clone. When genomic DNA from certain albino mice is subjected to similar analysis, there are no DNA fragments that contain complementary sequences to the same cDNA. This indicates that these mice lack the ability to produce tyrosinase because the DNA that encodes the enzyme must be deleted.

33. Plant 1 shows the typical inheritance for a dominant gene that is heterozygous. Assuming kanamycin resistance is dominant to kanamycin sensitivity, the cross can be outlined as follows:

$$kan^R/kan^S \times kan^S/kan^S$$
$$\downarrow$$
$$1/2\ kan^R/kan^S$$
$$1/2\ kan^S/kan^S$$

This would suggest that the gene of interest would be inserted once into the genome.

Plant 2 shows a 3:1 ratio in the progeny of the backcross. This suggests that there have been two unlinked insertions of the *kan^R* gene and presumably the gene of interest as well.

$$kan^{R1}/kan^{S1} \; ; kan^{R2}/kan^{S2} \; \times \; kan^{S1}/kan^{S1} \; ; kan^{S2}/kan^{S2}$$
$$\downarrow$$

$1/4$ $kan^{R1}/kan^{S1} \; ; kan^{R2}/kan^{S2}$

$1/4$ $kan^{R1}/kan^{S1} \; ; kan^{S2}/kan^{S2}$

$1/4$ $kan^{S1}/kan^{S1} \; ; kan^{R2}/kan^{S2}$

$1/4$ $kan^{S1}/kan^{S1} \; ; kan^{S2}/kan^{S2}$

34. Assuming that the DNA from this region is cloned, it could be used as a probe to detect this RFLP on Southern blots. DNA from individuals within this pedigree would be isolated (typically from blood samples containing white blood cells) and restricted with *Eco*RI, and Southern blots would be performed. Individuals with this mutant CF allele would have one band that would be larger (owing to the missing *Eco*RI site) when compared with wild type. Individuals that inherited this larger *Eco*RI fragment would, at minimum, be carriers for cystic fibrosis. In the specific case discussed in this problem, a woman that is heterozygous for this specific allele marries a man that is heterozygous for a different mutated CF allele. Just knowing that both are heterozygous, it is possible to predict that there is a 25 percent chance of their child's having CF. However, because the mother's allele is detectable on a Southern blot, it would be possible to test whether the fetus inherited this allele. DNA from the fetus (through either CVS or amniocentesis) could be isolated and tested for this specific *Eco*RI fragment. If the fetus did not inherit this allele, there would be a 0 percent chance of its having CF. On the other hand, if the fetus inherited this allele, there would be a 50 percent chance the child will have CF.

35. The promoter and control regions of the plant gene of interest must be cloned and joined in the correct orientation with the glucuronidase gene. This places the reporter gene under the same transcriptional control as the gene of interest. The companion text discusses the methodology used to create transgenic plants. Transform plant cells with the reporter gene construct, and as discussed in the text, grow into transgenic plants. The glucuronidase gene will now be expressed in the same developmental pattern as the gene of interest and its expression can easily be monitored by bathing the plant in an X-Gluc solution and assaying for the blue reaction product.

36. *Unpacking the Problem*

1. Hyphae in *Neurospora* are the threads of cells that grow out from the original ascospore. Therefore, hyphal extension refers to the pattern of these threads and the distance that they grow. Because this process is due to cell growth (and its control) and cell shape, anyone interested in a vast array of cell biological issues might be interested in the genes identified by such screens.

2. Mutational dissection is the attempt to identify all the genes and gene products involved in a particular process. In this experiment, the goal is to

identify all the genes that can mutate to a small-colony phenotype by random insertion of unrelated DNA (in this case a bacterial plasmid with a selectable marker).

3. *Neurospora* is a haploid organism, and this is relevant to this problem. What might otherwise be a recessive mutation in a diploid organism (typical of gene knockouts) would instead be immediately expressed in a haploid one.

4. The source of the DNA is not relevant to the problem as long as it contains a selectable marker for the organism being transformed. The ease of growing, manipulating, and isolating bacterial plasmids makes them an attractive choice.

5. Transformation, as originally discovered by Griffith, is the uptake of DNA from one organism by another organism and its ultimate expression. In this situation, *Neurospora* has been pretreated in such a way as to cause the uptake of the bacterial plasmid. This is a well-used technique in molecular genetics as a way of introducing genes from virtually any source into the organisms under study.

6. Plant and fungal cells are generally prepared for transformation by removal of their cell walls. The cell membranes are then exposed to a high salt concentration and the exogenous DNA. Studies indicate that the DNA enters the cell in two ways: (1) phagocytosis and (2) localized temporary dissolving of the membrane by the high salt concentration.

7. With successful transformation, the exogenous DNA passes through the cytoplasm and enters the nucleus, where it becomes integrated into a host chromosome.

8. Integration and Transformation

<u>Entry into Cell</u>

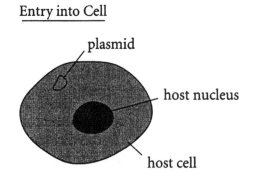

Integration and Transformation

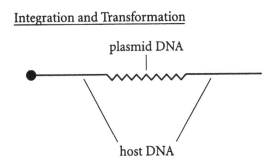

9. It is completely unnecessary to know what benomyl is. Its use simply allows for the selection of cells that received and integrated some exogenous DNA. Virtually any resistance marker could have been used. The choice of a resistance marker usually depends on what is easily available to the researcher, although questions of toxicity to humans may play a role in the choice.

10. Because hyphal extension occurs in colonies, not at the one-cell stage, the researcher must look for mutants that are expressed by a clone or colony. Therefore, he is looking for mutants that are "colonial." Mutations that produce an aberrant colony in size or shape are, by definition, involved with the extension of hyphae.

11. The "previous mutational analysis" could have been any random study. For example, in screens for specific auxotrophic mutants, experimenters would have noticed this abnormal phenotype also appearing.

12.

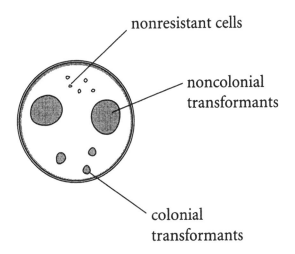

13. Tagging is a process to mutate and "mark" genes of interest by insertion of known DNA. In this case, the gene for benomyl resistance is being inserted randomly within the genome. Occasionally, insertion will occur within a gene involved with hyphal extension. These will cause the aberrant phenotype because the gene has been disrupted by the insertion of the

selectable DNA. The disruption causes a knockout mutation within the gene of interest and also supplies a "molecular handle" to later clone the DNA.

14. The orange-colored bread mold *Neurospora* is a multicellular haploid in which the cells are joined end to end to form hyphae. The hyphae grow through the substrate and also send up aerial branches that bud off haploid cells known as conidia (asexual spores). Conidia can detach and disperse to form new colonies, or alternatively they can act as paternal gametes and fuse with a maternal structure of a different individual. However, the different individual must be of the opposite mating type. In *Neurospora* there are two mating types, determined by the alleles *A* and *a*. A cross will succeed only if it is *A* × *a*. An asexual spore from the opposite mating type fuses with a receptive hair, and a nucleus travels down the hair to pair with a maternal gamete that waits inside a specialized knot of hyphae. The *A* and *a* pair then undergo synchronous mitoses, finally fusing to form diploid meiocytes. Meiosis occurs, and in each meiocyte four haploid nuclei are produced, which represent the four products of meiosis. For an unknown reason, these four nuclei divide mitotically, resulting in eight nuclei, which develop into eight football-shaped sexual spores, called ascospores. The ascospores are shot out of a flask-shaped fruiting body that has developed from the knot of hyphae that originally contained the maternal gametic cell. The ascospores can be isolated, each into a culture tube, where each ascospore will grow into a new culture by mitosis.

15. In order for the benomyl-resistance gene to be integrated within the host chromosome, it recombines with it. However, it is recombination between the *col* and *ben^R* genes that is interesting. It is either 0 percent (type 1), indicating that the insertion caused the small-colony phenotype, or 50 percent (type 2), showing the two events are unlinked.

16. Only two types are possible: integration into a "hyphal" gene (so the resistance and small-colony phenotype are linked) and ectopic integration and concurrent mutation of a gene, causing the small-colony phenotype where the two are unrelated and also unlinked. Which type is more likely depends on mutation rates and the number of genes that can mutate to the small-colony phenotype (i.e., the ease of generating spontaneous *col* mutants) and on the rate of transformation and integration.

17. A probe in experiments such as this one is usually a sequence of DNA that can be used to identify a specific DNA sequence within a genome or colony. The probe is labeled in some way to indicate its presence. In this experiment, the probe was probably the bacterial plasmid (although it might have been the benomyl-resistance gene), most likely radioactively labeled. A genomic library from each *col* mutant would be screened with the probe to identify those clones that contain complementary sequences (and, with luck, some sequences of the gene of interest).

18. A probe specific to the bacterial plasmid could be made by growing bacteria with the plasmid. The plasmids could be isolated through cesium chloride centrifugation and then labeled.

Solution to the Problem

a. Type 1 isolates behave genetically as if the benomyl-resistance and small-colony phenotype are completely linked. The progeny are all parental in phenotype (either *col ben-R* or *+ ben-S*). This would be the expected result if the insertion of the plasmid (and *ben-R* gene) caused the mutation that led to the *col* phenotype, which of course was the point of the experiment.

Type 2 isolates behave genetically as if the benomyl-resistance and small-colony phenotype are unlinked (i.e., the two markers are segregating randomly in the progeny). This is the expected result if the insertion of the *ben-R* gene was unrelated to the mutation that led to the *col* phenotype. In other words, the *col* mutation occurred randomly and separately from the insertion of the *ben-R* gene.

b. The type 1 isolates are the *col* genes that are mutated by the insertion of the plasmid, and therefore these are the genes that are tagged.

c. The type 1 isolates should be used to create genomic libraries. The libraries should be screened for clones that contain DNA adjacent to the insert by probing with known sequences from the plasmid. The identified clones will represent parts of the disrupted gene of interest. To recover the intact wild-type gene, a subclone of the disrupted gene sequence can be used to probe a wild-type genomic library.

d. All progeny that are benomyl resistant will also contain DNA from the bacterial plasmid integrated into their chromosome or chromosomes. Thus, all *ben-R* strains from this experiment will have DNA that hybridizes to a probe specific for the plasmid.

37. **a. and b.** During Ti plasmid transformation, the kanamycin gene will insert randomly into the plant chromosomes. Colony A, when selfed, has $3/4$ kanamycin-resistant progeny, and colony B, when selfed, has $15/16$ kanamycin-resistant progeny. This suggests that there was a single insertion into one chromosome in colony A and two independent insertions on separate chromosomes in colony B. This can be schematically represented by showing a single insertion within one of the pair of chromosome "A" for colony A.

Chromosomes "A"

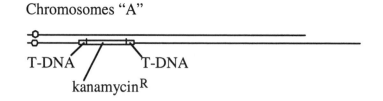

and two independent insertions into one of each of the pairs of chromosomes "B" and "C" for colony B.

Chromosomes "B"

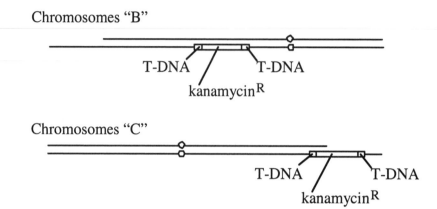

Chromosomes "C"

Genetically, this can be represented as

Colony A kan^{R_A}/kan^{S_A}
Colony B kan^{R_B}/kan^{S_B} ; kan^{R_C}/kan^{S_C}

When these are selfed kan^{R_A}/kan^{S_A} × kan^{R_A}/kan^{S_A}
$$\downarrow$$
$1/4$ kan^{R_A}/kan^{R_A}
$1/2$ kan^{R_A}/kan^{S_A}
$1/4$ kan^{S_A}/kan^{S_A}

kan^{R_B}/kan^{S_B} ; kan^{R_C}/kan^{S_C} × kan^{R_B}/kan^{S_B} ; kan^{R_C}/kan^{S_C}
$$\downarrow$$
$9/16$ $kan^{R_B}/-$; $kan^{R_C}/-$
$3/16$ $kan^{R_B}/-$; kan^{S_C}/kan^{S_C}
$3/16$ kan^{S_B}/kan^{S_B} ; $kan^{R_C}/-$
$1/16$ kan^{S_B}/kan^{S_B} ; kan^{S_C}/kan^{S_C}

38.   **a.**   Yeast plasmids can exist free in the cytoplasm or can integrate into a chromosome; however, the patterns of inheritance will differ for the two states. Free plasmids are present in multiple copies and will be distributed to all progeny. This is observed in crosses with YP1 transformed cells: YP1 leu^+ × leu^- all progeny are leu^+ and all have vector DNA. Crosses with

YP2 transformed cells show simple Mendelian inheritance and suggest that this plasmid has integrated into the yeast chromosome.

b. For YP1 transformed cells, the circular (and free) plasmid will be linearized by the single restriction cut, and when probed, a single band will be present on the Southern blot. Because the YP2 plasmid is integrated, a single cut within the plasmid will generate two fragments that contain plasmid DNA. However, the size of these fragments will depend on where in the genome other sites exist for this same restriction enzyme. This is schematically shown below:

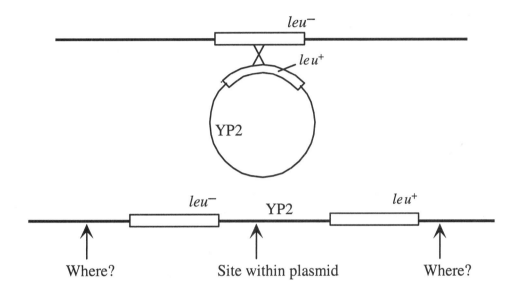

39. a. If the plasmid never integrates, the linear plasmid will be cut once by *Xba*I and two fragments will be generated that will both hybridize to the *Bgl*II probe. The autoradiogram will show two bands whose combined length will equal the full length of the plasmid.

b. If the plasmid integrates occasionally, most cells will still have free plasmids and these will be indicated by the two bands mentioned above. However, when the plasmid is integrated, two bands will still be generated, but their sizes will vary based on where other genomic *Xba*I sites are relative to the insertion point (see the figure for Problem 38b for similar logic). If integration is random, many other bands will be observed, but if its at a specific site, only two other bands will be detected.

12

Genomics

1. Contig describes a set of adjacent DNA sequences or clones assembled using overlapping sequences or restriction fragments. Because the pieces are assembled into a continuous whole, it makes sense that contig is derived from the word contiguous.

2. Because bacteria have relatively small genomes (roughly three megabase pairs) and essentially no repeating sequences, the whole genome shotgun approach would be used.

3. **a., b. and c.** Detection of hybridization at one end of every chromosome would not be indicative that the probe used is from a unique gene (as FISH indicates more than one region of homology) or from the telomere (as there is a telomere at each end of a chromosome and not just at one end). However, its possible that the probe is from the centromere as it hybridizes to one location on each chromosome.

4. Terminal sequence reads of cloned inserts. such as those generated during whole genome shotgun sequencing, are assembled into a scaffold by matching homologous sequences shared by reads from overlapping clones. In essence, the central sequence of any single clone will be generated from the terminal sequences of overlapping clones.

5. Assume the average BAC contains 200 kb. It would take 5000 BAC clones to hold a one gigabase genome. For five-fold coverage, you would be handling 25,000 clones.

6.

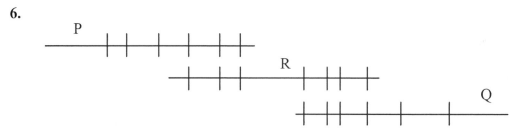

7. The minimal tiling path is the minimum number of clones that represents the entirety of the genome. For reasons of economy and efficiency, its desirable to sequence clones with as little overlap as possible.

8. BACs contain between 200–300 kb of cloned DNA. Because individual sequencing reactions provide about 600 bases of sequence, it is clear that the cloned DNA will have to be broken into much smaller pieces (subcloned) in order to determine its sequence. To subclone, you would cut the cloned DNA out of the BAC and then randomly shear the DNA to obtain the appropriately sized fragments. These 600 base-pair fragments would then be cloned into a vector appropriate for sequencing.

9. A scaffold is also called a supercontig. Contigs are sequences of overlapping reads assembled into units, and a scaffold is a collection of joined-together contigs.

10. Yes. If the gap is short, PCR fragments can be generated from primers based on the ends of your contigs, and these fragments can be directly sequenced without a cloning step. The process does not work if the gap is too long.

11. The data indicate that microsatellite locus and deletion are not linked. In essence, you see that segregation of M′ or M′′ is equally likely in deletion containing sperm. This is the expected result if the loci are unlinked.

12. The clone may contain DNA that hybridizes to a small family of repetitive DNA that is adjacent to the gene being studied or within one of its introns. Alternatively, the clone may share enough homology with members of a small gene family to hybridize to all or there may be four pseudogenes that are descendants of the unique gene being studied.

13. Yes. The clone hybridizes to and spans a translocation breakpoint that involves the X chromosome and an autosome. Because the DMD gene normally maps to the X chromosome, it is of interest that a translocation of the X is found in a patient with DMD. If the translocation is the cause of DMD mutation, the clone identifies a least a portion and/or location of the gene.

14. There are two approaches that can be attempted. One is to separately transform mutant *Neurospora* with each of the candidate genes and see if the transgene reverts the mutant phenotype (functional complementation). Alternatively, you can attempt to compare the phenotype of the mutant with the known functions of the candidate genes.

15. You still need to verify that the gene with the amino acid change is the correct candidate for the phenotype under study. (See question 14.)

16. A codon is the location that tRNAs will functionally bind through complementary base-pairing with their anticodons.

17. Yes. The operator is the location at which repressor functionally binds through interactions between the DNA sequence and the repressor protein.

18. There are numerous possible drawings that can be made. The goal is to indicate that at least one exon be present in all eight genomic fragments and that the ESTs define the 5′ and 3′ ends of the transcript.

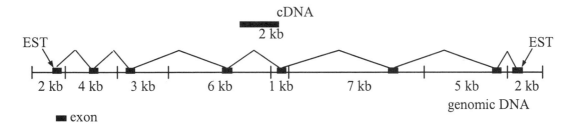

19. A level of 35 percent or more amino acid identity at comparable positions in two polypeptides is indicative of a common three-dimensional structure. It also suggests that the two polypeptides are likely to have at least some aspect of their function in common. However, it does not prove that your particular sequence does in fact encode a kinase.

20. The yeast two-hybrid test detects possible physical interactions between two proteins. These results indicate that gene A codes for a protein that interacts with proteins encoded by clones M and N, and further, that clone M encodes a protein that also interacts with proteins encoded by clones S and Q. For example:

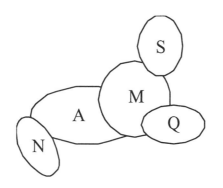

CHALLENGING PROBLEMS

21. The *Arabidopsis*-specific probe cross-hybridizes to DNA from cabbage. The number of bands observed is a function of where the specific restriction sites are relative to the region of DNA that hybridizes to the probe. Digestion with enzyme 2 results in a single band when hybridized to the probe, because there are no restriction sites within the sequence that hybridizes. Digestion with enzyme 1 results in three bands, because there are two restriction sites within the sequence that hybridizes. Similarly, digestion with enzyme 3 results in two bands when hybridized to the probe, because there is one restriction site within this sequence. The schematic below shows an example:

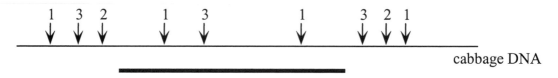

cabbage DNA

Arabidopsis probe

22. **a.** To determine the physical map showing the STS order, simply list the STSs that are positive, using parentheses if the order is unknown, and align them with one another to form a consistent order.

YAC A:			1	4	3	
YAC B:		5	1			
YAC C:				4	3	7
YAC D:	(6 2)	5				
YAC E:					3	7

b. Once the sequence of STSs is known, the YACs can be aligned as follows, although precise details of overlapping and the locations of ends are unknown:

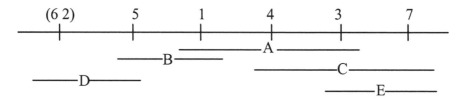

23. **a.** RAPDs are formed when regions of DNA are bracketed by two inverted copies of a "random" PCR primer sequence. Below, the primers are indicated by Xs, and the amplified regions would be the DNA between the brackets. For convenience, the two amplified regions are shown on the same lengthy piece of DNA for strain 1.

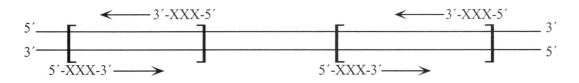

Strain 2 lacks one or two regions complementary to the primer.

b. Progeny 1 and 6 are identical with the strain 1 parent. Progeny 4 and 7 are identical with the strain 2 parent. Progeny 2 and 5 received the chromosome holding the upper band from the strain 1 parent and the chromosome holding the lower band from the strain 2 parent (resulting in no second band). Progeny 3 received the opposite: the chromosome holding the lower band from the strain 1 parent and the chromosome holding the upper band from the strain 2 parent (resulting in no second band). Therefore, bands 1 and 2 appear to be unlinked.

c. Recall that a nonparental ditype has two types only, both of which are recombinant. Therefore, the tetrad would be composed of two progeny like progeny 2 and two progeny like progeny 3.

24. a. The following stylized schematic of a reciprocal translocation between chromosome 3 and 21 is arbitrarily chosen to show the salient details. Band 3.1 of the q arm of chromosome 3, is split by the translocations that are correlated to the *N* disease allele. Probe c hybridizes to the region of 3q3.1 that remains with chromosome 3 and probes a, b, and d hybridize to the region of 3q3.1 that is translocated in this case to chromosome 21.

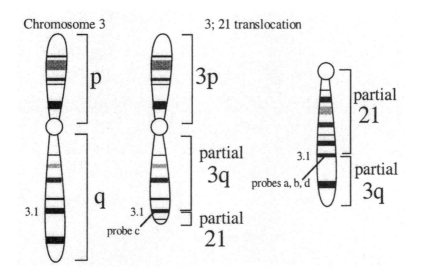

b. Because translocations of chromosome 3 that break band 3q3.1 are correlated to the disease, it is reasonable to assume that these rearrangements split the normal gene (*n*) in two, separating vital coding or regulatory regions. Therefore analysis and cloning of this specific region should be attempted.

In order to isolate and characterize the normal allele, chromosome walking from the known clones should be attempted in genomic libraries from individuals with the translocation and affected with the disease. Probe c is to one side of the breakpoint, while a, b, and d are on the other side. Also, translocation breakpoints serve as useful molecular landmarks, because they are easily identified on Southerns as "split bands" when probed with cloned DNA spanning the breakpoint. Once the breakpoint has been identified and cloned, the appropriate subclones would be used to clone the normal allele from a "normal" genomic library. This would be in conjunction with the usual techniques to identify a gene: sequencing, open reading frame analysis, Northern blots, etc.

   c.   Once *n* is cloned, it can be used to clone the various alleles from individuals who have the disease but not a translocation. The various alleles could then be compared with *n* by sequence, regulation, etc.

25.   From low to highest resolution the order would be:
f, c, (a, d), (e, h), b, g

26.   This is just a matter of aligning the sequences to determine their overlap.

Read 1:   TGGCCGTGATGGGCAGTTCCGGTG
Read 2:   TTCCGGTGCCGGAAAGA
Read 3: CTATCCGGGCGAACTTTTGGCCG
Read 4:   CGTGATGGGCAGTTCCGGTG
Read 5:   TTGGCCGTGATGGGCAGTT
Read 6:   CGAACTTTTGGCCGTGATGGGCAGTTCC

And this creates the contig:

CTATCCGGGCGAACTTTTGGCCGTGATGGGCAGTTCCGGTGCCGGAAAGA

27.

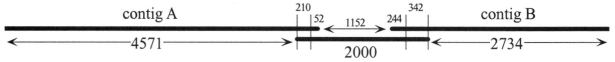

28.   You can determine whether the cDNA clone was a monster or not, by alignment of the cDNA sequence against the genomic sequence. (There are computer programs available to do this.) Is it derived from two different sites? Does the cDNA map within one [gene-sized] region in the genome or to two different regions? Of course, introns may complicate the issue.

29.   **a.**   Because the triplet code is redundant, changes in the DNA nucleotide sequence, (especially at those nucleotides coding for the third position of a codon) can occur without change to its encoded protein.

b. It can be expected that protein sequences will evolve and diverge more slowly than the genes that encode them.

30. *Unpacking the problem*

1. Two types of hybridizations that have already been discussed are hybridizations between strains of a species and hybridizations between species. A third type of hybridization is referred to in this problem: molecular hybridization. Molecular hybridization can involve either DNA-DNA hybridization or DNA-RNA hybridization. In both instances, it relies on the specificity of complementary pairing and can take place in solution, on a gel, on a filter, or on a slide. For example:

$$5'—U\,A\,C\,G\,G\,G\,A\,U—3'\ RNA$$
$$3'—A\,T\,G\,C\,C\,C\,T\,A—5'\ DNA$$

2. In situ hybridization is usually conducted on a slide so that the stained chromosomes can be observed and the specific portion of a chromosome to which the probe hybridizes can be identified.

3. A YAC is a yeast artificial chromosome. It contains a yeast centromere, autonomous replication sequences (origins of replication), telomeres, and DNA that has been attached between them.

4. Chromosome bands are dark regions along the length of a chromosome that occur in a characteristic pattern for each chromosome within an organism. They can occur naturally, as with *Drosophila* polytene chromosomes, or they can be induced by a number of chemical and physical agents, combined with staining to accentuate the bands and interbands.

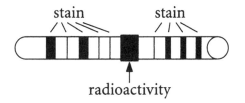

5. The five YACs could have been hybridized sequentially to the same chromosome preparation, which is, however, unlikely. Alternatively, the information could have been determined in five separate experiments. In either case, a YAC labeled with either radioactivity or fluorescence, and including the DNA of interest, was hybridized to a chromosome preparation. The chromosomes were properly treated to reveal the banding pattern, and the YACs were determined to hybridize to the same band.

6. A genomic fragment, by definition, contains a subportion of the genome being studied. In most instances, it actually contains a subportion of one

chromosome. Five randomly chosen YACs would not be expected to contain the same genomic fragment or even fragments from the same chromosome. The fragments could have been produced by either physical (X-irradiation, shearing) or chemical (digestion, restriction) means, but it does not matter how they were produced.

7.   A restriction enzyme is a naturally occurring bacterial enzyme that is capable of causing either single- or double-stranded breaks in DNA at specific DNA sequences.

8.   A long cutter is a restriction enzyme that produces very long fragments of DNA because the sequence it recognizes occurs infrequently within the genome.

9.   The YACs were radioactively labeled so that their location after hybridization could be detected through autoradiography. To radioactively label is to attach an isotope that emits energy through decay. Commonly used radioactive labels are tritium (^{3}H) in place of hydrogen and ^{32}P in place of phosphorus.

10.   An autoradiogram is a "self-picture" taken through radioactive decay from a labeled probe. When a gel or blot is used, the radioactive decay is captured by a piece of X-ray film. When in situ hybridization is performed on slides, the photographic emulsion coats the slide directly.

11.   Free choice. Be sure you truly know the meaning of each term.

12.   We are given a diagram of the composite autoradiographic results. The DNA from humans was isolated and subjected to digestion by a restriction enzyme that cuts very infrequently. Once the DNA was electrophoresed, it was Southern-blotted and then probed sequentially with radioactively labeled YACs, followed by sequential exposure to X-ray film. Between probings, the previous YAC hybrid was removed through denaturing of the DNA-DNA hybrid. Alternatively, five separate Southern blottings were done.

13.   The haploid human genome is thought to contain approximately 3.3×10^6 kilobases of DNA.

14.   Restriction digestion of human genomic DNA would be expected to produce hundreds of thousands of fragments.

15.   The fragments produced by restriction of human genomic DNA would be expected to be mostly different.

16.   When subjected to electrophoresis and then stained with a DNA stain, the digested human genome would produce a continuous "smear" of DNA,

from very large fragments (in excess of tens of thousands of base pairs in length) to fragments that are very small (under a hundred bases in length).

17. In this question, only two distinct bands are produced, at most, in any one probing. The difference between what is seen with a DNA stain and what is seen with probing lies in the specificity of the agent being used. DNA stain will detect any DNA, while a DNA probe will detect only DNA that is complementary to the probe.

18. Number them from top to bottom, 1–3, across the gel. Thus, YACs A–C contain band 1, YACs C–D contain band 2, and YACs A and E contain band 3.

19. There are no restriction fragments on the autoradiogram. The fragments are on the filter (nitrocellulose, nylon) used to blot the gel. The radioactivity of the probes is captured by the X-ray film as it decays, producing an exposed region of film.

20. YACs B, D, and E hybridize to one fragment, and YACs A and C hybridize to two fragments.

21. A YAC can hybridize to two fragments if the YAC contains continuous DNA and there is a restriction site within that region. A YAC can also hybridize to two fragments if it contains discontinuous DNA from two locations in the genome that either are on different chromosomes (this is analogous to a translocation) or are separated by at least two restriction sites if they are on the same chromosome (this is analogous to a deletion). In this case, the former makes more sense. Because the YACs were selected for their binding to one specific chromosome band, it is unlikely that the YACs are composed of discontinuous DNA sequences. A YAC could hybridize to more than two fragments because the continuous DNA could contain many restriction sites or the discontinuous DNA could be composed of DNA from a number of regions in the genome.

22. Cytogeneticists use the term *band* to designate a region of a chromosome that is dark staining. Molecular biologists use *band* to designate a region of dark appearing on an autoradiogram, which is produced by radioactive decay from a specific probe that reacted with a population of molecules localized by gel electrophoresis. In both cases, *band* refers to a localization.

Solution to the Problem

a. Note that fragments 1 and 3 occur together and fragments 1 and 2 occur together, but that fragments 2 and 3 do not occur together. This suggests that the sequence is 2 1 3 (or 3 1 2).

b. If the sequence of the fragments is 2 1 3, then the YACs can be shown in relation to these fragments. YAC A spans at least a portion of both 1 and 3. YAC B is within region 1. YAC C spans at least a portion of regions 1 and 2. YAC D is contained within region 2. YAC E is contained within region 3. A diagram of these results is shown below. In the diagram, there is no way to know the exact location of the ends of each YAC.

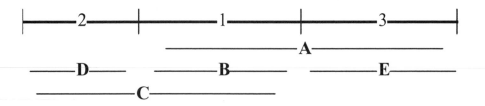

31. a. This is just a matter of aligning the sequences to determine their overlap.

```
Read 1:                                                   ATGCGATCTGTGAGCCGAGTCTTTA
Read 2:                    AACAAAAATGTTGTTATTTTTATTTCAGATG
Read 3:                                    TTCAGATGCGATCTGTGAGCCGAG
Read 4:  TGTCTGCCATTCTTAAAAACAAAAATGT
Read 5:                           TGTTATTTTTATTTCAGATGCGA
Read 6:                    AACAAAAATGTTGTTATT
```

And this creates the contig

```
TGTCTGCCATTCTTAAAAACAAAAATGTTGTTATTTTTATTTCAGATGCGATCTGTGAGCCGAGTCTTTA
```

and the transcript of

```
UGUCUGCCAUUCUUAAAAACAAAAAUGUUGUUAUUUUUAUUUCAGAUGCGAUCUGUGAGCCGAGUCUUUA
```

b. Translation of the contig starting at the first letter would give

CLPFLKTKMLLFLFQMRSVSRVF

Translation of the contig starting at the second letter would give

VCHS stop

Translation of the contig starting at the third letter would give

SAILKNKNVVIFISDAICRPSL

c. Using the nucleotide sequence of the contig and performing BLASTn or the possible translation products and performing tBLASTn, you will discover that this sequence and the translation product above listed first match perfectly with a region of exon 19 of the human CFTR gene.

32. The cross is

$$cys\text{-}1\ RFLP\text{-}1^O\ RFLP\text{-}2^O \times\ cys\text{-}1^+\ RFLP\text{-}1^M\ RFLP\text{-}2^M$$

Scoring the progeny, a parental type will have the genotype of either strain and, if the markers are all linked, be the most common. A recombinant type will have a mixed genotype and be less common. Clearly, the first two ascospore types are parental, with the remaining being recombinant.

a. The *cys-1* locus is in this region of chromosome 5. If it were not in this region, linkage to either of the RFLP loci would not be observed.

b. To calculate specific distances, you may need to review previous chapters. Here, it is assumed that you recall basic mapping strategies.

$$cys\text{-}1 \text{ to } RFLP\text{-}1 = {}^{(2\,+\,3)}/_{100} \times 100\% = 5 \text{ map units}$$
$$cys\text{-}1 \text{ to } RFLP\text{-}2 = {}^{(7\,+\,5)}/_{100} \times 100\% = 12 \text{ map units}$$
$$RFLP\text{-}1 \text{ to } RFLP\text{-}2 = {}^{(2\,+\,3\,+\,7\,+\,5)}/_{100} \times 100\% = 17 \text{ map units}$$

```
|—5 m.u.—|————12 m.u.————|
RFLP-1    cys-1           RFLP-2
```

c. A number of strategies could be tried. Because this is an auxotrophic mutant, functional complementation can be attempted. Positional cloning or chromosome walking from the RFLPs is also a very common strategy.

33. The correct assembly of large and nearly identical regions is problematic with either method of genomic sequencing. However, the whole genome shotgun method is less effective at finding these regions than the clone-based strategy. This method also has the added advantage of easy access to the suspect clone(s) for further analysis.

34. a. Of the regions of overlap for cosmids C, D, and E, region 5 is the only region in common. Thus, gene *x* is localized to region 5.

b. The common region of cosmids E and F, or the location of gene *y*, is region 8.

c. Both probes are able to hybridize with cosmid E because the cosmid is long enough to contain part of genes *x* and *y*.

35. a. DNA from each individual was obtained. It was restricted, electrophoresed, blotted, and then probed with the five probes. After each probing, an autoradiograph was produced.

b. First identify which chromosome came from the affected parent. This is easily determined by identifying which chromosome could not have come from the mother. For the first daughter, the chromosome with 2′ was inherited from the father. Likewise, 2″, 3″, and 2″ identify the paternal chromosome in the other children. In all cases, the chromosome drawn to the left in this problem is the one inherited from the mother.

Next compare the maternal chromosomes of affected offspring with unaffected offspring to determine which RFLP is most closely correlated to the disease. This analysis is based on the co-segregation of one of the RFLPs and the disease-causing gene. Notice that all of these chromosomes show evidence of recombination. For example, when compared with the mother's chromosomes, it can be deduced that the maternally inherited chromosome of the unaffected daughter is the result of a double crossover event.

Affected:	1º 2º 3′ 4º 5º
Unaffected:	1º 2º 3′ 4′ 5º
Unaffected:	1′ 2º 3′ 4′ 5º
Affected:	1′ 2º 3′ 4º 5′

The only RFLP that correlates to the disease and therefore is likely closest to the disease allele is 4º. It is present in both affected children and absent in both unaffected children.

c. It appears that RFLP 4 is the closest marker to the gene and could be used for positional cloning by chromosome walking. However, with only four offspring, the genetic distance between the gene and this marker could be quite large. The number of markers for each human chromosome is already large and increasing almost daily. If possible, it makes sense to analyze this family further (and as many other families with the same trait that can be found) to see if the gene can be further localized before the arduous task of "walking" is attempted.

36. a., b. and c. Cystic fibrosis (CF) is a recessive, autosomally inherited disease. Both parents in this pedigree must be carriers because some of their children are affected. Because the problem states that the three probes used are very closely linked to the *CF* gene, recombination will be ignored.

The data from three probes are presented, but only probes 1 and 3 detect RFLPs in this pedigree and are therefore informative. Both probes detect either one or two bands depending on the allele present. Calling the one-band pattern allele *A* and the two-band pattern allele *B*, the individuals of the pedigree are

Father	*RFLP-1^B RFLP-3^A*
Mother	*RFLP-1^A RFLP-3^B*
Child 1 (II-1)	*RFLP-1^B RFLP-3^A* (does not have CF)
Child 2 (II-2)	*RFLP-1^B RFLP-3^B* (does have CF)
Child 3 (II-3)	*RFLP-1^B RFLP-3^B* (does have CF)
Child 4 (II-4)	*RFLP-1^A RFLP-3^B* (does not have CF)
Child 5 (II-5)	*RFLP-1^B RFLP-3^A* (does not have CF)
Child 6 (II-6)	*RFLP 1^B RFLP-3^B* (does have CF)
Child 7 (II-7)	*RFLP-1^A RFLP-3^A* (does not have CF)

The first step is to determine which RFLP alleles are linked to the disease-causing *CF* alleles. The pattern of inheritance suggests that *RFLP-1^B* from the father and *RFLP-3^B* from the mother are both linked to *CF* alleles because all children that are *RFLP-1^B RFLP-3^B* also have CF.

The oldest son (II-1) is a carrier because he has inherited a *CF* allele (linked to *RFLP-1^B*)from his father. Similarly, II-4 has inherited a *CF* allele from his mother, II-5 has inherited a *CF* allele from his father, and II-7 is homozygous normal.

37. Assessing whether a short sequence constitutes an exon is difficult. The best way to determine if a suspected micro-exon is actually used is to look for a cDNA or an EST that includes it. Alternatively, identification of consensus donor and acceptor splice site sequences can be tried and also the use of comparative genomics, that is, the conservation of the predicted amino acid encoded by the micro-exon in the same or other genomes.

13

The Dynamic Genome:
Transposable Elements

BASIC PROBLEMS

1. Mutations in *gal* can be generated, and from these strains, λ*dgal* phage isolated. Through hybridization of denatured λ*dgal* DNA containing the mutation with wild-type λ*dgal* DNA, some of the molecules will be heteroduplexes between one mutant and one wild-type strand. If the mutation was caused by an insertion, the heteroduplexes will show a "looped out" section of single-stranded DNA, confirming that one DNA strand contains a sequence of DNA not present in the other.

The text also illustrates a method to compare the densities of *gal*⁺-carrying λ phage with *gal*⁻-carrying phage. In this experiment, the *gal*⁻-phage are denser, indicating that they contain a larger DNA molecule.

If the *gal* genes are cloned, direct comparison of the restriction maps or even the DNA sequence of mutants compared with wild type will give specific information about whether any are the results of insertions.

Using primers that flank the gene, PCR can be used to determine the size of the intervening fragment. Insertions would result in a larger-sized fragment compared to wild type.

2. In replicative transposition, transposable elements move to a new location by replicating into the target DNA, leaving behind a copy of the transposable element at the original site. If, on the other hand, the transposable element excises from its original position and inserts into a new position, this is called conservative transposition.

To test either mechanism, experiments must be designed so that both the "old" and "new" positions of the transposon can be assayed. If the transposon remains in the old site at the same time that a new copy is detected elsewhere, a replicative mechanism must be in use. If the transposon no longer exists in the old site when a copy is detected elsewhere, a conservative mechanism must be in use.

Figure 13-13 in the companion text describes how replicative transposition can be observed between two plasmids. The same general protocol could be used to detect conservative transposition, but of course the results would be different. Kleckner and co-workers actually demonstrated conservative transposition by following the movement of a transposon that contained a small heteroduplex within the *lacZ* gene. The DNA of two derivatives of Tn10 carrying different *lacZ* alleles (one being wild-type and the other being mutant) were denatured and allowed to reanneal. In some cases, the DNA molecules that reformed were actually heteroduplexes; one strand contained the *lacZ*$^+$ allele, and the other strand contained the *lacZ*$^-$ allele. Transpositions of theses heteroduplexes were then followed. Based on the mechanism of movement, two outcomes are possible. If replicative transposition occurred, the semiconservative nature of DNA replication would generate two genetically different transposons: instead of the heteroduplex *lacZ*$^+$/*lacZ*$^-$ DNA, one would now be all *lacZ*$^+$ and the other all *lacZ*$^-$. If transposition was conservative, the *lacZ* gene would still be heteroduplex *lacZ*$^+$/*lacZ*$^-$ after transposition and the first cell division would resolve the heteroduplex. This is what was observed. (The "sectored" colonies are the result of the original cell still having the *lacZ*$^+$/*lacZ*$^-$ heteroduplex after transposition. After the first division one cell was now *lacZ*$^+$ and the other was *lacZ*$^-$). The experiment is outlined below:

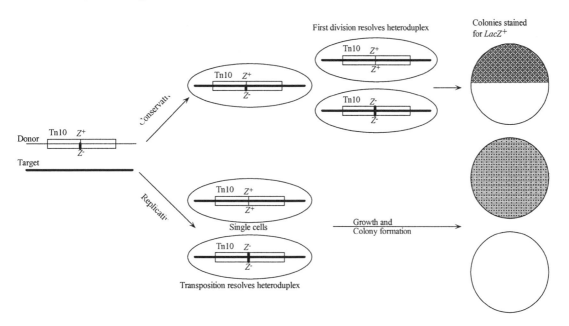

3. R plasmids are the main carriers of drug resistance. These plasmids are self-replicating and contain any number of genes for drug resistance, as well as the genes necessary for transfer by conjugation (called the RTF region). It is R plasmid's ability to transfer rapidly to other cells, even those of related species,

that allows drug resistance to spread so rapidly. R plasmids acquire drug-resistance genes through transposition. Drug-resistance genes are found flanked by IR (inverted repeat) sequences and as a unit are known as transposons. Many transposons have been identified, and as a set they encode a wide range of drug resistances (see the table in the companion text). Because transposons can "jump" between DNA molecules (e.g., from one plasmid to another or from a plasmid to the bacterial chromosome and vice versa), plasmids can continue to gain new drug-resistance genes as they mix and spread through different strains of cells. It is a classic example of evolution through natural selection. Those cells harboring R plasmids with multiple drug resistances survive to reproduce in the new environment of antibiotic use.

4. Boeke, Fink, and their co-workers demonstrated that transposition of the Ty element in yeast involved an RNA intermediate. They constructed a plasmid using a Ty element that had a promoter that could be activated by galactose, and an intron inserted into its coding region. First, the frequency of transposition was greatly increased by the addition of galactose, indicating that an increase in transcription (and production of RNA) was correlated to rates of transposition. More importantly, after transposition they found that the newly transposed Ty DNA lacked the intron sequence. Because intron splicing occurs only during RNA processing, there must have been an RNA intermediate in the transposition event.

5. P elements are transposable elements found in *Drosophila*. Under certain conditions they are highly mobile and can be used to generate new mutations by random insertion and gene knockout. As such, they are a valuable tool to tag and then clone any number of genes. P elements can also be manipulated and used to insert almost any DNA (or gene) into the *Drosophila* genome. P element-mediated gene transfer requires inserting the DNA of interest between the inverted repeats necessary for P element transposition. This recombinant DNA, along with helper intact P element DNA (to supply the transposase), are then co-injected into very early embryos. The progeny of these embryos are then screened for those that contain the randomly inserted DNA of interest.

6. Politics aside, Barbara McClintock was classically ahead of her time. She once wrote, "I stopped publishing detailed reports long ago when I realized, and acutely, the extent of disinterest and lack of confidence in the conclusions I was drawing from the studies." But beginning in the 1970's, transposition was discovered in bacteria and then in yeast. Transposable elements were subsequently found to be a significant component of most genomes, not just an oddity found only in maize. The importance of transposition in the transfer of drug resistance, cancer, and genetic engineering, just to name a few, was finally understood at the molecular level, and she was fully recognized for her lifetime of brilliant work.

CHALLENGING PROBLEMS

7. The staggered cut will lead to a nine base-pair target site duplication that flanks the inserted transposon.

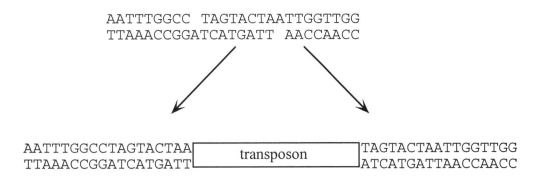

8. The sn^+ patches in an sn background and the occurrence of sn^+ progeny from an $sn \times sn$ mating indicate that sn^+ alleles are appearing at relatively high frequencies and that the sn alleles are unstable. High reversion rates suggest that the sn allele is the result of an insertion of a transposable element that is capable of (frequent) excision. Further, the transposable element must be capable of excision in the germline as well as in the soma.

9. In the *Ac-Ds* system, *Ac* can produce an unstable allele that is autonomous. *Ds* can revert only in the presence of *Ac* and is nonautonomous. In the following, Ac^+ indicates the absence of the *Ac* regulator gene.

Cross 1:

      P $C/c^{Ds}; Ac/Ac^+ \times c/c; Ac^+/Ac^+$

      F_1 $1/4$ $C/c; Ac/Ac^+$ (solid pigment)
              $1/4$ $C/c; Ac^+/Ac^+$ (solid pigment)
              $1/4$ $c^{Ds}/c; Ac/Ac^+$ (unstable colorless or spotted)
              $1/4$ $c^{Ds}/c; Ac^+/Ac^+$ (colorless)

      Overall: 2 solid:1 spotted:1 colorless

Cross 2:

      P $C/c^{Ac} \times c/c$

      F_1 $1/2$ C/c (solid pigment)
              $1/2$ C/c^{Ac} (unstable colorless or spotted)

      Overall: 1 solid:1 spotted

Cross 3:

P $C/c^{Ds}\,;Ac/Ac^+ \times C/c^{Ac}\,;Ac^+/Ac^+$

F_1 $\frac{1}{8}\ C/C\,;Ac/Ac^+$ (solid pigment)
$\frac{1}{8}\ C/c^{Ac}\,;Ac/Ac^+$ (solid pigment)
$\frac{1}{8}\ C/C\,;Ac^+/Ac^+$ (solid pigment)
$\frac{1}{8}\ C/c^{Ac}\,;Ac^+/Ac^+$ (solid pigment)
$\frac{1}{8}\ C/c^{Ds}\,;Ac^+/Ac^+$ (solid pigment)
$\frac{1}{8}\ C/c^{Ds}\,;Ac^+/Ac$ (solid pigment)
$\frac{1}{8}\ c^{Ds}/c^{Ac}\,;Ac^+/Ac^+$ (unstable colorless or spotted)
$\frac{1}{8}\ c^{Ds}/c^{Ac}\,;Ac^+/Ac$ (unstable colorless or spotted)

Overall: 3 solid:1 spotted

10. It would not be surprising to find a SINE element in an intron of a gene, rather than an exon. Processing of tbe pre-mRNA would remove the transposable element as part of the intron, and translation of the FB enzyme would not be effected.

11. There is no correct answer to this question. Yeast does have a very compact genome with almost 70 percent of its DNA representing exons, so it may not contain sufficient "safe havens" for DNA transposition. Transposons (and transposition events) that kill the host organism do not have a chance to spread. The *Ty* elements of yeast have evolved mechanisms that target their insertions so that they presumably do not harm their host. Perhaps it has only been chance that DNA transposons are not found in yeast but it is also possible that yeast has specific mechanisms that prevent their spread. One could look at related species of yeasts to determine if they have class 2 elements.

14

Mutation, Repair, and Recombination

BASIC PROBLEMS

1. You need to know the reading frame of the possible message.

2. The mutant has a deletion of one base and this will result in a frameshift (–1) mutation.

3. Proline can be coded for by CCN (N stands for any nucleotide) and histidine can be coded for by CAU or CAC. For a mutation to change a proline codon to a histidine codon requires a transverion (C to A) at the middle position. Therefore, a transition-causing mutagen cannot cause this change. Serine can be coded for by UCN. A change from C to U at the first position (a transition) would cause this missense mutation and would be possible with this mutagen.

4. Assuming single base-pair substitutions, then CGG can be changed to CGU, CGA, CGG, or AGG and still would code for arginine.

5. **a. and b.** By transversion, CGG (arginine) can be become AGG (arginine), GGG (glycine), CCG (proline), CUG (leucine), CGC (arginine), or CGU (arginine).

6. The enol form of thymine forms three hydrogen bonds with guanine.

7. **a. and b.** After the first round of replication, the enol form of thymine will be paired with guanine. At the next round, the guanine will pair with cytosine to result in the TA•CG transiton.

8. **a.** Acridine orange causes frameshift mutations and frameshift mutations often result in null alleles.

    b. A +1 frameshift mutation can be reverted by two further single insertions so that the reading frame is re-established.

9. Apurinic sites have lost either an A or G. If thymine is preferentially inserted opposite these sites, GC•AT transitions should be produced.

10. Misalignment of homologous chromosomes during recombination results in a duplication in one strand and a corresponding deletion in the other.

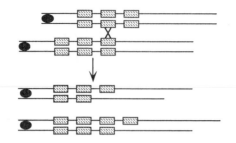

See figure 14-21 in the companion text for a figure representing slipped mispairing.

11. One of the mechanisms for repair of UV-induced photodimers is the enzyme photolyase. This enzyme requires visible light for function and would therefore be expected to be more active on bright sunny days.

12. The mismatched "T" would be corrected to C and the resulting ACG, after transcription, would be 5′ UGC 3′ and code for cysteine. Or if the other strand was corrected, ATG would be transcribed to 5′ UAC 3′ and code for tyrosine.

13. Gene conversion is the result of regions of heteroduplexes that form as part of the recombination process. Consequently, in meioses that produce aberrant ratios (gene conversions), there is a crossover between flanking genes at much higher frequencies than expected.

14. The formation of heteroduplex DNA is part of the double-strand break model. If the allelic mismatch is not repaired, a 5:3 ascus will result. However, if the mismatch is repaired prior to cell division, a 6:2 ascus results. If 6:2 asci are more frequent than 5:3 asci, it indicates that repair of the heteroduplex mismatch is more likely to happen than not.

15. Exonuclease is needed for the erosion of the ends of the DNA molecules after the double-strand break to form the regions of single-stranded DNA. Endonuclease is required to break the single-stranded DNA strands necessary to resolve the Holliday structures and also to form the double-stranded break that initiates recombination.

16. a. A transition mutation is the substitution of a purine for a purine or the substitution of a pyrimidine for a pyrimidine. A transversion mutation is the substitution of a purine for a pyrimidine, or vice versa.

b.   Both are base-pair substitutions. A synonymous mutation is one that does not alter the amino acid sequence of the protein product from the gene, because the new codon codes for the same amino acid as did the nonmutant codon. A neutral mutation results in a different amino acid that is functionally equivalent, and the mutation therefore has no known adaptive significance.

c.   A missense mutation results in a different amino acid in the protein product of the gene. A nonsense mutation causes premature termination of translation, resulting in a shortened protein.

d.   Frameshift mutations arise from addition or deletion of one or more bases in other than multiples of three, thus altering the reading frame for translation. Therefore, the amino acid sequence from the site of the mutation to the end of the protein product of the gene will be altered. Frameshift mutations can and often do result in premature stop codons in the new reading frame, leading to shortened protein products. A nonsense mutation causes premature termination of translation, resulting in a shortened protein.

17.   Frameshift mutations arise from addition or deletion of one or more bases in other than multiples of three. When translated, this will alter the reading frame and therefore the amino acid sequence from the site of the mutation to the end of the protein product. Also, frameshift mutations often result in premature stop codons in the new reading frame, leading to shortened protein products. A missense mutation changes only a single amino acid in the protein product.

18.   Misalignment of homologous chromosomes during recombination results in a duplication in one strand and a corresponding deletion in the other.

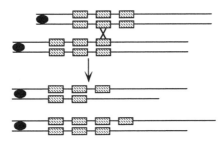

Recombination between two homologous repeats in a looped DNA molecule can lead to deletion.

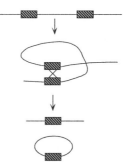

Also, "slipped mispairing" (see Figure 14-21 in the companion text) may result in a deletion in the newly synthesized DNA strand during replication.

All these mechanisms are supported by DNA sequencing results.

19. Depurination results in the loss of the adenine or guanine base from the DNA backbone. Because the resulting apurinic site cannot specify a complementary base, replication is blocked. Under certain conditions, replication proceeds with a near random insertion of a base opposite the apurinic site. In three-fourths of these insertions, a mutation will result.

   Deamination of cytosine yields uracil. If left unrepaired, the uracil will be paired with adenine during replication, ultimately resulting in a transition mutation.

   Deamination of 5-methylcytosine yields thymine and thus frequently leads to C to T transitions.

   Oxidatively damaged bases, such as 8-OxodG (8-oxo-7-hydrodeoxyguanosine) can pair with adenine, resulting in a transversion.

   Errors during DNA replication can lead to spontaneous indel mutations.

20. 5-BU is an analog of thymine. It undergoes tautomeric shifts at a higher frequency than does thymine and therefore is more likely to pair with G than thymine is during replication. At the next replication this will lead to a GC pair rather than the original AT pair. On the other hand, 5-BU can also be incorporated into DNA by mispairing with guanine. In this case it will convert a GC pair to an AT pair.

   EMS is an alkylating agent that produces O-6-ethylguanine. This alkylated guanine will mispair with thymine, which leads from a GC pair to an AT pair at the next replication.

21. An AP site is an apurinic or apyrimidinic site. AP endonucleases introduce chain breaks by cleaving the phosphodiester bonds at the AP sites. Some exonuclease activity follows, so that a number of bases are removed. The resulting gap is filled by DNA pol I and then sealed by DNA ligase.

General excision repair is used to remove damaged DNA including photodimers. It cleaves the phosphodiester backbone on either side of the damage, removing 12 or 13 nucleotides in bacteria and 27 to 29 nucleotides in eukaryotes. The resulting gap is filled by DNA pol I and then sealed by DNA ligase.

Photodimers can be also be repaired by photolyase (found in bacteria and lower eukaryotes), which binds to and splits the dimer in the presence of certain wavelengths of light. In bacteria the damaged DNA can also be bypassed by the SOS system, which enables DNA polymerase to "fill in" random bases where it encounters photodimers on the template strand. Finally, although photodimers on the template strand cause DNA pol III to stall, replication can restart downstream from the dimer, leaving a region of single-stranded DNA. The single-stranded DNA will attract single-stranded-binding protein and another protein, RecA, which can effect recombinational repair using DNA from the sister chromatid to patch the gap.

22. There are many repair systems that are available: direct reversal, excision repair, transcription-coupled repair, and non-homologous end-joining.

23. Yes. It will cause CG-to-TA transitions.

24. Depurination results in the formation of an AP site. AP endonucleases introduce chain breaks by cleaving the phosphodiester bonds at these sites. Some exonuclease activity follows, so that a number of bases are removed. The resulting gap is filled by DNA pol I and then sealed by DNA ligase.

Deamination of cytosine and adenine yields uracil and hypoxanthine, respectively. Specific glycosylases remove these bases, creating an AP site that is then repaired as above. Deamination of 5-methylcytosine yields thymine, and because this cannot be distinguished from any other thymine in the DNA, 5-methylcytosines represent mutational "hot spots."

25. The Streisinger model proposed that frameshifts arise when loops in single-stranded regions are stabilized by slipped mispairing of repeated sequences. In the *lac* gene of *E. coli*, a four-base-pair sequence is repeated three times in tandem, and this is the site of a hot spot.

The sequence is 5'-CTGG CTGG CTGG-3'. During replication the DNA must become single-stranded in short stretches for replication to occur. As the new strand is synthesized, transient disruptions of the hydrogen bonds holding the new and old strands together may be stabilized by the incorrect base-pairing of bases that are now out of register by the length of the repeat, or in this case, a total of four bases. Depending on which strand, new or template, loops out with respect to the other, there will be an addition or deletion of four bases, as diagrammed below:

```
                        T G
                  C        G
        5´-C T G G       C T G G-3´        ——> DNA synthesis
        3´-G A C C —— G A C C  G A C C -5´
```

In this diagram, the upper strand looped out as replication was occurring. The loop is stabilized by base pairing on either strand. As replication continues at the 3´ end, an additional copy of CTGG will be synthesized, leading to an addition of four bases. This will result in a frameshift mutation.

CHALLENGING PROBLEMS

26. **a.** Because 5´-UAA-3´ does not contain G or C, a transition to a GC pair in the DNA cannot result in 5´-UAA-3´. 5´-UGA-3´ and 5´-UAG-3´ have the DNA antisense-strand sequence of 3´-ACT-5´ and 3´-ATC-5´, respectively. A transition to either of these stop codons occurs from the nonmutant 3´-ATT-5´, respectively. However, a DNA sequence of 3´-ATT-5´ results in an RNA sequence of 5´-UAA-3´, itself a stop codon.

b. Yes. An example is 5´-UGG-3´, which codes for Trp, to 5´-UAG-3´.

c. No. In the three stop codons the only base that can be acted upon is G (in UAG, for instance). Replacing the G with an A would result in 5´-UAA-3´, a stop codon.

27. **a. and b.** Mutant 1: most likely a deletion. It could be caused by radiation.

Mutant 2: because proflavin causes either additions or deletions of bases and because spontaneous mutation can result in additions or deletions, the most probable cause was a frameshift mutation by an intercalating agent.

Mutant 3: 5-BU causes transitions, which means that the original mutation was most likely a transition. Because HA causes GC-to-AT transitions and HA cannot revert it, the original must have been a GC-to-AT transition. It could have been caused by base analogs.

Mutant 4: the chemical agents cause transitions or frameshift mutations. Because there is spontaneous reversion only, the original mutation must have been a transversion. X-irradiation or oxidizing agents could have caused the original mutation.

Mutant 5: HA causes transitions from GC-to-AT, as does 5-BU. The original mutation was most likely an AT-to-GC transition, which could be caused by base analogs.

   c.   The suggestion is a second-site reversion linked to the original mutant by 20 map units and therefore most likely in a second gene. Note that auxotrophs equal half the recombinants.

28. a.   A lack of revertants suggests either a deletion or an inversion within the gene.

   b.   To understand these data, recall that half the progeny should come from the wild-type parent.

   Prototroph A: because 100 percent of the progeny are prototrophic, a reversion at the original mutant site may have occurred.

   Prototroph B: half the progeny are parental prototrophs, and the remaining prototrophs, 28 percent, are the result of the new mutation. Notice that 28 percent is approximately equal to the 22 percent auxotrophs. The suggestion is that an unlinked suppressor mutation occurred, yielding independent assortment with the *nic* mutant.

   Prototroph C: there are 496 "revertant" prototrophs (the other 500 are parental prototrophs) and four auxotrophs. This suggests that a suppressor mutation occurred in a site very close [$100\%(4 \times 2)/1000 = 0.8$ m.u.] to the original mutation.

29. O^{-6}-methyl G leads predominantly to high levels of GC $\rightarrow$ AT transitions. 8-oxodG gives rise predominantly to high levels of G $\rightarrow$ T transversions. Finally, C-C photodimers will most often cause C $\rightarrow$ T transitions, but some transversions are also possible.

30.   Compare the original amino acid sequences to the mutant ones and list the changes.

   original: ile;   mutant: met
   original: asn;   mutant: ser
   original: stop; mutant: trp

Now compare the codons that must have been altered by this mutagen.

   original: ile AU<u>A</u>;   mutant: met AU<u>G</u>
   original: asn A<u>A</u>C or A<u>A</u>U;   mutant: ser A<u>G</u>C or A<u>G</u>U
   original: stop U<u>A</u>G or UG<u>A</u>; mutant: trp U<u>G</u>G

All these mutations can be the result of a T to C transition in the DNA. This would result in the A to G change in the mRNA that explains all three codon changes. This mutagen, then, might work by altering the base-pairing specificity of T so that it now base pairs with G.

31. Yes. Because DNA is a double-stranded molecule, replication of the DNA strand with a T to T* (altered T that base pairs with G) change produces an A to G transition in the newly replicated complementary DNA strand. If the mRNA is transcribed from the strand with the A to G change (the template strand), a U to C change is produced in the corresponding mRNA.

original: leu C<u>U</u>N;   mutant: pro C<u>C</u>N (where N = any base)

32. The new "wild-type" isolate contains an allele for a gene that increases the spontaneous reversion rate of *ad-3*. This allele appears to be unlinked to *ad-3* because independent assortment is observed. Call this gene *rev*. The cross becomes

$ad\text{-}3 ; rev^+$ (Strain A) $\times$ $ad\text{-}3^+ ; rev$ (wild-type isolate)

Of the *ad-3* progeny, half would be rev^+ (low reversion rate) and half would be *rev* (high reversion rate).

Further crosses to map *rev* could be performed as well as testing the allele and gene specificity of the high reversion rate to see if it is the result of increased spontaneous mutation rates (gene non-specific) or some other process (*ad-3* specific).

33. **a.** uracil DNA glycosylase

b. MutM glycosylase

c. general excision repair

d. methyl-directed mismatch repair

e. photolyase or general excision repair

f. endonuclease

g. alkyl transferase

34. We expect the linear asci to be 4:4 (4 *arg*:4 *arg*$^+$). Gene conversion is indicated by departures from this ratio: 6:2 (=2:6), 5:3 (=3:5), and 3:1:1:3 (known as aberrant 4:4). Asci 3, 4, and 6 show gene conversion.

35. The first ascus shows conversion of all three mutant sites. During recombination, the region of heteroduplex DNA can be of some length, resulting in the co-conversion of tightly-linked markers. This must be the case here as all three mutant sites are showing the same 2:6 ratio. For the second ascus, only mutant site 3 shows conversion. In this case, the region of heteroduplex did not

Chapter Fourteen   225

include mutant sites 1 and 2, or if it did, these mismatches were "repaired" back to the mutant sequences.

36. For the following, A and a will be used as the markers and for simplicity, only one possible starting combination will be shown.

Start, no repair	Repair of first heteroduplex		Repair of second heteroduplex		Repair of both heteroduplexes			
A	A	A	A	A	A	A	A	A
A	A	A	A	A	A	A	A	A
A	A	a	A	A	A	A	a	a
a	A	a	a	a	A	A	a	a
A	A	A	A	a	A	a	A	a
a	a	a	A	a	A	a	A	a
a	a	a	a	a	a	a	a	a
a	a	a	a	a	a	a	a	a

15

Large-Scale Chromosomal Changes

BASIC PROBLEMS

1. MM N OO would be classified as $2n-1$; MM NN OO would be classified as $2n$; and MMM NN PP would be classified as $2n+1$.

2. It would more likely be an autopolyploid. To make sure it was polyploid, you would need to microscopically examine stained chromosomes from mitotically dividing cells and count the chromosome number.

3. Aneuploid. Trisomic refers to three copies of one chromosome. Triploid refers to three copies of all chromosomes.

4. There would be one possible quadrivalent.

5. No. Amphidiploid means "doubled diploid." Because cauliflower has $n = 9$ chromosome, it could not have arose in this fashion. It has, however, contributed to other amphidiploid species.

6. The progenitor had nine chromosomes from a cabbage parent and nine chromosomes from a radish parent. These chromosomes were different enough that pairs did not synapse and segregate normally at meiosis. By doubling the chromosomes in the progenitor ($2n = 36$), all chromosomes now had homologous partners and meiosis could proceed normally.

7. In wheat, $2n = 6x = 42$ so B (or x) represents seven chromosomes.

8. Cross *T. tauschii* and Emmer to get ABD offspring. Treat the offspring with colchicine to double the chromosome number to AABBDD to get the hexaploid bread wheat.

9. Cells destined to become pollen grains can be induced by cold treatment to grow into embryoids. These embryoids can then be grown on agar to form monoploid plantlets.

10. The likely origin of a disomic is nondisjunction during meiosis. Depending whether the nondisjunction took place during the first or second division, you would expect one nullosomic, or two nullosomics and another disomic, respectively.

11. Yes. You would expect that one-sixth of the gametes would be *a*. Also, two-sixths would be *A*, two-sixths would be *Aa*, and one-sixth would be *AA*.

12. Both XXY and XXX would be fertile. XO (Turner) and XXY (Klinefelter) are both sterile.

13. Older mothers have an elevated risk of having a child with Down syndrome and other non-disjunctional events.

14. No. The DNA backbone has strict 5′ to 3′ polarity, and 5′ ends can only be joined to 3′ ends.

15. A crossover within a paracentric inversion heterozygote results in a dicentric bridge (and an acentric fragment).

16. An acentric fragment cannot be aligned or moved during meiosis (or mitosis) and consequently is lost.

17. possible translocation

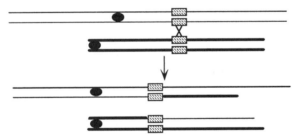

possible deletion (and duplication)

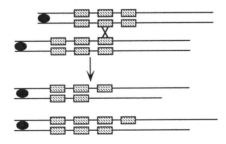

18. You could cross a strain with the appropriate 3; 4 reciprocal translocation to a wild-type strain to generate heterozygotes. One of the meiotic products of

adjacent-1 segregation in these heterozygotes will have a duplication of the translocated portion of chromosome 3 and a deletion for the translocated portion of chromosome 4.

19. Very large deletions tend to be lethal. This is likely due to genomic imbalance or the unmasking of recessive lethal genes. Therefore, the observed very large pairing loop is more likely to be from a heterozygous inversion.

20. Because the new mutant does not show pseudodominance with any of the deletions that span chromosome 2, it is likely that the mutation does not map to this chromosome.

21. Williams syndrome is the result of a deletion of the 7q11.23 region of chromosome 7. Cri du chat syndrome is the result of a deletion of a significant portion of the short arm of chromosome 5 (specifically bands 5p15.2 and 5p15.3). Both Turner syndrome (XO) and Down syndrome (trisomy 21) result from meiotic non-disjunction. The term syndrome is used to describe a set of phenotypes (often complex and varied) that generally occur together.

22. **a.** Cytologically, deletions lead to shorter chromosomes with missing bands (if banded) and an unpaired loop during meiotic pairing when heterozygous. Genetically, deletions are usually lethal when homozygous, do not revert, and when heterozygous, lower recombinational frequencies and can result in "pseudodominance" (the expression of recessive alleles on one homolog that are deleted on the other). Occasionally, heterozygous deletions express an abnormal (mutant) phenotype.

   b. Cytologically, duplications lead to longer chromosomes and, depending on the type, unique pairing structures during meiosis when heterozygous. These may be simple unpaired loops or more complicated twisted loop structures. Genetically, duplications can lead to asymmetric pairing and unequal crossing-over events during meiosis, and duplications of some regions can produce specific mutant phenotypes.

   c. Cytologically, inversions can be detected by banding, and when heterozygous, they show the typical twisted "inversion" loop during homologous pairing. Pericentric inversions can result in a change in the p:q ratio. Genetically, no viable crossover products are seen from recombination within the inversion when heterozygous, and as a result, flanking genes show a decrease in RF.

   d. Cytologically, reciprocal translocations may be detected by banding, or they may drastically change the size of the involved chromosomes as well as the positions of their centromeres. Genetically, they establish new linkage relationships. When heterozygous, they show the typical cross structure during meiotic pairing and cause a diagnostic 50 percent reduction of viable gamete production, leading to semisterility.

23. **a.** paracentric inversion

b. deletion

c. pericentric inversion

d. duplication

24. **a.** The products of crossing-over within the inversion will be inviable when the inversion is heterozygous. This paracentric inversion spans 25 percent of the region between the two loci and therefore will reduce the observed recombination between these genes by a similar percentage (i.e., 9 percent.) The observed RF will be 27 percent.

   b. When the inversion is homozygous, the products of crossing-over within the inversion will be viable, so the observed RF will be 36 percent.

25. The following represents the crosses that are described in this problem:

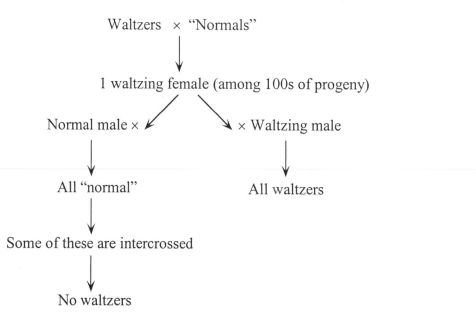

Waltzers × "Normals"

↓

1 waltzing female (among 100s of progeny)

Normal male × ↙   ↘ × Waltzing male

↓   ↓

All "normal"   All waltzers

↓

Some of these are intercrossed

↓

No waltzers

The single waltzing female that arose from a cross between waltzers and normals is expressing a recessive gene. It is possible that this represents a new "waltzer" mutation that was inherited from one of the "normal" mice, but given the cytological evidence (the presence of a shortened chromosome), it is more likely that this exceptional female inherited a deletion of the wild-type allele, which allowed expression of the mutant recessive phenotype.

When this exceptional female was mated to a waltzing male, all the progeny were waltzers; when mated to a normal male, all the progeny were normal. When some of these normal offspring were intercrossed, there were no progeny that were waltzers. If a "new" recessive waltzer allele had been inherited, all these "normal" progeny would have been w^+/w. Any intercross should have therefore produced 25 percent waltzers. On the other hand, if a deletion had occurred, half the progeny would be w^+/w and half would be $w^+/w^{deletion}$. If $w^+/w^{deletion}$ are intercrossed, 25 percent of the progeny would not develop (the homozygous deletion would likely be lethal), and no waltzers would be observed. This is consistent with the data.

26. This problem uses a known set of overlapping deletions to order a set of mutants. This is called deletion mapping and is based on the expression of the recessive mutant phenotype when heterozygous with a deletion of the corresponding allele on the other homolog. For example, mutants a, b, and c are all expressed when heterozygous with Del1. Thus it can be assumed that these genes are deleted in Del1. When these results are compared with the crosses with Del2 and it is discovered that these progeny are b^+, the location of gene b is mapped to the region deleted in Del1 that is not deleted in Del2. This logic can be applied in the following way:

Compare deletions 1 and 2: this places allele *b* more to the left than alleles *a* and *c*.
The order is *b* (*a*, *c*), where the parentheses indicate that the order is unknown.

Compare deletions 2 and 3: this places allele *e* more to the right than (*a*, *c*).
The order is *b* (*a*, *c*) *e*.

Compare deletions 3 and 4: allele *a* is more to the left than *c* and *e*, and *d* is more to the right than *e*. The order is *b a c e d*.

Compare deletions 4 and 5: allele *f* is more to the right than *d*. The order is *b a c e d f*.

Allele	Band
b	1
a	2
c	3
e	4
d	5
f	6

27. The data suggest that one or both breakpoints of the inversion are located within an essential gene, causing a recessive lethal mutation.

28. **a.** The Sumatra chromosome contains a pericentric inversion when compared with the Borneo chromosome.

   b.

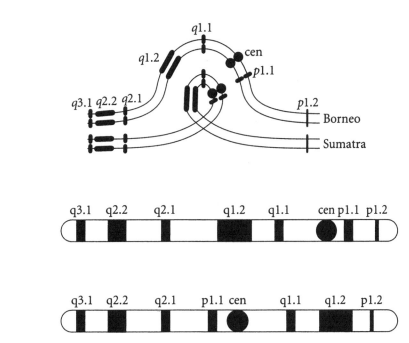

   c.

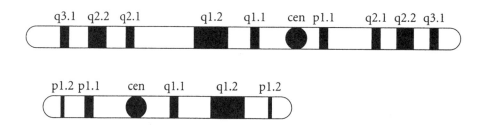

d. Recall that all single crossovers within the inverted region will lead to four meiotic products: two that will be viable, nonrecombinant (parental) types and two that will be extremely unbalanced, (most likely nonviable), recombinant types. In other words, if 30 percent of the meioses have a crossover in this region, 15 percent of the gametes will not lead to viable progeny. That means that 85 percent of the gametes should produce viable progeny.

29. *Unpacking the Problem*

1. A "gene for tassel length" means that there is a gene with at least two alleles (T and t) that controls the length of the tassel. A "gene for rust resistance" means that there is a gene that determines whether the corn plant is resistant to a rust infection or not (R and r).

2. The precise meaning of the allelic symbols for the two genes is irrelevant to solving the problem, because what is being investigated is the distance between the two genes.

3. A locus is the specific position occupied by a gene on a chromosome. It is implied that gene loci are the same on both homologous chromosomes. The gene pair can consist of identical or different alleles.

4. Evidence that the two genes are normally on separate chromosomes would have come from previous experiments showing that the two genes independently assort during meiosis.

5. Routine crosses could consist of F_1 crosses, F_2 crosses, backcrosses, and testcrosses.

6. The genotype T/t ; R/r is a double heterozygote, or dihybrid, or F_1 genotype.

7. The pollen parent is the "male" parent that contributes to the pollen tube nucleus, the endosperm nucleus, and the progeny.

8. Testcrosses are crosses that involve a genotypically unknown and a homozygous recessive organism. They are used to reveal the complete

genotype of the unknown organism and to study recombination during meiosis.

9. The breeder was expecting to observe 1 *T/t*; *R/r*:1 *T/t*; *r/r*:1 *t/t*; *R/r*:1 *t/t*; *r/r*.

10. Instead of a 1:1:1:1 ratio indicating independent assortment, the testcross indicated that the two genes were linked, with a genetic distance of 100%(3 + 5)/210 = 3.8 map units.

11. The equality and predominance of the first two classes indicate that the parentals were *TR/tr*.

12. The equality and lack of predominance of the second two classes indicate that they represent recombinants.

13. The gametes leading to this observation were:
    46.7% *TR*   1.4% *Tr*
    49.5% *tr*   2.4% *tR*

14. 46.7% *TR*
    49.5% *tr*

15. 1.4% *Tr*
    2.4% *tR*

16. *Tr* and *tR*

17. *T* and *R* are linked, as are *t* and *r*.

18. Two genes on separate chromosomes can become linked through a translocation.

19. One parent of the hybrid plant contained a translocation that linked the *T* and *R* alleles and the *t* and *r* alleles.

20. A corn cob is a structure that holds on its surface the progeny of the next generation.

21.

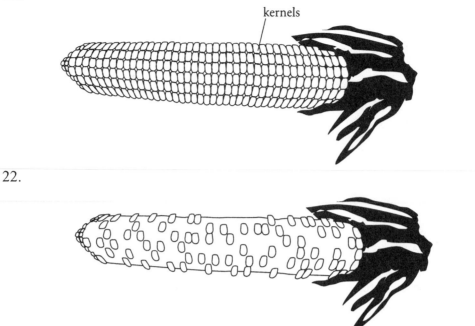

kernels

22.

23. A kernel is one progeny on a corn cob.

24. Absence of half the kernels, or 50 percent aborted progeny (semisterility), could result from the random segregation of one normal with one translocated chromosome (T1 + N2 and T2 + N1) during meioses in a parent that is heterozygous for a reciprocal translocation.

25. Approximately 50 percent of the progeny died. It was the "female" that was heterozygous for the translocation.

Solution to the Problem

a. The progeny are not in the 1:1:1:1 ratio expected for independent assortment; instead, the data indicate close linkage. And, half the progeny did not develop, indicating semisterility.

b. These observations are best explained by a translocation that brought the two loci close together.

c.   Parents: TR/tr × t/t; r/r

Progeny: 98 TR/t; r
104   tr/t; r
3   Tr/t; r
5   tR/t; r

d. Assume a translocation heterozygote in coupling. If pairing is as diagrammed below, then you would observe the following:

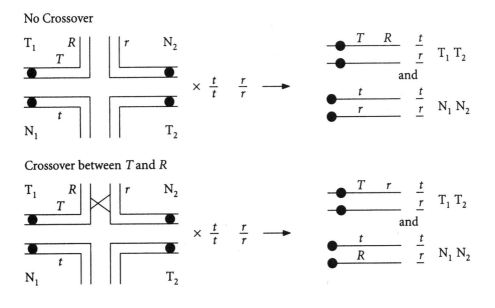

No Crossover

Crossover between *T* and *R*

e. The two recombinant classes result from a recombination event followed by proper segregation of chromosomes, as diagrammed above.

30. The cross was

P X^{e^+}/Y (irradiated) $\times$ X^e/X^e

F_1 Most X^e/Y yellow males

Two ? gray males

a. Gray male 1 was crossed with a yellow female, yielding yellow females and gray males, which is reversed sex linkage. If the e^+ allele was translocated to the Y chromosome, the gray male would be X^e/Y^{e^+} or gray. When crossed with yellow females, the results would be

X^e/Y^{e^+} gray males

X^e/X^e yellow females

b. Gray male 2 was crossed with a yellow female, yielding gray and yellow males and females in equal proportions. If the e^+ allele was translocated to an autosome, the progeny would be as below, where "A" indicates autosome:

P $A^{e^+}/A \, ; X^e/Y \times A/A \, X^e/X^e$

F_1 $A^{e^+}/A \, ; X^e/X^e$ gray female

$A^{e^+}/A \, ; X^e/Y$ gray male

$$A/A\,;X^e/X^e \quad \text{yellow female}$$
$$A/A\,;X^e/Y \quad \text{yellow male}$$

31. The break point can be treated as a gene with two "alleles," one for normal fertility and one for semisterility. The problem thus becomes a two-point cross.

Parentals	764	Semisterile *Pr*
	727	Normal *pr*
Recombinants	145	Semisterile *pr*
	186	Normal *Pr*
	1822	

$$100\%(145 + 186)/1822 = 18.17 \text{ m.u.}$$

32.
Klinefelter syndrome	XXY male
Down syndrome	Trisomy 21
Turner syndrome	XO female

33. Create a hybrid by crossing the two plants and then double the chromosomes with a treatment that disrupts mitosis, such as colchicine treatment. Alternatively, diploid somatic cells from the two plants could be fused and then grown into plants through various culture techniques.

34. **a.**

$$b^+/b/b \quad \times \quad b/b$$
$$\downarrow \qquad\qquad \downarrow$$

Gametes: $\frac{1}{6}\ b^+$ b

$\frac{1}{3}\ b$

$\frac{1}{3}\ b^+/b$

$\frac{1}{6}\ b/b$

Among the progeny of this cross, the phenotypic ratio will be 1 wild-type (b^+) : 1 *b*.

b.

$$b^+/b^+/b \quad \times \quad b/b$$
$$\downarrow \qquad\qquad \downarrow$$

Gametes: $\frac{1}{6}\ b$ b

$\frac{1}{3}\ b^+$

$\frac{1}{3}\ b^+/b$

$\frac{1}{6}\ b^+/b^+$

Among the progeny of this cross, the phenotypic ratio will be 5 wild-type (b^+) : 1 *b*.

c.
$$b^+/b^+/b \quad \times \quad b^+/b$$
$$\downarrow \qquad\qquad \downarrow$$

Gametes: $\quad$ $\frac{1}{6}\ b$ $\qquad$ $\frac{1}{2}\ b$

$\qquad\qquad\qquad$ $\frac{1}{3}\ b^+$ $\qquad$ $\frac{1}{2}\ b^+$

$\qquad\qquad\qquad$ $\frac{1}{3}\ b^+/b$

$\qquad\qquad\qquad$ $\frac{1}{6}\ b^+/b^+$

Among the progeny of this cross, the phenotypic ratio will be 11 wild-type (b^+) : 1 b.

35. **a., b. and c.** One of the parents of the woman with Turner syndrome (XO) must have been a carrier for colorblindness, an X-linked recessive disorder. Because her father has normal vision, she could not have obtained her only X from him. Therefore, nondisjunction occurred in her father. A sperm lacking a sex chromosome fertilized an egg with the X chromosome carrying the colorblindness allele. The nondisjunctive event could have occurred during either meiotic division.

d. If the colorblind patient had Klinefelter syndrome (XXY), then both Xs must carry the allele for colorblindness. Therefore, nondisjunction had to occur in the mother. Remember that during meiosis I, given no crossover between the gene and the centromere, allelic alternatives separate from each other. During meiosis II, identical alleles on sister chromatids separate. Therefore, the nondisjunctive event had to occur during meiosis II because both alleles are identical.

36. **a.** If a 6x were crossed with a 4x, the result would be 5x.

b. Cross *A/A* with *a/a/a/a* to obtain *A/a/a*.

c. The easiest way is to expose the *A/a** plant cells to colchicine for one cell division. This will result in a doubling of chromosomes to yield *A/A/a*/a**.

d. Cross 6x (*a/a/a/a/a/a*) with 2x (*A/A*) to obtain *A/a/a/a*.

e. In culture, expose haploid plants cells to the herbicide and select for resistant colonies. Treat cells that grow with colchicine to obtain diploid cells.

37. Type a: the extra chromosome must be from the mother. Because the chromosomes are identical, nondisjunction had to have occurred at M_{II}.

Type b: the extra chromosome must be from the mother. Because the chromosomes are not identical, nondisjunction had to have occurred at M_I.

Type c: the mother correctly contributed one chromosome, but the father did not contribute any chromosome 4. Therefore, nondisjunction occurred in the male during either meiotic division.

38. a. The cross is $P/P/p \times p/p$.

The gametes from the trisomic parent will occur in the following proportions:

$1/6$ p
$2/6$ P
$1/6$ P/P
$2/6$ P/p

Only gametes that are p can give rise to potato leaves, because potato is recessive. Therefore, the ratio of normal to potato will be 5:1.

b. If the gene is not on chromosome 6, there should be a 1:1 ratio of normal to potato.

39. The generalized cross is $A/A/A \times a/a$, from which $A/A/a$ progeny were selected. These progeny were crossed with a/a individuals, yielding the results given. Assume for a moment that each allele can be distinguished from the other, and let $1 = A$, $2 = A$ and $3 = a$. The gametic combinations possible are

1-2 (A/A) and 3 (a)
1-3 (A/a) and 2 (A)
2-3 (A/a) and 1 (A)

Because only diploid progeny were examined in the cross with a/a, the progeny ratio should be 2 wild type:1 mutant if the gene is on the trisomic chromosome. With this in mind, the table indicates that y is on chromosome 1, cot is on chromosome 7, and h is on chromosome 10. Genes d and c do not map to any of these chromosomes.

40. a.

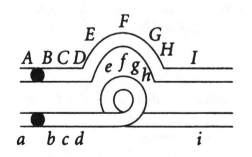

b.

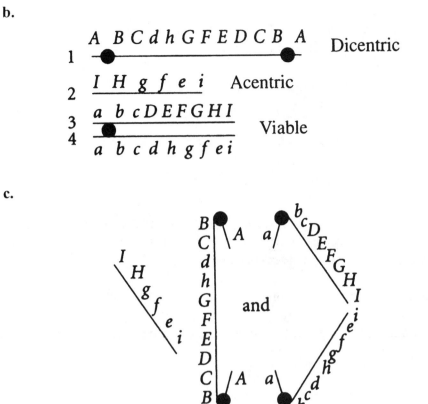

c.

d. The chromosomes numbered 3 and 4 will give rise to viable progeny. The genotypes of those progeny will be *A B C D E F G H I/a b c D E F G H I* and *A B C D E F G H I/a b c d h g f e i*.

41. a. Two crosses show 28 map units between the loci for body color and eye shape in a testcross of the F$_1$: California × California and Chile × Chile. The third type of cross, California × Chile, leads to only four map units between the two genes when the hybrid is testcrossed. This indicates that the genetic distance has decreased by 24 map units, or 100%(24/28) = 85.7%. A deletion cannot be used to explain this finding, nor can a translocation. Most likely the two lines are inverted with respect to each other for 85.7 percent of the distance between the two genes.

b.

A single crossover in either region would result in 4% crossing-over between *B* and *R*. The products are

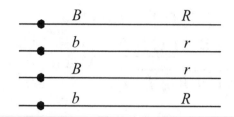

42. **a.** The aberrant plant is semisterile, which suggests an inversion. Because the *d–f* and *y–p* frequencies of recombination in the aberrant plant are normal, the inversion must involve *b* through *x*.

b. To obtain recombinant progeny when an inversion is involved, either a double crossover occurred within the inverted region or single crossovers occurred between *f* and the inversion, which occurred someplace between *f* and *y*.

43. The cross is

P   *c bz wx sh d/c bz wx sh d* × *C Bz Wx Sh D/C Bz Wx Sh D*

F₁   *C Bz Wx Sh D/c bz wx sh d*

Backcross   *C Bz Wx Sh D/c bz wx sh d* × *c bz wx sh d/c bz wx sh d*

a. The total number of progeny is 1000. Classify the progeny as to where a crossover occurred for each type. Then, total the number of crossovers between each pair of genes. Calculate the observed map units.

Region	#COs	M.U. observed	M.U. expected
C–Bz	103	10.3	12
Bz–Wx	13	1.3	8
Wx–Sh	13	1.3	10
A–D	186	18.6	20

Notice that a reduction of map units, or crossing-over, is seen in two intervals. Results like this are suggestive of an inversion. The inversion most likely involves the Bz, Wx, and Sh genes.

Further notice that all those instances in which crossing-over occurred in the proposed inverted region involved a double crossover. This is the expected pattern.

b. A number of possible classes are missing: four single-crossover classes resulting from crossing-over in the inverted region, eight double-crossover classes involving the inverted region and the noninverted region, and triple crossovers and higher. The 10 classes detected were the only classes that were viable. They involved a single crossover outside the inverted region or a double crossover within the inverted region.

c. Class 1: parental; increased due to nonviability of some crossovers

Class 2: parental; increased due to nonviability of some crossovers

Class 3: crossing-over between *C* and *Bz*; approximately expected frequency

Class 4: crossing-over between *C* and *Bz*, approximately expected frequency

Class 5: crossing-over between *Sh* and *D*; approximately expected frequency

Class 6: crossing-over between *A* and *D*; approximately expected frequency

Class 7: double crossover between *C* and *Bz* and between *Sh* and *D*; approximately expected frequency

Class 8: double crossover between *C* and *Bz* and between *A* and *D*; approximately expected frequency

Class 9: double crossover between *Bz* and *Wx* and between *Wx* and *Sh*; approximately expected frequency

Class 10: double crossover between *Bz* and *Wx* and between *Wx* and *Sh*; approximately expected frequency

d. Cytological verification could be obtained by looking at chromosomes during meiotic pairing. Genetic verification could be achieved by mapping these genes in the wild-type strain and observing their altered relationships.

44. The original plant had two reciprocal translocation chromosomes that brought genes *P* and *S* very close together. Because of the close linkage, a ratio suggesting a monohybrid cross, instead of a dihybrid cross, was observed, both with selfing and with a testcross. All gametes are fertile because of homozygosity.

original plant: *P S/p s*
tester: *p s/p s*

F₁ progeny: heterozygous for the translocation:

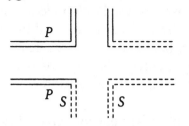

The easiest way to test this is to look at the chromosomes of heterozygotes during meiosis I.

45. The percent degeneration seen in the progeny of the exceptional rat is roughly 50 percent larger than that seen in the progeny from the normal male. Semisterility is an important diagnostic for translocation heterozygotes. This could be verified by cytological observation of the meiotic cells from the exceptional male.

46. **a.**   The cross is *Fr/Fr × fr/fr/fr*.

Trisomic progeny are then crossed to a diploid wild-type plant.

$$Fr/fr/fr \times fr/fr$$

Because only diploid progeny of this cross are evaluated, the ratio of fast- to slow-ripening plants will be 1:2.

b.   If *Fr* is not located on the trisomic chromosome, the crosses are

$$Fr/Fr \times fr/fr$$
and
$$Fr/fr \times fr/fr$$

Therefore the ratio of fast- to slow-ripening plants will be 1:1.

CHALLENGING PROBLEMS

47. **a.**   Single crossovers between a gene and its centromere lead to a tetratype (second-division segregation.) Thus a total of 20 percent of the asci should show second division segregation, and 80 percent will show first-division segregation. The following are representative asci

un3⁺ ad3⁺	un3⁺ ad3⁺	un3⁺ ad3⁺	un3⁺ ad3	un3⁺ ad3
un3⁺ ad3⁺	un3⁺ ad3⁺	un3⁺ ad3⁺	un3⁺ ad3	un3⁺ ad3
un3⁺ ad3⁺	un3⁺ ad3	un3⁺ ad3	un3⁺ ad3⁺	un3⁺ ad3⁺

$un3^+\ ad3^+$	$un3^+\ ad3$	$un3^+\ ad3$	$un3^+\ ad3^+$	$un3^+\ ad3^+$
$un3\ ad3$	$un3\ ad3^+$	$un3\ ad3$	$un3\ ad3^+$	$un3\ ad3$
$un3\ ad3$	$un3\ ad3^+$	$un3\ ad3$	$un3\ ad3^+$	$un3\ ad3$
$un3\ ad3$	$un3\ ad3$	$un3\ ad3^+$	$un3\ ad3$	$un3\ ad3^+$
$un3\ ad3$	$un3\ ad3$	$un3\ ad3^+$	$un3\ ad3$	$un3\ ad3^+$
80%	5%	5%	5%	5%

In all cases, the "upside-down" version would be equally likely.

b. The aborted spores could result from a crossing-over event within an inversion of the wild type, compared with the standard strain. Crossing-over within heterozygous inversions leads to unbalanced chromosomes and nonviable spores. This could be tested by using the wild type from Hawaii in mapping experiments of other markers on chromosome 1 in crosses with the standard strain and looking for altered map distances.

48. Cross 1: independent assortment of the 2 genes (expected for genes on separate chromosomes).

Cross 2: the 2 genes now appear to be linked (the observed RF is 1%); also, half of the progeny are inviable. These data suggest a reciprocal translocation occurred and both genes are very close to the breakpoints.

Cross 3: the viable spores are of 2 types: half contain the normal (nontranslocated chromosomes) and half contain the translocated chromosomes.

49. a.
$$a^+/a^+/a/a \times a/a/a/a$$
Gametes: $\frac{1}{6}\ a^+/a^+$ a/a
$\frac{2}{3}\ a^+/a$
$\frac{1}{6}\ a/a$

Among the progeny of this cross, the phenotypic ratio will be 5 wild-type (a^+) : 1 a.

b.
$$a^+/a/a/a \times a/a/a/a$$
Gametes: $\frac{1}{2}\ a^+/a$ a/a
$\frac{1}{2}\ a/a$

Among the progeny of this cross, the phenotypic ratio will be 1 wild-type (a^+) : 1 a.

c.

$$a^+/a/a/a \;\times\; a^+/a/a/a$$
$$\downarrow \qquad\qquad \downarrow$$

Gametes: $\;\tfrac{1}{2}\;a^+/a \qquad \tfrac{1}{2}\;a^+/a$

$\qquad\qquad\;\; \tfrac{1}{2}\;a/a \qquad\;\; \tfrac{1}{2}\;a/a$

Among the progeny of this cross, the phenotypic ratio will be 3 wild-type (a^+) : 1 a.

d.

$$a^+/a^+/a/a \;\times\; a^+/a/a/a$$
$$\downarrow \qquad\qquad\quad \downarrow$$

Gametes: $\;\tfrac{1}{6}\;a^+/a^+ \qquad \tfrac{1}{2}\;a^+/a$

$\qquad\qquad\;\; \tfrac{2}{3}\;a^+/a \qquad\;\; \tfrac{1}{2}\;a/a$

$\qquad\qquad\;\; \tfrac{1}{6}\;a/a$

Among the progeny of this cross, the phenotypic ratio will be 11 wild-type (a^+) : 1 a.

50. Consider the following table, in which "L" and "S" stand for 13 large and 13 small chromosomes, respectively:

Hybrid	Chromosomes
G. hirsutum × *G. thurberi*	S, S, L
G. hirsutum × *G. herbaceum*	S, L, L
G. thurberi × *G. herbaceum*	S, L

Each parent in the cross must contribute half its chromosomes to the hybrid offspring. It is known that *G. hirsutum* has twice as many chromosomes as the other two species. Furthermore, its chromosomes are composed of chromosomes donated by the other two species. Therefore, the genome of *G. hirsutum* must consist of one large and one small set of chromosomes. Once this is realized, the rest of the problem essentially solves itself. In the first hybrid, the genome of *G. thurberi* must consist of one set of small chromosomes. In the second hybrid, the genome of *G. herbaceum* must consist of one set of large chromosomes. The third hybrid confirms the conclusions reached from the first two hybrids.

The original parents must have had the following chromosome constitution:

G. hirsutum	26 large, 26 small
G. thurberi	26 small
G. herbaceum	26 large

G. hirsutum is a polyploid derivative of a cross between the two Old World species. This could easily be checked by looking at the chromosomes.

51. **a.** *B. campestris* was crossed with *B. napus*, and the hybrid had 29 chromosomes consisting of 10 bivalents; and 9 univalents. *B. napus* had to have contributed a total of 19 chromosomes to the hybrid. Therefore, *B. campestris* had to have contributed 10 chromosomes. The $2n$ number of *B. campestris* is 20.

When *B. nigra* was crossed with *B. napus*, *B. nigra* had to have contributed 8 chromosomes to the hybrid. The $2n$ number of *B. nigra* is 16.

B. oleracea had to have contributed 9 chromosomes to the hybrid formed with *B. juncea*. The $2n$ number in *B. oleracea* is 18.

b. First list the haploid and diploid number for each species:

Species	Haploid	Diploid
B. nigra	8	16
B. oleracea	9	18
B. campestris	10	20
B. carinata	17	34
B. juncea	18	36
B. napus	19	38

Now, recall that a bivalent in a hybrid indicates that the chromosomes are essentially identical. Therefore, the more bivalents formed in a hybrid, the closer the two parent species. Three crosses result in no bivalents, suggesting that the parents of each set of hybrids are not closely related:

Cross	Haploid number
B. juncea × *B. oleracea*	18 vs. 9
B. carinata × *B. campestris*	17 vs. 10
B. napus × *B. nigra*	19 vs. 8

Three additional crosses resulted in bivalents, suggesting a closer relationship among the parents:

Cross	Haploid #	Bivalents	Univalents
B. juncea × *B. nigra*	18 vs. 8	8	10
B. napus × *B. campestris*	19 vs. 10	10	9
B. carinata × *B. oleracea*	17 vs. 9	9	8

Note that in each cross the number of bivalents is equal to the haploid number of one species. This suggests that the species with the larger haploid number is a hybrid composed of the second species and some other species. In each case, the haploid number of the unknown species is the number of univalents. Therefore, the following relationships can be deduced:

B. juncea is an amphidiploid formed by the cross of *B. nigra* and *B. campestris*.

B. napus is an amphidiploid formed by the cross of *B. campestris* and *B. oleracea*.

B. carinata is an amphidiploid formed by the cross of *B. nigra* and *B. oleracea*.

These conclusions are in accord with the three crosses that did not yield bivalents:

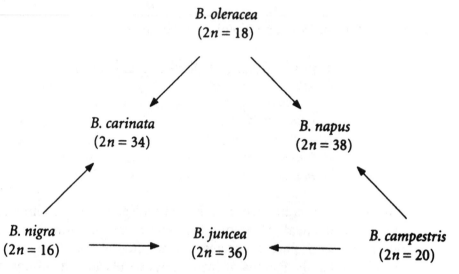

52. **a.** Loss of one X in the developing fetus after the two-cell stage

    b. Nondisjunction leading to Klinefelter syndrome (XXY), followed by a nondisjunctive event in one cell for the Y chromosome after the two-cell stage, resulting in XX and XXYY

    c. Nondisjunction of the X at the one-cell stage

    d. Fused XX and XY zygotes (from the separate fertilizations either of two eggs or of an egg and a polar body by one X-bearing and one Y-bearing sperm)

    e. Nondisjunction of the X at the two-cell stage or later

53. *Unpacking the Problem*

    1. *Homozygous* means that an organism has two identical alleles.
    A *mutation* is any deviation from wild type.
    An *allele* is one particular form of a gene.

Closely linked means that two genes are physically very close to each other on the same chromosome.

Recessive refers to a type of allele that is expressed only when it is the sole type of allele for that gene found in an individual.

Wild type is the most frequent type found in a laboratory population or in a population in the "wild."

Crossing-over refers to the physical exchange of alleles between homologous chromosomes.

Nondisjunction is the failure of separation of either homologous chromosomes or sister chromatids in the two meiotic divisions.

A *testcross* is a cross to a homozygous recessive organism for the trait or traits being studied.

Phenotype is the appearance of an organism.

Genotype is the genetic constitution of an organism.

2. No. The genes in question are on an autosome, specifically, number 4.

3. The most common lab species, *Drosophila melanogaster*, has eight chromosomes.

4.

P $b\,e^+/b\,e^+ \times b^+\,e/b^+\,e$ (cross 1)

F_1 $b\,e^+/b^+\,e \times b\,e/b\,e$ (cross 2)

Progeny	$b\,e^+/b\,e$	expected parental
	$b^+\,e/b\,e$	expected parental
	$b^+\,e^+/b\,e$	unexpected recombinant ("very closely linked" so rare)
	$b\,e/b\,e$	unexpected recombinant ("very closely linked" so rare)

rare wild type $\times$ $b\,e/b\,e$ (cross 3)

Progeny		
$1/6$	wild type
$1/6$	bent, eyeless
$1/3$	bent
$1/3$	eyeless

5. $b\,e^+$ and $b^+\,e$

6. $b\,e^+/b^+\,e$

7. It is not at all surprising that the F_1 are wild type. This means that both mutations are recessive and complement (are in different genes).

8. $b\,e/b\,e \rightarrow$ gametes: $b\,e$

9.   The two common gametes are $b\,e^+$ and $b^+\,e$. The two rare gametes are $b^+\,e^+$ and $b\,e$.

10. Normal

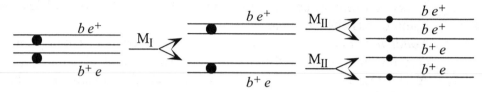

First-Division Nondisjunction: all gametes are aneuploid

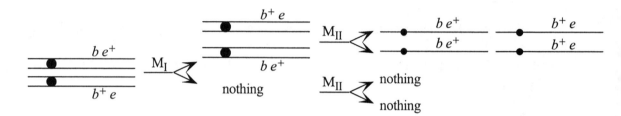

Second-Division Nondisjunction: half the gametes are aneuploid

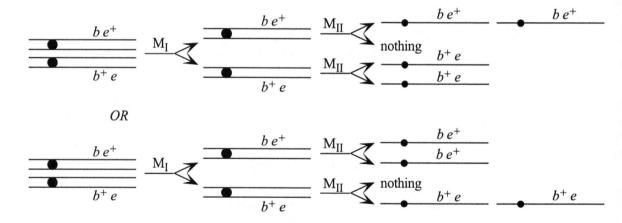

11. This is answered in 10, above.

12. Viable progeny may be able to arise from aneuploidic gametes because chromosome 4 is very small and, percentage-wise, contributes little to the genome. The progeny would be monosomic and trisomic.

13. Listed below are the gametes from 9 and 10 above, the contribution of the male parent, and the phenotype of the progeny.

Female gamete	Male gamete	Phenotype
$b^+ e$	$b\,e$	eyeless
$b\,e^+$	$b\,e$	bent
$b^+ e^+$	$b\,e$	wild type
$b\,e$	$b\,e$	bent, eyeless
$b^+ e/b\,e^+$	$b\,e$	wild type
—	$b\,e$	bent, eyeless
$b\,e^+/b\,e^+$	$b\,e$	bent
$b^+ e/b^+ e$	$b\,e$	eyeless

14. The ratio points to meiosis of a trisomic.

15. Research with artificial chromosomes has indicated that extremely small chromosomes segregate improperly at higher rates than longer chromosomes. It is suspected that the chromatids from homologous chromosomes need to intertwine in order to remain together until the onset of anaphase. Very short chromosomes are thought to have some difficulty in doing this and therefore have a higher rate of nondisjunction. In this instance, which deals with natural chromosomes as opposed to artificial chromosomes, very small chromosomes would be expected to have very little genetic material in them, and therefore their loss or gain may not be of too much importance during development.

16. rare wild type $\times$ $e\,b/e\,b$ (cross 3)

If the rare wild type is from recombination, then the cross becomes

$$b^+ e^+/b\,e \times b\,e/b\,e$$

Progeny $b^+ e^+/b\,e$ parental: wild type
         $b\,e/b\,e$ parental: bent, eyeless
         $b^+ e/b\,e$ rare recombinant: eyeless
         $b\,e^+/b\,e$ rare recombinant: bent

If the rare wild type is from nondisjunction, then the cross becomes

$$b\,e^+/b^+ e/b\,e \times b\,e/b\,e$$

Progeny $b\,e^+/b^+ e/b\,e$ wild type
         $b\,e^+/b\,e/b\,e$ bent
         $b^+ e/b\,e/b\,e$ eyeless
         $b\,e^+/b\,e$ bent
         $b^+ e/b\,e$ eyeless
         $b\,e/b\,e$ bent, eyeless

Solution to the Problem

Cross 1:   P $b\,e^+/b\,e^+$ × $b^+\,e/b^+\,e$
             F$_1$ $b^+\,e/b\,e^+$

Cross 2:   P   X/X ; $b^+\,e/b\,e^+$ ×   X/Y ; $b\,e/b\,e$
             F$_1$ expect 1 $b\,e^+/b\,e$: 1 $b^+\,e/b\,e$, X/X and X/Y
                 one rare observed X/X ; $b^+\,e^+$

a.   The common progeny are $b^+\,e/b\,e$ and $b\,e^+/b\,e$.

b.   The rare female could have come from crossing-over, which would have resulted in a gamete that was $b^+\,e^+$. The rare female could also have come from nondisjunction that gave a gamete that was $b\,e^+/b^+\,e$. Such a gamete might give rise to viable progeny.

c.   If the female had been wild-type ($b^+\,e^+/b\,e$) as a result of crossing-over, her progeny would have been as follows:

    Parental: $b^+\,e^+/b\,e$ wild type (common)
                   $b\,e/b\,e$ bent, eyeless (common)

    Recombinant: $b\,e^+/b\,e$ bent (rare)
                   $b^+\,e/b\,e$ eyeless (rare)

These expected results are very far from what was observed, so the rare female was not the result of recombination.

If the female had been the product of nondisjunction ($b\,e^+/b^+\,e/b\,e$), her progeny when crossed to $b\,e/b\,e$ would be as follows:

    $1/6$ $b^+\,e/b\,e$ eyeless
    $1/6$ $b\,e^+/b\,e/b\,e$ bent
    $1/6$ $b^+\,e/b\,e/b\,e$ eyeless
    $1/6$ $b\,e^+/b\,e$ bent
    $1/6$ $b\,e/b\,e$ bent, eyeless
    $1/6$ $b\,e^+/b^+\,e/b\,e$ wild type

Overall, 2 bent:2 eyeless:1 bent eyeless:1 wild type

These results are in accord with the observed results, indicating that the female was a product of nondisjunction.

54. Recall that ascospores are haploid. The normal genotype associated with the phenotype of each spore is given below:

1	2	3
$b^+ f^+$	$b f^+$	$b^+ f$
$b^+ f^+$	$b f^+$	$b^+ f$
$b^+ f^+$	abort	$b^+ f^+$
$b^+ f^+$	abort	$b^+ f^+$
abort	$b^+ f$	$b f$
abort	$b^+ f$	$b f$
abort	$b^+ f$	$b f^+$
abort	$b^+ f$	$b f^+$

    a. For the first ascus, the most reasonable explanation is that nondisjunction occurred at the first meiotic division. Second-division nondisjunction or chromosome loss are two explanations of the second ascus. Crossing-over best explains the third ascus.

    b.

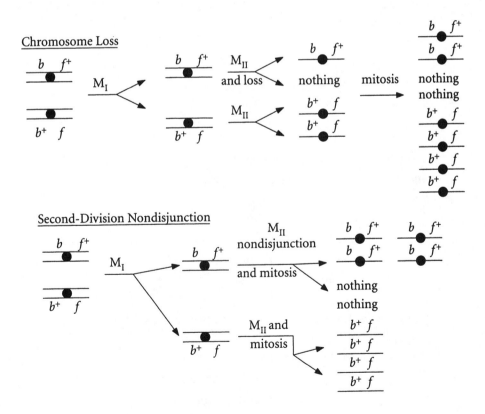

55.   **a.** Each mutant is crossed with wild type, or

$$m \times m^+$$

The resulting tetrads (octads) show 1:1 segregation, indicating that each mutant is the result of a mutation in a single gene.

b. The results from crossing the two mutant strains indicate either both strains are mutant for the same gene,

$$m_1 \times m_2$$

or, that they are mutant in different but closely linked genes

$$m_1 \, m_2^+ \times m_1^+ m_2$$

c. and d. Because phenotypically black offspring can result from nondisjunction (notice that in Case C and Case D, black appears in conjunction with aborted spores), it is likely that mutant 1 and mutant 2 are mutant in different but closely linked genes. The cross is therefore

$$m_1 \, m_2^+ \times m_1^+ m_2$$

Case A is an NPD tetrad and would be the result of a four-strand double cross-over.

$m_1^+ \, m_2^+$	black
$m_1^+ \, m_2^+$	black
$m_1 \, m_2$	fawn
$m_1 \, m_2$	fawn

Case B is a tetratype and would be the result of a single cross-over between one of the genes and the centromere.

$m_1^+ \, m_2^+$	black
$m_1^+ \, m_2$	fawn
$m_1 \, m_2^+$	fawn
$m_1 \, m_2$	fawn

Case C is a the result of nondisjunction during meiosis I.

$m_1^+ \, m_2 \, ; m_1 \, m_2^+$	black
$m_1^+ \, m_2 \, ; m_1 \, m_2^+$	black
no chromosome	abort
no chromosome	abort

Case D is a the result of recombination between one of the genes and the centromere followed by nondisjunction during meiosis II. For example:

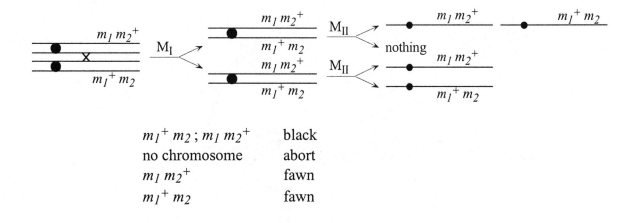

$m_1{}^+ m_2$; $m_1 m_2{}^+$ black

no chromosome   abort

$m_1 m_2{}^+$ fawn

$m_1{}^+ m_2$ fawn

16

Dissection of Gene Function

BASIC PROBLEMS

1. This would be considered a screen for auxotrophs. If it were a selection, only the desired mutant phenotype would survive.

2. *Chlamydomonas* is a photosynthetic organism. At moderate light intensities, they swim towards the light (positive phototaxis). Therefore, suspend the progeny of mutagenized organisms in a long column with light at the top. Collect those that do not swim to the top. This would select for those organisms with defective flagellar function (or defective response to light).
    After mutagenesis, visually screen individual organisms using a microscope.

3. You can screen for new *ad-3* mutants by mutagenizing the *ad-3$^+$*/Δ heterokaryons and then growing them on minimal medium supplemented with adenine. Colonies would then be replica-plated unto plates with minimal medium only. Those colonies that do not grow are likely candidates for new *ad-3* mutants. Alternatively, mutagenized *ad-3$^+$*/Δ heterokaryons may be grown on minimal medium supplemented with adenine. New *ad-3* mutants will be purple, so large numbers of colonies can be screened rapidly. This test relies on the identification of purple (mutant) colonies among a large number of white (wild-type) colonies.

    As for selection of mutants, filtration enrichment can be used for finding those heterokaryons that cannot grow in the absence of adenine after mutagenesis.

4. Mutagenize the fungus and then grow them on plates containing red food coloring. Those colonies that do not become red are defective in this uptake process. These "white" colonies should be crossed to each other to determine the number of genes affected (complementation test). Representatives of each gene should then be tested for pathogenicity to see if the defect in the uptake of red dye also affects nutrient uptake.

5. Growth on these plates is a control to check viability of the progeny.

6. Wild-type flies migrate towards the light. Any mutation that affects the detection of light or the behavior of positive phototaxis will cause flies to randomly follow the dark rather than the light pathway.

7. *cdc* mutants affect the cell cycle. Comparative genomics has shown these same genes are at work in the cell cycle of humans, and further study has shown that many of these genes are defective in cancers.

8. *nim* (never in mitosis) would be expected to be epistatic to *bim* (block in mitosis).
    nim would be expected to be epistatic to *nud* (nuclear distribution).
    bim would be expected to be epistatic to *nud*.

9. Translational. Translation of fused "hybrid" proteins is useful for studies in which the fused protein (because of the LacZ) can be followed in cellular processes. A vector that contains the *lacZ* gene that has no translational start sites is randomly inserted into the genome. Expression of *lacZ* is determined by both the transcription and translation of the mutant gene. Gene fusions produce a hybrid protein with the N-terminus from the mutant gene covalently attached to LacZ at the C-terminus.

10. These phrases are used as "pace holders" so that protocols, experiments, and results may be discussed generically without concern for any specific gene or unique situations.

11. Parental males are treated with mutagen and then crossed to untreated females to produce the F_1 generation. In zebrafish, UV-irradiated sperm are able to fuse with eggs and trigger the development of haploid embryos with the DNA derived only from the egg. If the F_1 mother is heterozygous for a particular gene, half of the embryos will have the desired phenotype. Once found, inbreeding the originally mutagenized parental male will produce a line homozygous for the mutation.

12. The LacZ construct is carried on a transposon. Crosses are made that mobilize the construct so that it is transposed to various, random sites in the genome. Flies with the inserted transgene are then screened for LacZ expression. Those flies with the desired expression pattern have insertions near genes that are candidates for involvement in some aspect of head development.

13. Targeted gene knockout replaces the resident wild-type copy of the gene with a transgenic segment containing a null version of the gene. The mutated gene inserts into the chromosome by a mechanism resembling homologous recombination, replacing the normal sequence with the mutant. Therefore, transform the cells using the clone with the hygromicin resistance gene spliced into the middle of the PLC gene and then select for resistance. Screen through the resistance colonies for those where the clone replaced the PLC gene, and did not insert ectopically.

For knocking out the PLC gene by RIP (repeat-induced point mutation), introduce the wild-type clone into haploid cells. The transgene will insert into the genome randomly. When these cells are crossed, both copies of the gene will be mutated (multiple GC→AT transitions) and rendered null.

14. The goal of PCR mutagenesis is to amplify the DNA sequence of a gene with an altered base within its sequence. The long primer is necessary as it contains both the "mutant" base and all the sequence that goes from that point to one end of the gene. (The other primer of this reaction only has to define the other end of the gene but does not have to stretch to the mutant base.)

15. DICER cuts the introduced double-stranded RNA into segments called interfering RNAs. These segments are then bound by RISC (RNA-induced silencing complex). An RNA component of RISC, called the guide RNA, then helps the complex find the mRNA that has complementarity to the interfering RNAs. Once bound, the RISC degrades the mRNA.

16. Gene position is not a necessary requirement for using RNAi. What is needed is the sequence of the gene that you intend to inactivate, as you need to start with double-stranded RNA that is made with sequences that are homologous to part of the gene.

17. The mutations on chromosome 1 all fail to complement each other and therefore represent one gene. The four mutations that map to chromosome 5 represent two genes, as there are two complementation groups. Mutants 1 and 4 fail to complement each other but both complement 2 and 3. Mutants 2 and 3 fail to complement each other but both complement 1 and 4.

18. Generally, mutations that directly affect the active site are more likely to destroy enzyme function than those that map elsewhere in the polypeptide. Therefore, site-directed mutagenesis of this region of the gene should be attempted and the resultant mutants assayed for enzyme function.

19. Based on the results, the gene is haplosufficient. The phenotype of the heterozygote is wild type and the mutant is recessive.

20. Because the mutant phenotype is insensitive to the dosage of the wild-type allele, the mutation would be classified as a neomorph.

21. The new mutation is a recessive hypomorph or leaky mutation. When homozygous, a mutant phenotype results but when paired with a deletion, the single copy of the hypomorphic allele is not sufficient for survival.

22. All cells derived from the cell in which the reversion took place will now be w^+/w. Depending on when during development this took place, the petal will now be blue, either in part or in whole. Because the petal is part of the plant's soma, this reversion would not be inherited.

23. Starting with a yeast strain that is *pro-1*, plate the cells on medium lacking proline. Only those cells that are able to synthesize proline will form colonies. Most of these will be revertants; however, some will have second-site suppressors. (Treating the cells with a mutagen prior to plating them would significantly increase the yield.)

24. There are many ways to test a chemical for mutagenicity. For example, the text discusses a detection system for recessive somatic mutations in mice. Mice bred to be heterozygous for seven genes involved in coat color are exposed to a potential mutagen by injecting it into the uterus of their pregnant mother. Any somatic mutation from wild-type to mutant at one of the seven loci will result in a patch (or mutant sector) of differently colored fur. The number of mutant sectors later found on these chemically treated mice would be compared with the number found on genetically identical, but chemically untreated, mice (the control mice). (A proper control would expose the control mice to exactly the same experimental protocol except for the caffeine. This would include injection of whatever solvent was used into the uterus of their mother at the same developmental time as for the experimental mice.)

    Much larger screens or selections could be done with fungi. For example, haploid *Neurospora* auxotrophic for the amino acid leucine could be exposed to caffeine and then plated onto minimal medium selecting for *leu*⁺ colonies. Only those cells in which a reverse mutation (from *leu*⁻ to *leu*⁺) occurred would grow. Reversion rates of treated and untreated (control) cells would be compared to see if the caffeine was mutagenic. Alternatively, yeast cells could be exposed to caffeine and mutations in the gene *ade-3* could be scored and numbers compared between treated and untreated populations. (Mutations in this gene actually cause the yeast to be red instead of white, so large numbers of colonies can be screened rapidly.)

    Although this question asks specifically for mutations in higher organisms, a rapid and widely used mutagen-detection system using bacteria was developed by Bruce Ames in the 1970s. Using a genetically modified bacterium (*Salmonella typhimurium*) that is auxotrophic for histidine and defective in DNA repair, the Ames test quickly ascertains the mutagenicity of various chemicals. Because the basic properties of DNA and mutation are the same in prokaryotes and eukaryotes, this test does have relevance. The Ames test has been further enhanced by using rat liver extracts to modify the tested chemicals to simulate human (and mammalian) metabolism. This is important because although the liver is responsible for most of the detoxification and metabolism of ingested chemicals (hence the connection between alcohol and liver disease!), some chemicals are modified in ways that actually make them toxic or mutagenic.

25. The commission was looking for induced recessive X-linked lethal mutations, which would show up as a shift in the sex ratio. A shift in the sex ratio is the first indication that a population has sustained lethal genetic damage. Other

recessive mutations might have occurred, of course, but they would not be homozygous and therefore would go undetected. All dominant mutations would be immediately visible, unless they were lethal. If they were lethal, there would be lowered fertility, an increase in detected abortions, or both, but the sex ratio would not shift as dramatically.

26. The mutants can be categorized as follows:

Mutant 1: an auxotrophic mutant
Mutant 2: a non-nutritional, temperature-sensitive mutant
Mutant 3: a leaky, auxotrophic mutant
Mutant 4: a leaky, non-nutritional, temperature-sensitive mutant
Mutant 5: a non-nutritional, temperature-sensitive, auxotrophic mutant

27. **a.** To select for a nerve mutation that blocks flying, place *Drosophila* at the bottom of a cage and place a poisoned food source at the top of the cage.

b. Make antibodies against flagellar protein and expose mutagenized cultures to the antibodies.

c. Do filtration through membranes with variously sized pores.

d. Screen visually.

e. Go to a large shopping mall and set up a rotating polarized disk. Ask the passersby to look through the disk for a free evaluation of their vision and their need for sunglasses. People with normal vision will see light with a constant intensity through the disk. Those with polarized vision will see alternating dark and light.

f. Set up a Y tube (a tube with a fork giving the choice of two pathways) and observe whether the flies or unicellular algae move to the light or the dark pathway.

g. Set up replica cultures and expose one of the two plates to low doses of UV.

28. The allele for NF must have arisen spontaneously in one of the parents' germ lines. Depending on when this mutation happened (the size of the mutant clone of germ-line tissue), their chance of having another affected child would range from 0 to 50 percent (the latter number if the entire germ line was mutant).

29. Phenocopying is the mimicking of a mutant phenotype by inactivating the gene product rather than the gene itself. It can be used regardless of how well-developed the genetic technology has been elaborated for a particular organism and can provide meaningful results when gene knockout or site-directed mutagenesis is not possible. Three methods of producing phenocopies are: the

prevention of gene-specific translation by the introduction of anti-sense RNA (RNA complementary to the gene-specific mRNA); the introduction of double-stranded RNA with sequences homologous to part of a specific mRNA resulting in major reduction in the level of the mRNA (called double-stranded RNA interference); and inhibition of a specific protein through high-affinity binding of compound(s) identified through chemical genetics.

30. In cross-fertilizing species, F$_3$ individuals must be analyzed in order to recover homozygous autosomal recessive mutations after mutagenesis. A balancer chromosome typically contains a dominant morphological marker and a series of inversions that prevent crossing over between it and its homolog. These features make mutant screens much more efficient as any F$_3$ individual that lacks the balancer's dominant morhpological marker will be homozygous for the homologous mutagenized chromosome.

31. A forward mutation is any change away from the wild-type allele while any change back to the wild-type allele is called a reverse mutation.

32. Genetic selections permit highly efficient recovery of mutations because individuals die if they lack a newly induced mutation affecting the phenotype in question. This allows for rapid identification of even very rare events as only desired mutants grow, so large numbers of organisms can be tested easily. Genetic screens are much less efficient (often similar to searching for a needle in haystack) but are much more flexible in allowing the investigator to recover mutations with essentially any kind of phenotype—even those that were unexpected!

CHALLENGING PROBLEMS

33. *Unpacking the Problem*

1.

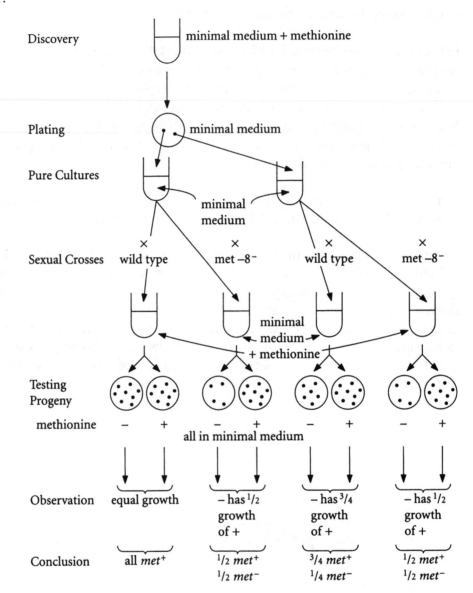

2. Haploid refers to possessing a single genome.

Auxotrophic means that an organism requires dietary provision of some substance that is not normally required by members of its species.

Methionine is an amino acid.

Asexual spores are a mode of propagation used by some species in which the spores are derived from an organism without a genetic contribution from another organism. Of necessity, the spores have the same number of genomes as the organism from which they are derived.

Prototrophic means that an organism does not have any special dietary requirements beyond those normal for the species.

A colony is a collection of cells or organisms all derived mitotically, from a single cell or organism and all possessing the same genotype.

A mutation is the process that generates alternative forms of genes, and it results in an inherited difference between parent and progeny.

3.   The "8" in *met-8* refers to the eighth locus found that leads to a methionine requirement. It is unnecessary to know the specifics of the mutation in order to work the problem.

4.   The following crosses were made in this problem:

>   Cross 1: prototroph 1 × wild type
>   Cross 2: prototroph 1 × backcross to *met-8*
>   Cross 3: prototroph 2 × wild type
>   Cross 4: prototroph 2 × backcross to *met-8*

5.   Use *met-8** to indicate the prototroph derived from the *met-8* strain.

>   Cross 1: *met-8** 1 × *met-8*$^+$
>   Cross 2: *met-8** I × *met-8*
>   Cross 3: *met-8** 2 × *met-8*$^+$
>   Cross 4: *met-8** 2 × *met-8*

6.   In this organism, asexual spores give rise to an organism that is capable of forming sexual spores following a mating. Therefore, the original mutation occurred in somatic tissue that subsequently gave rise to germinal tissue.

7.   Because the trait being selected is the ability to grow in the absence of methionine, a reversion is being studied.

8.   Only two revertants were observed, because reversion occurs at a much lower frequency than forward mutation.

9.   The millions of asexual spores did not grow because they required methionine and the medium used did not contain methionine.

10. A low percentage of the millions of spores that did not grow would be expected to have other mutations that rendered them incapable of growth. In addition, a low percentage would be expected to have chromosome abnormalities that would lead to death.

11. The wild type used in this experiment was prototrophic, by definition; that is, *wild type* refers to the norm for a species, which means "prototrophic."

12. It is highly unlikely that visual inspection could distinguish between wild type and prototrophic revertants.

13. One way to select for a *met-8* mutation is to grow a large number of spores on a medium that lacks methionine. Filtration will separate those spores capable of growth from those incapable of growth. Once spores have been isolated that are incapable of growth in a medium lacking methionine, they can be tested for a methionine mutation by plating them on medium containing methionine. If they are capable of growth on this second medium, they are methionine auxotrophs.

14. The starting auxotrophic spores were haploid. Both mitotic crossing-over and haploidization require diploids. Therefore, it is unlikely that either process is involved with producing the observed results.

15.   Cross 1: *met-8* × wild type → 1 *met-8*:1 wild type
                  Cross 2: *met-8* × *met-8* → all *met-8*
                  Cross 3: wild type × wild type → all wild type

16. While the analysis could have been conducted using tetrad analysis, it is more likely that random selection of progeny was used.

17.

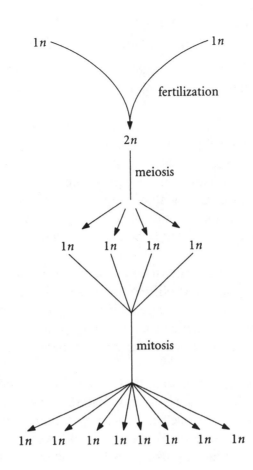

18. If a 3:1 ratio is obtained in haploids, then two genes must be segregating.

Solution to the Problem

a. and b. The pattern of growth for prototroph 1 suggests that it is a reversion of the original mutation. When crossed with wild type, a reversion would be expected to produce all *met*$^+$ progeny, and when backcrossed, it would be expected to produce a 1:1 ratio. Let the reversion be symbolized by *met-8**. The crosses are

P   *met-8** × *met-8*$^+$ P   *met-8** × *met-8*
           ↓   ↓
F_1 $^1/_2$ *met-8** prototrophic F_1 $^1/_2$ *met-8** prototrophic
        $^1/_2$ *met-8*$^+$ prototrophic $^1/_2$ *met-8* auxotrophic

The pattern of growth for prototroph 2 suggests that a suppressor at another, unlinked locus is responsible for its prototrophic growth. Let *met-s* symbolize this suppressor. Then the crosses are

P   *met-8*; *met-s* × *met-8*$^+$; *met-s*$^+$ P   *met-8*; *met-s* × *met-8*; *met-s*$^+$
                  ↓   ↓

F_1	$^{1}/_{4}$ *met-8*; *met-s*	Prototrophic	F_1	$^{1}/_{4}$ *met-8*; *met-s*	Prototrophic
	$^{1}/_{4}$ *met-8*$^{+}$; *met-s*$^{+}$	Prototrophic		$^{1}/_{4}$ *met-8*; *met-s*$^{+}$	Auxotrophic
	$^{1}/_{4}$ *met-8*$^{+}$; *met-s*	Prototrophic		$^{1}/_{4}$ *met-8*; *met-s*	Prototrophic
	$^{1}/_{4}$ *met-8*; *met-s*$^{+}$	Auxotrophic		$^{1}/_{4}$ *met-8*; *met-s*$^{+}$	Auxotrophic

34. **a. and b.** Hypomorphic mutations result in less gene product activity than in the wild type. It would be expected that these would be more frequent among mutations caused by base substitution than frameshift mutagens, because the former will likely change just a single amino acid while the latter will alter the entire amino acid sequence coded downstream of the lesion. Amorphic mutations result in the complete loss of gene product activity. These would be expected to be more frequent among frameshift mutations.

35. Neomorphic mutations result in novel gene activity and are dominant. Reversion of the dominant phenotype will commonly be the result of introducing another mutation in the already mutant gene that now eliminates its function completely. Most gene knockout mutations are recessive so it is likely that most "revertants" will actually be recessive loss-of-function mutations.

36. **a.** No, because the mutational target size will vary considerably. Genes come in different physical sizes and that may play a role. More difficult to calculate are issues such as what proportion of mutations make the gene product sufficiently abnormal or what proportion of non-protein-coding mutations alter the expression of the gene sufficiently to cause the desired mutant phenotype.

   b. Not likely. The dorsal fin is a complex structure and it would be expected that more than five genes would contribute to its development. Of course, although it represents a lot of work, 40 independently isolated mutations is not sufficient to "saturate" the genome for mutations

   Certainly, the original screen could be extended to search for more mutations. Screening after mutagenesis with different mutagens would also be of use. Because some genes that affect fin development might also be required for other developmental programs or might otherwise be lethal to the organism when absent, screening for conditional alleles might also identify genes missed by nonconditional screens.

37. **a.** Forward genetics identifies heritable differences by their phenotypes and map locations, and precedes the molecular analysis of the gene products. Reverse genetics starts with an identified protein or RNA and works towards mutating the gene that encodes it (and in the process, discovers the phenotype when the gene is mutated). Because you have known proteins and want to determine the phenotypes associated with loss-of-function mutations in the genes that encoded them, a reverse genetic approach is the answer.

b. The two general approaches would be either directed mutations in the gene of interest or the generation of phenocopies of the mutant phenotype by inactivating the gene product rather than the gene itself.

38. Leaky mutants are mutants with an altered protein product that retains a low level of function. Enzyme activity may, for instance, be reduced rather than abolished by a mutation.

39. a. You are told that all mutants are simple Mendelian recessives, so you need not worry about mutations that map to the chloroplast's genome. To determine the number of genes involved, a complementation test (conducting pairwise crosses) is performed. All combinations complement (are mutant for different genes) except for 1×4. In this cross, the F_1 is still mutant, the mutations fail to complement and are in the same gene. Thus, 3 genes are represented.

b. As stated above, 1 and 4 map to the same gene. Of the other combinations, only 2 and 3 show linkage. In this case, the testcross of the 2×3 F_1 produces 10 percent wild-type progeny, not the expected 25 percent if the genes were unlinked. You can also use these data to determine the map distance between genes 2 and 3. The percentage of wild-type progeny from the testcross will be equal to half that of the recombinants (the other half will be mutant for both genes). Thus genes 2 and 3 are 20 map units apart.

c. 1×2:

$$m_1/m_1 \,;\, +/+ \quad \times \quad +/+ \,;\, m_2/m_2$$
$$\downarrow$$
$$m_1/+ \,;\, m_2/+ \quad \times \quad m_1/m_1 \,;\, m_2/m_2 \text{ (testcross)}$$
$$\downarrow$$

25% $m_1/m_1 \,;\, m_2/+$
25% $m_1/+ \,;\, m_2/m_2$
25% $m_1/m_1 \,;\, m_2/m_2$
25% $m_1/+ \,;\, m_2/+$ (wild type)

1×3:

$$m_1/m_1 \,;\, +/+ \quad \times \quad +/+ \,;\, m_3/m_3$$
$$\downarrow$$
$$m_1/+ \,;\, m_3/+ \quad \times \quad m_1/m_1 \,;\, m_3/m_3 \text{ (testcross)}$$
$$\downarrow$$

25% $m_1/m_1 \,;\, m_3/+$
25% $m_1/+ \,;\, m_3/m_3$
25% $m_1/m_1 \,;\, m_3/m_3$
25% $m_1/+ \,;\, m_3/+$ (wild type)

1 × 4: m_1/m_1 × m_4/m_4

                            ↓

                 m_1/m_4 × m_1/m_1 or m_4/m_4 (testcross)

                            ↓

                 50% m_1/m_4
                 50% m_1/m_1

2 × 3: $m_2 +/m_2 +$ × $+ m_3/+ m_3$

                            ↓

                 $m_2 +/+ m_3$ × $m_2 m_3/m_2 m_3$ (testcross)

                            ↓

                 40% $m_2 +/m_2 m_3$
                 40% $+ m_3/m_2 m_3$
                 10% $m_2 m_3/m_2 m_3$
                 10% $+ +/m_2 m_3$ (wild type)

2 × 4: as in 1 × 2
3 × 4: as in 1 × 2

40. 15 percent are essential gene functions (such as enzymes required for DNA replication or protein synthesis).

25 percent are auxotrophs (enzymes required for the synthesis of amino acids or the metabolism of sugars, etc.).

60 percent are redundant or pathways not tested (genes for histones, tubulin, ribosomal RNAs, etc., are present in multiple copies; the yeast may require many genes under only unique or special situations or in other ways that are not necessary for life in the "lab").

41. a. and b. It is likely that the observed abnormalities are the result of mitotic recombination. A cross-over between the genes of interest and the centromere in this case will lead to "twin spots" (the adjacent patches of stubby and ebony body observed). On the other hand, a cross-over that occurs between the two genes will lead to "single spots" (the solitary patches of ebony). The position of the two genes with respect to the centromere determines the phenotype of the single spot. When recombination occurs in the region between the genes, the gene more distal to the centromere becomes homozygous, while the more proximal gene remains heterozygous. In this problem, because the single patches are ebony, e is more distal than s. The other product of this event will appear normal and will not be detected among the other "normal" cells.

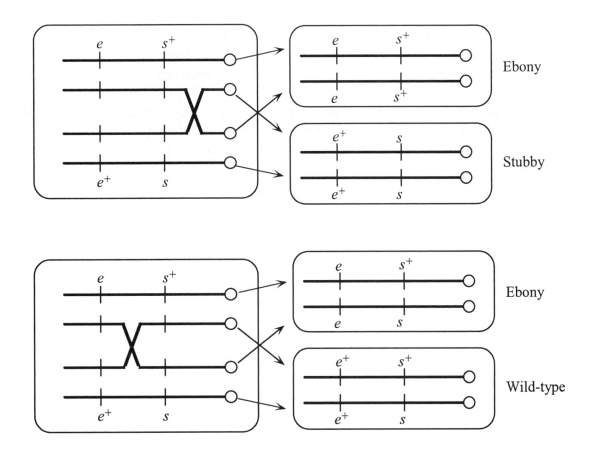

42. The cross is *trp^r* × *trp^+*, where *trp^r* is the revertant. However, it might also be *trp^-* · *su* × *trp^+* · *su^+*, where *su^+* has no effect on the *trp* gene and *su* is a suppressor of *trp–*.

    a. If the revertant is a precise reversal of the original change that produced the mutant allele, 100 percent of the progeny would be tryptophan-independent.

    b. If a suppressor mutation on a different chromosome is involved, then the cross is *trp^-* ; *su* × *trp^+* ; *su^+*. Independent assortment would lead to the following:

        1 *trp^-* ; *su*   tryptophan independent
        1 *trp^+* ; *su^+*   tryptophan independent
        1 *trp^-* ; *su^+*   tryptophan dependent
        1 *trp^+* ; *su*   tryptophan independent

    c. If a suppressor mutation 24 map units from the *trp* locus occurred, then the cross is *trp^-* *su* × *trp^+* *su^+* and the diploid intermediate is *trp^-* *su* /*trp^+* *su^+*. The two parental types would occur 76 percent of the time, and the two recombinant types would occur 24 percent of the time. The progeny would be

38%	$trp^-\ su$	tryptophan independent
38%	$trp^+\ su^+$	tryptophan independent
12%	$trp^-\ su^+$	tryptophan dependent
12%	$trp^+\ su$	tryptophan independent

d. A common reason that a mutant is nonreverting is because it is the result of a deletion. However, because only "one" revertant was isolated from mutant B, it is possible that not enough cells were looked at to find an A revertant.

17

Genetic Regulation of Cell Number: Normal and Cancer Cells

BASIC PROBLEMS

1. (1) Certain cancers are inherited as highly penetrant simple Mendelian traits.
   (2) Most carcinogenic agents are also mutagenic.
   (3) Various oncogenes have been isolated from tumor viruses.
   (4) A number of genes that lead to the susceptibility to particular types of cancer have been mapped, isolated, and studied.
   (5) Dominant oncogenes have been isolated from tumor cells.
   (6) Certain cancers are highly correlated to specific chromosomal rearrangements.

2. The activities of proteins controlling the cell cycle or cell-death pathways are controlled by phosphorylation/dephosphorylation, protein/protein interactions, and proteolysis.

3. Cyclins complex with cyclin-dependent protein kinases (CDKs) and activate their kinase activity. They also tether target proteins so that they can be phosphorylated by the complexed CDK.

   The levels of cyclins are regulated by rapid inactivation of the required transcription factors, the high degree of instability of the mRNA encoding the cyclin, and the high degree of instability of the cyclin itself.

4. Oncoproteins may be the result of point mutations, gene fusions, or deletions of key regulatory domains.

5. Apaf is a positive regulator of apoptosis; Bcl is a negative one. Apaf binds to cytoplasmic cytochrome c protein. This complex then binds to and activates the initiator caspase. Bcl blocks the release of cytochrome c from the mitochondria and also binds to Apaf, preventing its interaction with the initiator caspase.

6. Translocations can cause the fusion of the enhancer of one gene to the transcription unit of another, leading to the misexpression or overexpression of a nonmutated protein, or they can cause the fusion of two genes resulting in an abnormal chimeric protein.

7. As caspases (cysteine-containing aspartate-specific proteases) are first translated, they contain sequences that prevent their activity. These inactive versions are called zymogens. Part of the polypeptide then must be removed by enzymatic cleavage for the caspase to become active.

8. Translocation and fusion of an immunoglobin enhancer to the *bcl2* gene causes large amounts of Bcl2 to be expressed in B lymphocytes. Because Bcl2 is a negative regulator of apoptosism, this overexpression essentially blocks apoptosis in these mutant B lymphocytes and allows them to accumulate cell proliferation-promoting mutations over their unusually long lifetime.

   Another example would be the point mutation in the gene encoding the Ras protein. The mutant version of this G-protein is always bound to GTP and thus remains active even in the absence of the correct signal.

9. Tumor-promoting mutant alleles of tumor-suppressor genes inactivate the proteins that they encode. Because only when both copies of one of these genes are inactivated are tumors observed, these loss-of-function mutations are by definition recessive. Examples of such genes are the *p53* gene and the *RB* gene.

10. a. The v-*erbB* oncogene encodes a mutated form of the epidermal growth factor receptor (EGFR). Compared to wild type, the mutant EGFR oncoprotein lacks key regulatory regions as well as the extracellular ligand-binding domain. These deletions allow the mutant protein to constitutively dimerize, leading to continuous autophosphorylation, which results in continuous transduction of a signal from the receptor.

    b. The *ras* oncogene is due to a missense mutation that allows the mutated Ras protein to always bind GTP, even in the absence of the signals normally required for such binding by wild-type Ras protein. As a result, the Ras oncoprotein continually promotes cell proliferation.

    c. The Philadelphia chromosome is a translocation between chromosomes 9 and 22 that results in the fusion of two genes, *bcr1* and *abl*. The Bcr1-Abl fusion protein has an activated kinase activity that is responsible for causing chronic myelogenous leukemia.

CHALLENGING PROBLEMS

11. **a.** Many growth-factor receptors are receptor tyrosine kinases (RTKs) and are activated by dimerization caused by binding their ligand (the growth factor). Structural aberrations of such a protein can lead to activation in the absence of ligand or dimerization. In the case of the v-*erbB* oncoprotein, loss of the extracellular ligand-binding domain as well as some regulatory components of the cytoplasmic domain leads to dimerization and continuous activation in the absence of its ligand.

   b. Changes in transcriptional regulation that lead to the overproduction or misexpression of an otherwise normal protein may lead to cancer. For example, overproduction of Bcl2 protein (a negative regulator of apoptosis) in B lymphocytes, as the result of a translocation that fuses the enhancer of one of the immunoglobulin genes to the transcriptional unit of the *bcl2* gene, is the major cause of follicular lymphoma.

   c. G proteins cycle between being bound to GDP (inactive state) and being bound to GTP (active state). Mutations that alter the protein preventing its GTPase activity or allowing it to bind to GTP even in the absence of proper signals, may cause cancer (as in the case of Ras oncoprotein).

12. **a.** Cyclins bind to, activate, and direct CDKs to phosphorylate specific cellular targets, and by doing so, control the cell cycle. Normal cell division requires the sequential production and then removal of different cyclins. The activity of the cyclin/CDK heterodimer is also regulated through p21. Overproduction of one of the cyclins could disrupt the orderly process of cell division, but it would be limited by the amount of CDK present as well as the state of the p21 "brake."

   b. A nonsense mutation would lead to a decrease of the normal protein product. If that protein were part of the receptor for a growth factor, which stimulates cell proliferation, then cell division could not be triggered. This would likely be recessive and would slow cell proliferation, not accelerate it.

   c. Overproduction of FasL will signal adjacent cells through their Fas cell-surface receptors, which in turn leads to Apaf activation. This in turns causes proteolysis and activation of the initiator caspase, ultimately leading to apoptosis of the cell. Although this would be dominant, it would lead to excess cell death, not proliferation.

   d. Cytoplasmic tyrosine-specific protein kinase phosphorylates proteins in response to signals received by the cell. These phosphorylations lead to activation of the transcription factors for the next step in the cell cycle. If the active site is disrupted, then phosphorylations will not occur and transcription factors for the next step will not be activated. This would likely be recessive and would slow cell proliferation, not accelerate it.

e. If the enhancer causes large amounts of the apoptosis inhibitor to be expressed in the liver, the normal pathway of cell death will be blocked. These liver cells (the enhancer is liver-specific) will have an unusually long lifetime in which to accumulate various mutations that could lead to cancer. This chromosomal rearrangement would be dominant.

13. Once apoptosis is initiated, a self-destruct switch has been thrown: endonucleases and proteases are released, DNA is fragmented, and organelles are disrupted. This obviously is a "terminal" state from which the cell will not have a need or chance to reuse the machinery of destruction. On the other hand, the various proteins needed for the regulation and execution of the cell cycle will be needed again if the cell continues to divide. By recycling many of these, the cell obviously conserves its resources (proteins are energetically expensive to make) and recycling also allows for more rapid divisions, because the cell does not have to spend time remaking all the pieces.

14. Gain-of-function mutations affect mitogenic pathways to increase cell proliferation or cell survival pathways to decrease apoptosis. Loss-of-function mutations promote tumors by loss of growth inhibitor pathways, p53 pathways, or apoptosis activator pathways.

15. a. This would be dominant. The misexpression of FasL from one allele would be dominant to the normal expression of the wild-type FasL allele. In this case, each liver cell would signal its neighboring cells to undergo apoptosis.

b. No. It would lead to excess cell death, not proliferation.

16. Normal Ras is a G-protein that activates a protein kinase, which in turn phosphorylates a transcription factor. If it were simply deleted, no cancer could develop, because cell division would not occur. If it were simply duplicated, an excess of the G-protein could not cause cancer, because it must be activated before it can activate the protein kinase, and presumably the enzyme that activates normal Ras is closely regulated and would not activate too many copies. However, if it were to have a point mutation, it might now bind GTP, even in the absence of normal control signals and be in a state of permanent activation. As a positive regulator of cell growth, this mutant Ras would continually promote cell proliferation.

In contrast, normal *c-myc* is a transcription factor. If the gene were to be duplicated, too much transcription factor could lead to malignancy.

17. Inhibition of apoptosis can contribute to tumor formation by allowing cells to have an unusually long lifetime in which to accumulate various mutations that lead to cancer. Also, the normal role of apoptosis in removing abnormal cells

and, through p53, killing cells that have "damaged" DNA would also be inhibited.

18. **a.** Mutations in a tumor suppressor gene are recessive and result in its loss of function. That function can be restored by the introduction of a wild-type allele.

    b. Mutations in an oncogene are dominant and due to gain of function (overexpression or misexpression). The normal function will not inhibit these mutants, and the introduced gene would be ineffective in restoring the normal phenotype.

19. **a.** Type A diabetes is most likely due to a defect in the pancreas. The pancreas normally makes insulin, and type A diabetes can be treated by supplying insulin. Type B diabetes is most likely due to a target cell defect because type B is unresponsive to exogenous insulin.

    b. Type B diabetes appears to be caused by a defect in the target cell. A number of genes are responsible for the receptor and the subsequent cascade of changes that occur in leading to a change in transcription. Any of these genes could have a mutant form.

20. p53 detects and is activated by DNA damage. When activated, p53 activates p21, an inhibitor of the cyclin/CDK complex necessary for the progression of the cell cycle. If the DNA damage is repairable, this system will eventually deactivate p53 and allow cell division. However, if the damage is irreparable, p53 would stay active and would activate the apoptosis pathway, ultimately leading to cell death. It is for this reason that the "loss" of p53 is often associated with cancer.

21. Tumors form from cells in which both copies of the *RB* gene are inactivated. Patients with HBR are heterozygous (*RB/rb*) for a mutant copy of the *RB* gene. For these individuals, the mutation rate (and number of retinal cells) makes it virtually certain that at least some of their retinal cells will acquire a mutation in the remaining normal *RB* gene, thereby producing cells with no functional Rb protein. Sporadic retinoblastoma, on the other hand, is the result of a single cell acquiring mutations in both copies of the *RB* gene. Because these mutations are independent events, the occurrence of cells without functional Rb protein will be much rarer.

22. **a.** In the absence of functional Rb protein, E2F will be in the nucleus.

    b. No. The absence of functional E2F would be epistatic to the absence of functional Rb protein.

    c. A mutation that causes permanent sequestering of E2F in the cytoplasm will likely inhibit the cell cycle.

18

The Genetic Basis of Development

BASIC PROBLEMS

1. Both the cytoskeleton and body skeleton give shape and structure, and both are involved in the generation of movements, but in very different ways. The cytoskeleton consists of three types of filaments: microfilaments, intermediate filaments, and microtubules. They have a strict polar organization and for microfilaments and microtubules, they also act as tracks along which molecules and organelles may be moved.

2. Microfilaments are linear polymers of the protein actin. The arrangement of the actin proteins in the polymer gives a polarity similar to that observed in polypeptide chains of nucleic acids. Some proteins that interact with microfilaments are able to distinguish the two ends (called "+" and "−") or distinguish which direction they travel along the filament.

3. Often, microtubules are said to be like railroad tracks but perhaps the term monorail might be a better analogy.

4. From the zygote (called P_0) to the cell that becomes the precursor of the germ line (called P_4) there are four divisions. There is no specific reason that four divisions rather than say three or five are required, but precursors to other tissues need to be formed and possibly other cellular determinants need to be segregated prior to committing a cell to a single developmental fate.

5. The term syncitium is used to describe a multinucleate cell. During early *Drosophila* embryogenesis, the first 13 mitoses are nuclear divisions without any cell division. In essence, the embryo at this stage is a single cell with 1000's of nuclei. Eventually, each of these nuclei will be surrounded by membrane and become individual cells. So in this sense, these future cells share one common cytoplasm at the syncitial stage of development.

6.

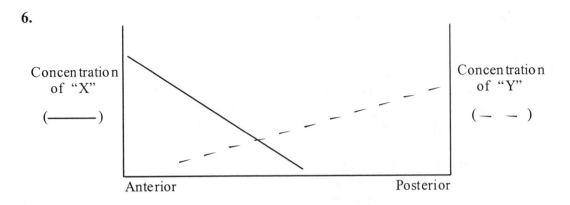

Concentration of "X" (———)

Concentration of "Y" (— —)

Anterior   Posterior

7. There are virtually an unlimited number of experiments one can design to demonstrate a gradient in solution. One very simple approach would be to carefully pipette several drops of food dye into the bottom of test tube filled with water. As the dye slowly diffuses, a very visible gradient will be easily observed.

8. A DNA probe with sequence complementarity to *bicoid* mRNA was used. This probe needs to be labeled (fluorescence, radioactivity) so that after it hybridizes to the *bicoid* mRNA, the probe's localization can be observed and by inference, the localization of the mRNA can be known.

9. The correct localization of *bicoid* and *nanos* is dependent on specific microtubule-association sequences located within the 3′ UTR. Therefore, swapping the 5′ UTRs of these genes will not affect their gradients.

10. The cells that invaginate to form the cephalic furrow are determined by a specific concentration of BCD protein. Thus, the cephalic furrow is a marker for the BCD gradient within the embryo. As the furrow forms more posteriorly, you can infer that the increase in BCD gene dosage has affected its gradient, and that anterior fates are forming more posteriorly than normal.

11. The DL protein is found in the cytoplasm of cells in the dorsal region of the blastoderm. The DL protein is bound to CACT protein in these cells and therefore remains sequestered in the cytoplasm.

12. The DL-CACT complex must be phosphorylated. The resulting conformational changes breaks apart the complex and the now free and phosphorylated DL protein is able to migrate into the nucleus.

13. These are names of genes that are required for normal *Drosophila* development. The mutant phenotypes associated with these genes provide clues to their different roles: *knirps* is an A-P cardinal gene (also called a gap gene); *runt* is a pair-rule gene; *gooseberry* is a segment polarity gene; and *antennapedia* is a segment identity gene.

14. The primary pair-rule gene *eve* (*even-skipped*) would be expressed in seven stripes along the A-P axis of the late blastoderm.

15. Pair-rule genes encode transcription factors expressed in repeating patterns of seven stripes. They are necessary to divide the embryo into the correct number of segments. Homeotic genes are also transcription factors, but they are essential for segment identity and do not affect segment number.

16. Geneticists say cells "communicate" because many cell-cell interactions have been demonstrated. Cells "talk" to each other through the various molecules secreted as signals that then bind to receptors on or in other cells and in doing so, deliver the "message."

17. The anchor cell secretes a ligand that binds to a specific receptor (RTK) on certain neighboring cells. This is an inductive interaction, as the cell that receives the highest level of signal is "told" to develop as a primary vulval cell. The primary cell then secretes another signal that prevents its neighbors from also adopting a primary fate. This is called lateral inhibition. Comparing the two, induction acts to promote a particular cell fate while inhibition acts to prevent adoption of a particular fate.

18. By sequence comparison, the ancestor of the Hox gene B6 is the *Antp* gene in the *Drosophila* Hom set.

19. SRY appears to be expressed constitutively. Its presence leads to male development while its absence allows the default female developmental pathway.

20. The B gene transcripts are required to differentiate the sepal and petal fates and also the stamen and carpel fates. In the absence of B transcripts, the flower whorls would be sepal, sepal, carpel, and carpel.

21. In humans, a single copy of the Y chromosome is sufficient to shift development toward normal male phenotype. The extra copy of the X chromosome is simply inactivated. Both mechanisms seem to be all-or-none, rather than based on concentration levels.

22. Because maleness is based on the presence of androgens produced by the developing testes and femaleness is based on the absence of those androgens, what seems to be crucial here is whether the migrating germ cells organize testes. Although what determines this is unknown, it may be that a minimal number of XY cells are required to organize a testis. If, in the mosaic, not enough of these cells exist, then development will be female. If a sufficient number exist, development will be male.

23. The *bcd* mRNA is tethered to the minus ends of microtubules that are located at the anterior pole of the egg. Upon translation, BCD protein diffuses in the common cytoplasm creating the observed gradient.

The *hb-m* mRNA is uniformly distributed throughout the oocyte. However, translation of *hb-m* mRNA is blocked by the NOS protein product. The *nos* mRNA is localized to the posterior pole through its association with the plus ends of microtubules and upon translation, a posterior to anterior gradient of NOS is generated.

24. CACT and DL form an inactive complex in the cytoplasm. Phosphorylation of the sequestered DL and CACT proteins causes conformational changes that break apart the cytoplasmic complex. The now free phosphorylated DL protein is able to migrate to the nucleus, where it acts as a transcription factor.

Rb and E2F proteins are also combined in a cytoplasmic complex that is inactive. Phosphorylation of the Rb protein alters its shape and causes it to dissociate from the E2F protein. The free E2F protein is then able to diffuse into the nucleus and promote transcription.

25. a. There must be a diffusible substance produced by the anchor cell that affects development of the six cells. The 1° has the strongest response to the substance, and the 3° represents a lack of response due to a low concentration or absence of the diffusible substance.

b. Remove the anchor cell and the six equivalent cells. Arrange the six cells in a circle around the anchor cell. All six cells will develop the same phenotype, which will depend on the distance from the anchor cell.

26. a. The results suggest that ABa and ABp are not determined at this point in their development. Also, future determination and differentiation of these cells are dependent on their position within the developing organism.

b. Because an absence of EMS cells leads to a lack of determination and differentiation of AB cells, the EMS cells must be, at least in part, responsible for AB-cell development, either through direct contact or by the production of a diffusible substance.

c. Most descendants of the AB cells do not become muscle cells when P2 is present; all descendants of the AB cells become muscle cells when P2 is absent. Therefore, P2 must prevent some AB descendants from becoming muscle cells.

27. Proper *ftz* expression requires *Kr* in the fourth and fifth segments, and *kni* in the fifth and sixth segments.

28. The anterior-posterior polarity of the *Drosophila* embryo is developed through the action of maternal effect genes. The anterior determinant is the product of the *bcd* gene. *bcd* mRNA is localized to the posterior pole and translation of this localized mRNA creates the anterior-to-posterior gradient of BCD protein. *hb-m* mRNA is uniformly distributed throughout the oocyte. Its translation is inhibited by the NOS protein product. *nos* mRNA is localized to the posterior pole and translation of this localized mRNA creates the posterior-to-anterior gradient of NOS protein. The NOS translational repressor gradient produces the shallow anterior-to-posterior gradient of HB-M protein. In embryos lacking functional NOS protein, *hb-m* mRNA will be translated throughout the embryo and posterior segments will be lost. The absence of functional BCD protein allows anterior segments to develop as posterior ones.

29. A number of experiments could be devised. A comparison of amino acid sequence between mammalian gene products and insect gene products would indicate which genes are most similar to each other. Using cloned cDNA sequences from mammalian genes for hybridization to insect DNA would also indicate which genes are most similar to each other.

CHALLENGING PROBLEMS

30. a. The anterior 20 percent of the embryo is normally devoted to the head and thorax regions. The bicaudal phenotype results in the loss of these regions and in the loss of A1 through A3. The gap proteins are responsible for the induction of the pair-rule proteins, which ultimately set the number of segments, and the homeotic proteins, which set the identity of the segments. Obviously, the gap proteins are improperly regulated to produce the bicaudal phenotype.

Normal regulation of the gap proteins is accomplished by differential sensitivity to the differing concentrations of the maternally derived morphogens. Because the anterior portion of the embryo has been removed, high concentrations of the morphogens in these regions have also been removed. This results in the abnormal segment number and identity that are observed.

b. The *oskar* mutation results in the loss of the posterior localization of the *nos* mRNA and protein. Therefore, there is no repression of HB-M translation. The lack of repression of the HB-M transcription factor results in an excess of the HB-M protein. The normal shallow gradient, A to P, is therefore lost. Because the gap genes respond differentially to the BCD:HB-M ratio, no induction of the gap genes occurs, which leads to reduced segmentation. This results simply in a broader head and thorax, and no mirror-image phenotype is possible.

31. If you diagram these results, you will see that deletion of a gene that functions posteriorly allows the next-most anterior segments to extend in a posterior direction. Deletion of an anterior gene does not allow extension of the next-most posterior segment in an anterior direction. The gap genes activate *Ubx* in both thoracic and abdominal segments, whereas the *abd-A* and *Abd-B* genes are activated only in the middle and posterior abdominal segments. The functioning of the *abd-A* and *Abd-B* genes in those segments somehow prevents *Ubx* expression. However, if the *abd-A* and *Abd-B* genes are deleted, *Ubx* can be expressed in these regions.

32. It may be that the wild-type allele in the embryo produces a gene product that can inhibit the gene product of the rescuable maternal-effect lethal mutations, while the nonrescuable maternal-effect lethal mutations produce a product that cannot be inhibited.

Alternatively, the nonrescuable maternal-effect lethal mutations may produce a product that is required very early in development, before the developing fly is producing any proteins, while the rescuable maternal-effect lethal mutations may act later in development when embryo protein production can compensate for the maternal mutation.

33. The anterior-posterior axis would be reversed.

34. **a.** A pair-rule gene.

  b. Look for expression of the mRNA from the candidate gene in repeating pattern of seven stripes along the A-P axis of developing embryo.

  c. No. An embryo mutant for the gap gene *Krüppel* would be missing many anterior segments. This effect would be epistatic to the expression of a pair-rule gene.

35. There are a number of ways of approaching this problem, as various combinations of transgenic strains will give a satisfactory solution. The table below gives all possible combinations and results. Systematically create ten transgenic *Drosophila* strains by introducing each of the five plasmids into each of wild type and *bicoid* mutant strains. Observe where the BCD protein is spatially localized in the transgenic embryos. The abbreviations A, P, B, or N are used for anterior, posterior, both, or neither, respectively.

	Transgenic wild type	Transgenic *bicoid* mutant
plasmid 1	A	N
plasmid 2	A	A
plasmid 3	B	P
plasmid 4	A	A
plasmid 5	B	P

36. **a.** *Arabidopsis* strains that are mutant for both class A and class B genes would only have carpels, as expression of class C genes only, specifies carpel fate. Strains that are mutant for class C genes will be lacking carpels (and these strains will also be missing stamens).

19

Population Genetics

BASIC PROBLEMS

1. The frequency of an allele in a population can be altered by natural selection, mutation, migration, nonrandom mating, and genetic drift (sampling errors).

2. There are a total of $(2)(384) + (2)(210) + (2)(260) = 1708$ alleles in the population. Of those, $(2)(384) + 210 = 978$ are A_1 and $210 + (2)(260) = 730$ are A_2. The frequency of A_1 is $978/1708 = 0.57$, and the frequency of A_2 is $730/1708 = 0.43$.

3. The given data are $q^2 = 0.04$ and $p^2 + 2pq = 0.96$. Assuming Hardy–Weinberg equilibrium, if $q^2 = 0.04$, $q = 0.2$, and $p = 0.8$. The frequency of B/B is $p^2 = 0.64$, and the frequency of B/b is $2pq = 0.32$.

4. The frequency of a phenotype in a population is a function of the frequency of alleles that lead to that phenotype in the population. To determine dominance and recessiveness, do standard Mendelian crosses.

5. **a.** The needed equations are

$$p' = p\frac{pW_{AA} + qW_{Aa}}{\overline{W}}$$

and

$$\overline{W} = p^2W_{AA} + 2pqW_{Aa} + q^2W_{aa}$$

$p' = 0.5\,[(0.5)(0.9) + 0.5(1.0)]/[(0.25)(0.9) + (0.5)(1.0) + (0.25)(0.7)] = 0.53$

b. The needed equation is

$$\hat{p} = \frac{W_{a/a} - W_{A/a}}{(W_{a/a} - W_{A/a}) + (W_{A/A} - W_{A/a})}$$

$$= \frac{0.7 - 1.0}{(0.7 - 1.0) + (0.9 - 1.0)}$$

$$= 0.75$$

6. The needed equation is

$$q^2 = \mu/s$$

or

$$s = \mu/q^2 = 10^{-5}/10^{-3} = 0.01$$

7. Albinos appear to have had greater opportunity to mate. They may have been considered lucky and encouraged to breed at very high levels in comparison with nonalbinos. They may also have been encouraged to mate with each other. Alternatively, in the tribes with a very low frequency, albinos may have been considered very unlucky and destroyed at birth or prevented from marriage.

CHALLENGING PROBLEMS

8. This problem assumes that there is no backward mutation. Use the following equation:

$$p_n = p_0 e^{-n\mu}$$

That is, $p_{50,000} = (0.8)e^{-(5 \times 10^4)(4 \times 10^{-6})} = (0.8)e^{-0.2} = 0.65$

9. **a.** If the variants represent different alleles of gene X, a cross between any two variants should result in a 1:1 progeny ratio (because the organism is haploid.) All the variants should map to the same locus. Amino acid sequencing of the variants should reveal differences of one to just a few amino acids.

   b. There could be another gene (gene Y), with five variants, that modifies the gene X product post-transcriptionally. If so, the easiest way to distinguish between the two explanations would be to find another mutation in X and do a dihybrid cross. For example, if there is independent assortment,

P $X^1; Y^1 \times X^2; Y^2$
F_1 1 $X^1; Y^1$:1 $X^1; Y^2$:1 $X^2; Y^1$:1 $X^2; Y^2$

If the new mutation in X led to no enzyme activity, the ratio would be

2 no activity : 1 variant one activity : 1 variant two activity

The same mutant in a one-gene situation would yield 1 active : 1 inactive.

10. a. If the population is in equilibrium, $p^2 + 2pq + q^2 = 1$. Calculate the actual frequencies of p and q in the population and compare their genotypic distribution to the predicted values. For this population:

$$p = [406 + \tfrac{1}{2}(744)]/1482 = 0.52$$
$$q = [332 + \tfrac{1}{2}(744)]/1482 = 0.48$$

The genotypes should be distributed as follows, if the population is in equilibrium:

$L^M/L^M = p^2(1482) = 401$	Actual: 406	
$L^M/L^N = 2pq(1482) = 740$	Actual: 744	
$L^N/L^N = q^2(1482) = 341$	Actual: 332	

This compares well with the actual data, so the population is in equilibrium.

b. If mating is random with respect to blood type, then the following frequency of matings should occur:

$L^M/L^M \times L^M/L^M = (p^2)(p^2)(741) = 54$ Actual: 58
$L^M/L^M \times L^M/L^N$ or $L^M/L^N \times L^M/L^M = (2)(p^2)(2pq)(741) = 200$ Actual: 202
$L^M/L^N \times L^M/L^N = (2pq)(2pq)(741) = 185$ Actual: 190
$L^M/L^M \times L^N/L^N$ or $L^N/L^N \times L^M/L^M = (2)(p^2)(q^2)(741) = 92$ Actual: 88
$L^M/L^N \times L^N/L^N$ or $L^N/L^N \times L^M/L^N = 2(2pq)(q^2)(741) = 170$ Actual: 162
$L^N/L^N \times L^N/L^N = (q^2)(q^2)(741) = 39$ Actual: 41

Again, this compares nicely with the actual data, so the mating is random with respect to blood type.

11. a. and b. For each, p and q must be calculated and then compared with the predicted genotypic frequencies of $p^2 + 2pq + q^2 = 1$.

Population	p	q	Equilibrium?
1	1.0	0.0	Yes
2	0.5	0.5	No
3	0.0	1.0	Yes
4	0.625	0.375	No
5	0.375	0.625	No
6	0.5	0.5	Yes
7	0.5	0.5	No
8	0.2	0.8	Yes
9	0.8	0.2	Yes
10	0.993	0.007	Yes

c. The formulas to use are $q^2 = \mu/s$ and $s = 1 - W$.

$$4.9 \times 10^{-5} = 5 \times 10^{-6}/s \; ; \; s = 0.102, \text{ so } W = 0.898$$

d. For simplicity, assume that the differences in survivorship occur prior to reproduction. Thus, each genotype's fitness can be used to determine the relative percentage each contributes to the next generation.

Genotype	Frequency	Fitness	Contribution	A	a
A/A	0.25	1.0	0.25	0.25	0.0
A/a	0.50	0.8	0.40	0.20	0.20
a/a	0.25	0.6	0.15	0.0	0.15
				0.45	0.35

$$p' = 0.45/(0.45 + 0.35) = 0.56$$
$$q' = 0.35/(0.45 + 0.35) = 0.44$$

Alternatively, the formulas to use are

$$p' = p\frac{pW_{AA} + qW_{Aa}}{\overline{W}}$$

and

$$\overline{W} = p^2 W_{AA} + 2pq W_{Aa} + q^2 W_{aa}$$

$$p' = (0.5)[(0.5)(1.0) + (0.5)(0.8)]/[(0.25)(1.0) + (0.5)(0.8) + (0.25)(0.6)]$$
$$= (0.5)(0.9)/(0.8) = 0.56$$

12. a. Assuming the population is in Hardy-Weinberg equilibrium and that the allelic frequency is the same in both sexes, we can directly calculate the frequency of the colorblind allele as $q = 0.1$. (Because this trait is sex-linked, q is equal to the frequency of affected males.) Colorblind females must be homozygous for this X-linked recessive trait, so their frequency in the population is equal to $q^2 = 0.01$.

b. There would be 10 colorblind men for every colorblind woman (q/q^2).

c. For this condition to be true, the mothers must be heterozygous for the trait and the fathers must be colorblind ($X^C/X^c \times X^c/Y$). The frequency of heterozygous women in the population will be $2pq$, and the frequency of colorblind men will be q. Therefore, the frequency of such random marriages will be $(2pq)(q) = 0.018$.

d. All children will be phenotypically normal only if the mother is homozygous for the noncolorblind allele ($p^2 = 0.81$). The father's genotype does not matter and therefore can be ignored.

e. There are several ways of approaching this problem. One way to visualize the data, however, is to construct the following

Mother \ Father	$0.4\ X^C$	$0.6\ X^c$	Y
$0.8\ X^C$	$0.32\ X^C/X^C$	$0.48\ X^C/X^c$	$0.8\ X^C/Y$
$0.2\ X^c$	$0.08\ X^C/X^c$	$0.12\ X^c/X^c$	$0.2\ X^c/Y$

As can be seen, the frequency of colorblind females will be 0.12 and colorblind males 0.2.

f. From analysis of the results in (e), the frequency of the colorblind allele will be 0.2 in males (the same as in the females of the previous generation) and $^1/_2(0.08 + 0.48) + 0.12 = 0.4$ in females.

13. Assume that proper function results from the right gene products in the proper ratio to all other gene products. A mutation will change the gene product, eliminate the gene product, or change the ratio of it to all other gene products. All three outcomes upset a previously balanced system. While a new and "better" balance may be achieved, this is less likely than being deleterious.

14. Wild-type alleles are usually dominant because most mutations result in lowered or eliminated function. To be dominant, the heterozygote has approximately the same phenotype as the dominant homozygote. This will typically be true when the wild-type allele produces a product and the mutant allele does not.

Chromosomal rearrangements are often dominant mutations because they can cause gross changes in gene regulation or even cause fusions of several gene products. Novel activities, overproduction of gene products, etc., are typical of dominant mutations.

15. Prior to migration, $q^A = 0.1$ and $q^B = 0.3$ in the two populations. Because the two populations are equal in number, immediately after migration, $q^{A + B} = {}^1/_2(q^A + q^B) = {}^1/_2(0.1 + 0.3) = 0.2$. At the new equilibrium, the frequency of affected males is $q = 0.2$, and the frequency of affected females is $q^2 = (0.2)^2 = 0.04$. (Colorblindness is an X-linked trait.)

16. For a population in equilibrium, the probability of individuals being homozygous for a recessive allele is q^2. Thus for small values of q, few individuals in a randomly mating population will express the trait. However, if two individuals share a close common ancestor, there is an increased chance of homozygosity by descent, because only one "progenitor" need be heterozygous.

For the following, it is assumed that the allele in question is rare. Thus the chance of both "progenitors" being heterozygous will be ignored.

a. For a parent-sib mating, the pedigree can be represented as follows:

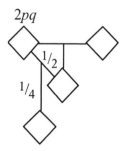

In this example, it is only the chance of the incestuous parent's being heterozygous that matters. Thus, the chance of the descendant's being homozygous is

$$2pq(^1/_2)(^1/_4) = pq/4$$

If q is very small, then p is nearly 1.0 and the chance of an affected child can be represented as approximately $q/4$. (Again, this should be compared to the expected random-mating frequency of q^2.)

b. For a mating of first cousins, the pedigree can be represented as follows:

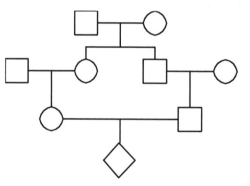

The probability of inheriting the recessive allele if *either* grandparent is heterozygous can be represented as follows:

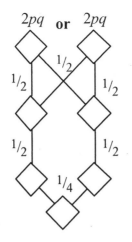

Thus the chance of this child's being affected is

$$2pq \ (^1/_2)(^1/_2)(^1/_2)(^1/_2)(^1/_4) + 2pq \ (^1/_2)(^1/_2)(^1/_2)(^1/_2)(^1/_4) = {}^{pq}/_{16}$$

Again, if q is rare, p is nearly 1.0, so the chance of homozygosity by descent is approximately $^q/_{16}$.

c. An aunt-nephew (or uncle-niece) mating can be represented as:

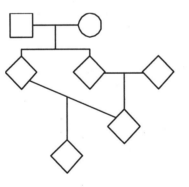

Following the possible inheritance of the recessive allele from either grandparent,

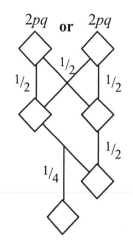

The chance of this child's being homozygous is

$$2pq(^1/_2)(^1/_2)(^1/_2)(^1/_4) + 2pq(^1/_2)(^1/_2)(^1/_2)(^1/_4) = pq/_8,$$

or for rare alleles approximately $q/_8$.

17. The allele frequencies are
$$f(A) = 0.2 + ^1/_2(0.60) = 50\%$$
$$f(a) = ^1/_2(0.60) + 0.2 = 50\%$$

Positive assortative mating: the alleles will randomly unite within the same phenotype. For *A/–*, the mating population is 0.2 *A/A* + 0.6 *A/a*. The allelic frequencies within this subpopulation are

$$f(A) = [0.2 + ^1/_2(0.6)]/0.8 = 0.625$$
$$f(a) = ^1/_2(0.6)/0.8 = 0.375$$

The phenotypic frequencies that result are

A/–: $p^2 + 2pq = (0.625)^2 + 2(0.625)(0.375) = 0.39 + 0.47 = 0.86$
a/a: $q^2 = (0.375)^2 = 0.14$

However, assuming that all contribute equally to the next generation and this subpopulation represents 0.8 of the total population, these figures must be adjusted to reflect this weighting:

A/–: $(0.86)(0.8) = 0.69$
a/a: $(0.14)(0.8) = 0.11$

The *a/a* contribution from the other subpopulation will remain unchanged because there is only one genotype, *a/a*. Their weighted contribution to the total phenotypic frequency is 0.20. Therefore, after one generation, the phenotypic frequencies will be $A/- = 0.69$ and $a/a = 0.20 + 0.11 = 0.31$, and the genotypic frequencies will be $f(A) = 0.5$ and $f(a) = 0.5$. Over time, these allelic frequencies will stay the same, but the frequency of heterozygotes will continue to decrease until there are two separate populations, A/A and a/a, which will not interbreed. *Negative assortative mating*: mating is between unlike phenotypes. The two types of progeny will be A/a and a/a. A/A will not exist. A/a will result from all $A/A \times a/a$ matings and half the $A/a \times a/a$ matings. These matings will occur with the following relative frequencies

$$A/A \times a/a = (0.2)(0.2) = 0.04$$
$$A/a \times a/a = (0.6)(0.2) = 0.12$$

Because these are the only matings that will occur, they must be put on a 100 percent basis by dividing by the total frequency of matings that occur:

$A/A \times a/a$:   0.04/0.16 = 0.25, all of which will be A/a
$A/a \times a/a$:   0.12/0.16 = 0.75, half A/a and half a/a

The phenotypic frequencies in this generation will be

A/a: 0.25 + 0.75/2 = 0.625
a/a: 0.75/2 = 0.375

In the next generation, because all matings are now between heterozygotes and homozygous recessives, the final allelic frequencies of $f(A) = 0.25$ and $f(a) = 0.75$ will be obtained and the population will be 50 percent A/a and 50 percent a/a.

18. Many genes affect bristle number in *Drosophila*. The artificial selection resulted in lines with mostly high-bristle-number alleles. Some mutations may have occurred during the 20 generations of selective breeding, but most of the response was due to alleles present in the original population. Assortment and recombination generated lines with more high-bristle-number alleles.

Fixation of some alleles causing high bristle number would prevent complete reversal. Some high-bristle-number alleles would have no negative effects on fitness, so there would be no force pushing bristle number back down because of those loci.

The low fertility in the high-bristle-number line could have been due to pleiotropy or linkage. Some alleles that caused high bristle number may also have caused low fertility (pleiotropy). Chromosomes with high-bristle-number alleles may also carry alleles at different loci that caused low fertility (linkage). After artificial selection was relaxed, the low-fertility alleles would have been selected against by natural selection. A few generations of relaxed selection would have allowed low-fertility-linked alleles to recombine away, producing high-bristle-number chromosomes that did not contain low-fertility alleles. When selection was reapplied, the low-fertility alleles had been reduced in frequency or separated from the high-bristle loci, so this time there was much less of a fertility problem.

19.   Affected individuals $= B/b = 2pq = 4 \times 10^{-6}$. Because q is almost equal to 1.0, $2p = 4 \times 10^{-6}$. Therefore, $p = 2 \times 10^{-6}$.

$$\mu = hsp = (1.0)(0.7)(2 \times 10^{-6}) = 1.4 \times 10^{-6}$$

where h = degree of dominance of the deleterious allele

20.   The probability of not getting a recessive lethal genotype for one gene is $1 - 1/8 = 7/8$. If there are n lethal genes, the probability of not being homozygous for any of them is $(7/8)^n = 13/31$. Solving for n, an average of 6.5 recessive lethals are predicted. If the actual percentage of "normal" children is less owing to missed in utero fatalities, the average number of recessive lethals would be higher.

21. **a.**   The formula needed is

$$\hat{q} = \sqrt{\mu/_s}$$
$$= 4.47 \times 10^{-3}$$

   so,   Genetic cost $= sq^2 = 0.5(4.47 \times 10^{-3})^2 = 10^{-5}$

   b.   Using the same formulas as part a,

$$\hat{q} = 6.32 \times 10^{-3}$$

Genetic cost $= sq^2 = 0.5(6.32 \times 10^{-3})^2 = 2 \times 10^{-5}$

   c. $\hat{q} = 5.77 \times 10^{-3}$

Genetic cost $= sq^2 = 0.3(5.77 \times 10^{-3})^2 = 10^{-5}$

20

Quantitative Genetics

BASIC PROBLEMS

1. There are many traits that vary more or less continuously over a wide range. For example, height, weight, shape, color, reproductive rate, metabolic activity, etc., vary quantitatively rather than qualitatively. Continuous variation can often be represented by a bell-shaped curve, where the "average" phenotype is more common than the extremes. Discontinuous variation describes the easily classifiable, discrete phenotypes of simple Mendelian genetics: seed shape, auxotrophic mutants, sickle-cell anemia, etc. These traits show a simple relationship between genotype and phenotype.

2. The mean (or average) is calculated by dividing the sum of all measurements by the total number of measurements, or in this case, the total number of bristles divided by the number of individuals.

$$\text{mean} = \bar{x} = \frac{[1 + 4(2) + 7(3) + 31(4) + 56(5) + 17(6) + 4(7)]}{(1 + 4 + 7 + 31 + 56 + 17 + 4)}$$

$$= {}^{564}/_{120} = 4.7 \text{ average number of bristle/individual}$$

The variance is useful for studying the distribution of measurements around the mean and is defined in this example as

$$\text{variance} = s^2 = \text{average of the (actual bristle count} - \text{mean})^2$$

$$s^2 = {}^1/_N \Sigma (x_i - \bar{x})^2$$
$$= {}^1/_{120} \Sigma[(1 - 4.7)^2 + (2 - 4.7)^2 + (3 - 4.7)^2 + (4 - 4.7)^2 + (5 - 4.7)^2 +$$
$$(6 - 4.7)^2 + (7 - 4.7)^2]$$
$$= 0.26$$

The standard deviation, another measurement of the distribution, is simply calculated as the square root of the variance.

$$\text{standard deviation} = s = \sqrt{0.26} = 0.51$$

3. **a.** H^2 has meaning only with respect to the population that was studied in the environment in which it was studied. Even if a trait shows high heritability, it does not imply the trait is unaffected by its environment. The only acceptable analysis is to study directly the norms of reaction of the various genotypes in the population over the range of projected environments. Because it is so difficult to fully replicate a human genotype so that it might be tested in different environments, there is no known norm of reaction for any human quantitative trait.

    b. Neither H^2 nor h^2 is a reliable measure that can be used to generalize from a particular sample to a "universe" of the human population. They certainly should not be used in social decision making (as implied by the terms eugenics and dysgenics).

    c. Again, H^2 and h^2 are not reliable measures, and they should not be used in any decision making with regard to social problems.

4. The following are unknown: (1) norms of reaction for the genotypes affecting IQ, (2) the environmental distribution in which the individuals developed, and (3) the genotypic distributions in the populations. Even if the above were known, because heritability is specific to a specific population and its environment, the difference between two different populations cannot be given a value of heritability.

CHALLENGING PROBLEMS

5. **a.** Broad heritability measures that portion of the total variance that is due to genetic variance. The equation to use is

    $H^2 = $ the genetic variance/phenotypic variance

    where genetic variance = phenotypic variance – environmental variance

    $$H^2 = \frac{s_p^2 - s_e^2}{s_p^2}$$

    Narrow heritability measures that portion of the total variance that is due to the additive genetic variation. The equation to use is

$$h^2 = \frac{\text{additive genetic variance}}{\text{additive genetic variance} + \text{dominance variance} + \text{environmental variance}}$$

$$h^2 = \frac{s_a^2}{s_a^2 + s_d^2 + s_e^2}$$

Shank length:

$H^2 = (310.2 - 248.1)/(310.2) = 0.200$

$h^2 = 46.5/(46.5 + 15.6 + 248.1) = 0.150$

Neck length:

$H^2 = (730.4 - 292.2/(730.4) = 0.600$

$h^2 = 73.0/(73.0 + 365.2 + 292.2) = 0.010$

Fat content:

$H^2 = (106.0 - 53.0)/(106.0) = 0.500$

$h^2 = 42.4/(42.4 + 10.6 + 53.0) = 0.400$

b. The larger the value of h^2, the greater the difference between selected parents and the population as a whole, and the more that characteristic will respond to selection. Therefore, fat content would respond best to selection.

c. The formula needed is

selection response $= h^2 \times$ selection differential

Therefore, selection response $= (0.400)(10.5\% - 6.5\%) = 1.6\%$ decrease in fat content, or 8.9% fat content.

6. **a.** The probability of any gene being homozygous is $1/2$ (e.g., for A: A/A or a/a), and the probability of being heterozygous (or not homozygous) is also $1/2$. Thus, the probability for any one gene being homozygous while the other two are heterozygous is $(1/2)^3$. Because there are three ways for this to happen (homozygosity at A or at B or at C), the total probability is

p(homozygous at 1 locus) $= 3(1/2)^3 = 3/8$

The same logic can be applied to any two genes being homozygous

p(homozygous at 2 loci) $= 3(1/2)^3 = 3/8$

There are two ways for all three genes to be homozygous, so

$$p(\text{homozygous at 3 loci}) = 2(^{1}/_{2})^{3} = {}^{2}/_{8}$$

b. $p(\text{0 capital letters}) = p(\text{all homozygous recessive}) = (^{1}/_{4})^{3} = {}^{1}/_{64}$

$p(\text{1 capital letter}) = p(\text{1 heterozygote and 2 homozygous recessive})$
$= 3(^{1}/_{2})(^{1}/_{4})(^{1}/_{4}) = {}^{3}/_{32}$

$p(\text{2 capital letters}) = p(\text{1 homozygous dominant and 2 homozygous recessive})$

or

$p(\text{2 heterozygotes and 1 homozygous recessive})$

$= 3(^{1}/_{4})^{3} + 3(^{1}/_{4})(^{1}/_{2})^{2} = {}^{15}/_{64}$

$p(\text{3 capital letters}) = p(\text{all heterozygous})$

or

$p(\text{1 homozygous dominant, 1 heterozygous, and 1 homozygous recessive})$

$= (^{1}/_{2})^{3} + 6(^{1}/_{4})(^{1}/_{2})(^{1}/_{4}) = {}^{10}/_{32}$

$p(\text{4 capital letters}) = p(\text{2 homozygous dominant and 1 homozygous recessive})$

or

$p(\text{1 homozygous dominant and 2 heterozygous})$

$= 3(^{1}/_{4})^{3} + 3(^{1}/_{4})(^{1}/_{2})^{2} = {}^{15}/_{64}$

$p(\text{5 capital letters}) = p(\text{2 homozygous dominant and 1 heterozygote})$

$= 3(^{1}/_{4})^{2}(^{1}/_{2}) = {}^{3}/_{32}$

$p(\text{6 capital letters}) = p(\text{all homozygous dominant}) = (^{1}/_{4})^{3} = {}^{1}/_{64}$

7. For three genes there are a total of 27 genotypes that will occur in predictable proportions. For example, there are three genotypes that have two genes that are heterozygous and one gene that is homozygous recessive (*A/a* ; *B/b* ; *c/c*, *A/a* ; *b/b* ; *C/c*, *a/a* ; *B/b* ; *C/c*). The frequency of this combination is $3(^{1}/_{2})(^{1}/_{2})(^{1}/_{4}) = {}^{3}/_{16}$, and the phenotypic score is $3 + 3 + 1 = 7$. For all the genotypes possible, the total distribution of phenotypic scores is as follows:

Score	Proportion
3	$1/64$
5	$3/32$
6	$3/64$
7	$3/16$
8	$3/16$
9	$11/64$
10	$3/16$
11	$3/32$
12	$1/64$

And the plot of these data will be

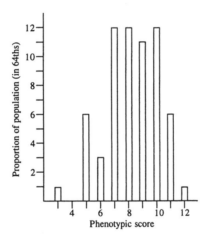

8. The population described would be distributed as follows:

3 bristles	$19/64$
2 bristles	$44/64$
1 bristle	$1/64$

The 3-bristle class would contain seven different genotypes, the 2-bristle class would contain 19 different genotypes, and the 1-bristle class would contain only one genotype. It would be very difficult to determine the underlying genetic situation by doing controlled crosses and determining progeny frequencies.

9.   a.   Solving the formula for values of x over the stated range for each genotype gives the following data:

x	1	2	3
0.03	0.90		
0.04	0.91		
0.05	0.93		
0.06	0.93		
0.07	0.94		
0.08	0.95		
0.09	0.96		
0.10	0.97		0.90
0.11	0.97		0.92
0.12	0.98	0.90	0.93
0.13	0.98	0.92	0.94
0.14	0.99	0.94	0.95
0.15	0.99	0.95	0.96
0.16	0.99	0.96	0.97
0.17	1.00	0.98	0.98
0.18	1.00	0.98	0.98
0.19	1.00	0.99	0.99
0.20	1.00	1.00	0.99
0.21	1.00	1.00	1.00
0.22	1.00	1.00	1.00
0.23	1.00	1.00	1.00
0.24	0.99	1.00	1.00
0.25	0.99		1.00
0.26	0.99		1.00
0.27	0.98		1.00
0.28	0.98		0.99
0.29	0.97		0.99
0.30	0.97		0.98
0.31	0.96		0.98
0.32	0.95		0.97
0.33	0.94		0.96
0.34	0.93		0.95
0.35	0.93		0.94
0.36	0.91		0.93
0.37	0.90		0.92
0.38			0.90

Plotting these data give the following curves

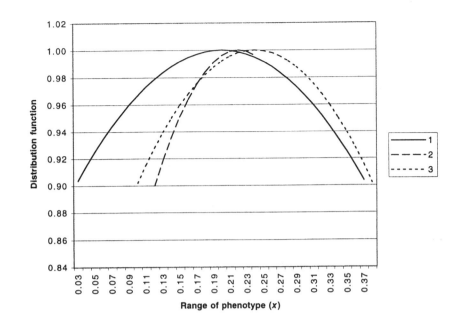

b. Because the three genotypes are equally frequent, the average distribution across the entire range of phenotypes will be

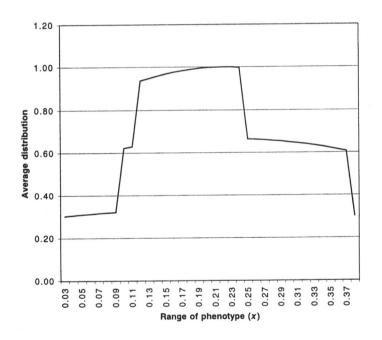

There are regions of the overall phenotypic distribution where the variation within a given genotype does not overlap, and this gives sharp steps to the distribution. On the other hand, any individual whose phenotype lies between values of 0.12 to 0.24 could have any of the three genotypes.

10. **a.**

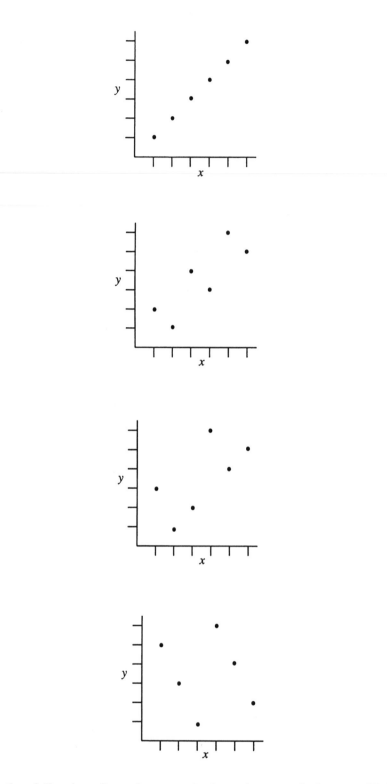

b.

c.

d.

Use the following formulas to calculate the correlation coefficient (r_{xy}) between x and y:

$$r_{xy} = \frac{\text{cov } xy}{s_x s_y}$$

and

$$\text{cov } xy = {}^{1}\!/_{N} \Sigma \, x_i \, y_i - \overline{xy}$$

a. $\text{cov } xy = {}^{1}\!/_{6} \, [(1)(1) + (2)(2) + (3)(3) + (4)(4) + (5)(5) + (6)(6)]$
$- ({}^{21}\!/_{6})({}^{21}\!/_{6}) = 2.92$

Standard deviation $x = s_x = \sqrt{{}^{1}\!/_{N} \Sigma \, (x_i - \bar{x})^2} = 1.71$

Standard deviation $y = s_y = \sqrt{{}^{1}\!/_{N} \Sigma \, (y_i - \bar{y})^2} = 1.71$

Therefore, $r_{xy} = 2.92/(1.71)(1.71) = 1.0$. The other correlation coefficients are calculated in a like manner.

b. 0.83

c. 0.66

d. −0.20

11. First, define alcoholism in behavioral terms. Next, realize that all observations must be limited to the behavior you used in the definition and all conclusions from your observations are applicable only to that behavior. In order to do your data gathering, you must work with a population in which familiality is distinguished from heritability. In practical terms, this means using individuals who are genetically close but who are found in all environments possible.

12. Before beginning, it is necessary to understand the data. The first entry, h/h h/h, refers to the II and III chromosomes, respectively. Thus, there are four h (high bristle number) sets of alleles in two or more genes on separate chromosomes. The next entry is h/l h/h. Chromosome II is now heterozygous and chromosome III is still homozygous, etc.

The effect of substituting one l chromosome II for an h chromosome II, and therefore going from homozygous h/h to heterozygous h/l, can be seen in the differences along the rows in the first two columns. The average change is (2.9 + 3.1 + 2.7)/3 = 2.9. When chromosome II goes from heterozygous h/l to homozygous l/l, the average change is (3.2 + 5.2 + 6.8)/3 = 5.1.

The effect of substituting one l chromosome III for an h chromosome III can be seen in the differences between rows: 25.1 − 23.0 = 2. 1; 22.2 − 19.9 = 2.3; and 19.0 − 14.7 = 4.3 (average change 2.9). And going from h/l to l/l for chromosome III gives an average change of (11.2 + 10.8 + 12.4)/3 = 11.5 bristles.

Here is a summary of these results:

	Chromosome II	Chromosome III
h/h to h/l	2.9	2.9
h/l to l/l	5.1	11.5

Each set of alleles for both chromosomes is expressed in the phenotype, but that expression varies with the chromosome. Chromosome III appears to have a stronger effect on the phenotype than does chromosome II. (Compare h/h h/h with both l/l h/h and h/h l/l. The difference in the first case is 6.1, and in the second case, 13.3.) Finally, there is partial dominance of h over l for both chromosomes. The change from h/h to h/l is less than the change from h/l to l/l.

13. **a.**

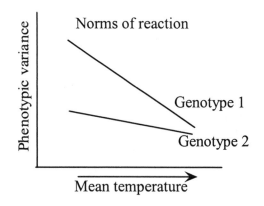

b. Broad heritability is defined as

$$H^2 = \frac{s_g^2}{s_p^2}$$

Assuming that the genetic variance stays the same, phenotypic variance must decrease if H^2 increases. Therefore, the same plot as (a) will satisfy the conditions.

c. To satisfy the conditions that genetic variance increases as H^2 decreases, phenotypic variance must also increase. Therefore the plot will be

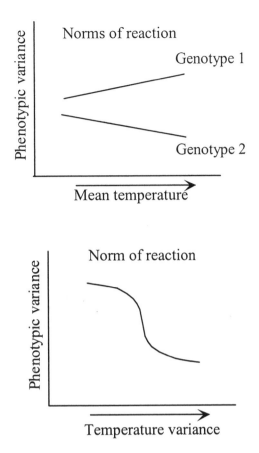

d.

14. **a.** The regression line shows the relationship between the two variables. It attempts to predict one (the son's height) from the other (the father's height). If the relationship is perfectly correlated, the slope of the regression line should approximate 1. If you assume that individuals at the extreme of any spectrum are homozygous for the genes responsible for these phenotypes, then their offspring are more likely to be heterozygous than are the original individuals. That is, they will be less extreme. Also, there is no attempt to include the maternal contribution to this phenotype.

b. For Galton's data, regression is an estimate of heritability (h^2), *assuming* that there were few environmental differences between all fathers and all sons both, individually and as a group. However, there is no evidence given to determine if the traits are familial but not heritable. This data would indicate genetic variation only if the relatives do not share common environments more than nonrelatives do.

21

Evolutionary Genetics

BASIC PROBLEMS

1. A transformational scheme of evolution predicts that all members of a species will change over time. By sharing similar environments and life experiences, each member is altered in its lifetime and its progeny inherit these alterations. For example, if all giraffes stretch their necks during their lifetimes to reach the ever higher foliage, all offspring will inherit this acquired "stretched" neck and begin their lives where their parents have left off. Over time, longer- and longer-necked giraffes would evolve.

   A variational scheme of evolution predicts that not all members of a species contribute equally to future generations. Heritable differences among members of the same species result in some being more "fit" and able to produce more offspring than others. Over time, one type of individual is replaced by another. For example, using an economic metaphor, during the 1990s, there was an explosion of internet and "dot.com" companies. A variation scheme of evolution would have suggested that some of these companies would grow, evolve, and prosper while others, based on intrinsic (heritable) differences (management, capitalization, etc.) would become extinct—a prediction that certainly came true.

2. The three principles are: (1) organisms within a species vary from one another, (2) the variation is heritable, and (3) different types leave different numbers of offspring in future generations.

3. The variation required by Darwin's proposed mechanism of evolution can not exist or be maintained if "blending" or nonrandom segregation occurs. In either case, populations will become homogeneous and variation would be rapidly lost. The "particulate" nature of genes described by Mendel allows for segregation of traits generation after generation. In this way, even currently detrimental (and recessive) alleles can be retested as the environment and circumstances change.

4.   A geographical race is a population that is genetically distinguishable from other local populations but is capable of exchanging genes with those other local populations.

A species is a group of organisms that exchanges genes within the group but cannot do so with other groups.

Populations that are geographically separated will diverge from each other as a consequence of a combination of unique mutations, selection, and genetic drift. For populations to diverge enough to become reproductively isolated, spatial separation sufficient to prevent any effective migration is usually necessary.

CHALLENGING PROBLEMS

5.   A population will not differentiate from other populations by local inbreeding if

$$\mu \geq 1/N$$

so

$$N \geq 1/\mu$$
$$N \geq 10^5$$

6.   The rate of loss of heterozygosity in a closed population is $1/(2N)$ per generation.

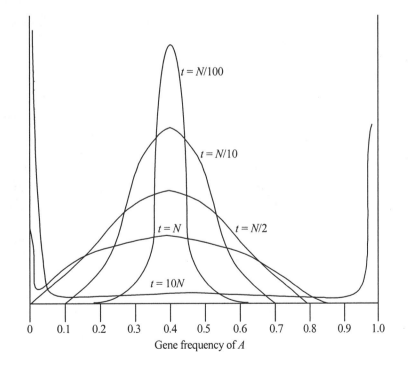

7. A population will not differentiate from other populations by local inbreeding if

   the number of migrant individuals ≥ 1 per generation

   For (a), migration is not sufficient to prevent local inbreeding so the results are roughly the same as seen in problem 6. In the case of (b), there is one migrant per generation so the populations will not differentiate and allelic frequencies will remain the same in all populations.

8. The mean fitness is calculated by summing the frequency of each progeny class times its fitness. For example, the frequency of $A/A \cdot B/B$ is $p(A)^2 \times p(B)^2$ or $(0.64)(0.81) = 0.52$. This frequency is multiplied by its fitness to give $(0.52 \times 0.95) = 0.49$. Summing for all classes, the mean fitness of population 1 $[p(A) = 0.8, p(B) = 0.9]$ is 0.92. Because selection acts to increase the mean fitness, the frequency of both A and B should increase in the next generation (the $A/A \cdot BB$ class has a fitness of 0.95).

   The mean fitness of population 2 $[p(A) = 0.2, p(B) = 0.2]$ is 0.73. Again, the frequency of both A and B should increase.

   There is a single adaptive peak at $A/A \cdot B/B$. By inspection, the fitness is lowest at $a/a \cdot b/b$ and highest at $A/A \cdot B/B$. The allelic frequencies at the peak is 1.0 for both A and B.

9. The mean fitness for population 1 $[p(A) = 0.5, p(B) = 0.5]$ is 0.825. The mean fitness for population 2 $[p(A) = 0.1, p(B) = 0.1]$ is 0.856. There are four adaptive peaks: at $p(A) = 0.0$ or 1.0 and $p(B) = 0.0$ or 1.0. (The mean fitness will be 0.90 and any of these points.) With population 1, the direction of change for both $p(A)$ and $p(B)$ will be random. Both higher or lower frequencies of either allele can result in increased mean fitness (although there are some combinations that would lower fitness). Because population 2 is already near an adaptive peak of $p(A) = 0.0$ and $p(B) = 0.0$, both $p(A)$ and $p(B)$ should decrease to increase the mean fitness.

10. The text discusses the frequency distribution of haploid chromosome numbers among dicots. Above a chromosome number of 12, even numbers are much more common than odd numbers. This is evidence of frequent polyploidization during plant evolution. For example, if a specie of plant with an odd haploid chromosome number undergoes a "doubling" event, the chromosome number becomes even.

11. The α and β gene families show remarkable amino acid sequence similarities (see table in the companion text). Within each gene family, sequence similarities are greater and in some cases, member genes have identical intron-exon structure.

12. All human populations have high *i*, intermediate I^A, and low I^B frequencies. The variations that do exist among the different geographical populations is most likely due to genetic drift. There is no evidence that selection plays any role regarding these alleles.

13. To test the species distinctness, it is necessary to be able to manipulate and culture *D. pseudoobscura* and *D. persimilis* in captivity. If they cannot be cultured in the laboratory, their species distinctness cannot be established. The mating behavior compatibility of the different *Drosophila* can be tested by placing a mixture of males of both forms with females of one of the forms to see whether there are any female mating preferences. The same experiment can then be repeated with mixed females and one sort of male and with mixtures of males and females of both forms. From such experiments, patterns of mating preference can be observed. Even if there is some small amount of mating of different forms, this may occur only because of the unnatural conditions in which the test is being carried out. On the other hand, no mating of any kind may occur, even between the same forms, because the necessary cues for mating are missing, in which case nothing can be concluded.

If matings between different forms do occur, the survivorship of the interpopulation hybrids can be compared with that of the intrapopulation matings. If hybrids survive, their fertility can be tested by attempting to back-cross them to the two different parental strains. As in the case of the mating tests, under the unnatural conditions of the laboratory, some survivorship or fertility of species hybrids is possible even though the isolation in nature is complete. Any clear reduction in observed survivorship or fertility of the hybrids is strong presumptive evidence that they belong to different species.

14.

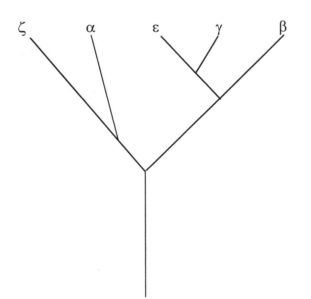

15. For polymorphic sites within a species, let nonsynonymous = a and synonymous = b. For polymorphic sites between the species, let nonsynonymous = c and synonymous = d. If divergence is due to neutral evolution, then

$$a/b = c/d$$

If divergence is due to selection, then

$$a/b < c/d$$

However, in this example, $a/b = 20/50 > c/d = 2/18$, which fits neither expectation.

Because the ratio of nonsynonymous to synonymous polymorphisms (a/b) is relatively high, the gene being studied may encode a protein tolerant of substitution (like fibrinopeptides, dscussed in the companion text). The relatively fewer species differences may suggest that speciation was a recent event. so few polymorphisms have been fixed in one species that are not variants in the other.

Interactive Genetics

Mendelian Analysis

Goals for Mendelian Analysis:

1. Describe the mode of inheritance of a phenotypic difference between two strains:
   - Distinguish dominance and recessiveness in traits
   - Determine whether the phenotype difference is due to a single gene with two alleles
   - Write genotypes and predict phenotypes
   - Predict the number of possible gamete types, their kinds and ratios
2. Distinguish self-crosses from test crosses.
3. Apply these concepts to simple human pedigrees for select traits:
   - Calculate probabilities using product and sum rules
   - Use the binomial expansion to calculate the combinations of possible outcomes

Mendelian Problems
Problem 1

Let's begin with a simple cross between two pure breeding peas, a purple flowered pea and a white flowered pea. How many traits or characters are different between the purple and white parents?

In this cross the pollen from the purple parent is used to fertilize the ovules of the white parent. How many genetically distinct gametes are produced from the purple parent?

How many genetically distinct gametes does the white flowered parent produce?

All the progeny of the parental cross are purple flowered pea plants. However, when these peas are self-crossed, both purple and white peas appear in the next or F_2 generation. Which phenotype is the dominant phenotype?

Using A and a to designate the different forms of the inherited trait (alleles), assign genotypes to the three generations of the cross described above. How many genetically distinct gametes are produced from the F_1? In the simple Punnett square shown below, fill in the genotype of the progeny using A/a.

gamete types

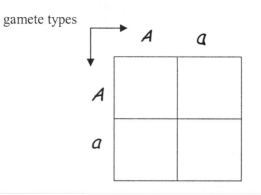

Indicate the phenotype expected for each genotype in the Punnett square above. What is the ratio of purple flowered progeny to white flowered progeny in the F_2?

What genotypic ratios would you predict?

Problem 2

The grass tree is the common name for the Australian genus *Xanthorrhoea* in the lily family, *Liliaceae*. Grass trees grow to a height of about 4.5m (15 feet) and bear long, narrow leaves in a tuft at the top of the trunk. White flowers or yellow flowers can be produced in a dense spike above the leaves in a given tree. A tree producing yellow flowers was self-crossed. The seeds were collected and planted to determine flower color of each progeny tree. Twenty-eight trees were found to produce yellow flowers and eight trees were found to produce white flowers.

What is the expected genetic (phenotypic) ratio of yellow flowering grass trees to white flowering grass trees among the progeny?

Which phenotype is dominant, white or yellow flowering trees?

Using the symbols of *Y/y*, write the genotypes of the original yellow flowering tree and the progeny.

Yellow × Yellow

Yellow   White

Predict the phenotypic ratios for the crosses shown below:

Cross	Phenotypic Ratios	
	Yellow	White
White tree × white tree		
Original yellow tree × white tree		
Pure breeding yellow tree × white tree		
Original yellow tree self-crossed		

Problem 3

Two pure breeding strains of peas, one giving wrinkled, yellow seeds and the other round, green seeds, were crossed and all the resulting peas were round and yellow. These round seeds were then planted and the flowers self-fertilized. The peas produced from this selfing contained four phenotypic classes.

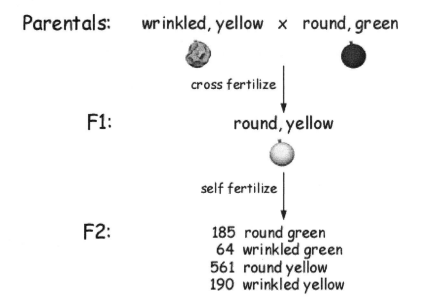

Parentals: wrinkled, yellow × round, green

cross fertilize

F1:   round, yellow

self fertilize

F2:   185 round green
             64 wrinkled green
            561 round yellow
            190 wrinkled yellow

How many traits or characters are different between the pure breeding parents?

Using A and a to designate pea shape and B and b to designate pea color, assign genotypes to the three generations in the diagram of the cross above.

How many genetically distinct gametes are produced from the F_1?
Fill out the Punnett square shown below.

gamete types

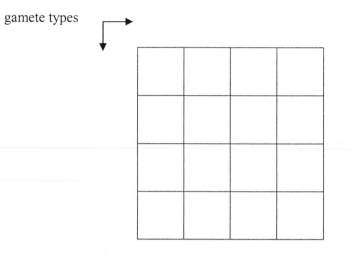

How many genotypes above correspond to the round and yellow phenotype?

How many genotypes above correspond to the wrinkled yellow phenotype?

How many genotypes above correspond to the round green phenotype?

How many genotypes above give rise to a green wrinkled pea?

Using the Punnett square above, what is the ratio of round peas to wrinkled peas (ignore the color)?

What is the ratio for yellow versus green?

Determine the ratio of round yellow peas to round green peas to wrinkled yellow peas to wrinkled green peas expected from this dihybrid cross.

Problem 4

You have been sent to Planet TATACCA to conduct basic genetic experiments on a new species: **X**. The two traits that you are observing are the shape and attachment of the ear. Pointy ears are dominant to round, and attached ear lobes are dominant to detached.

Assuming that genes responsible for these phenotypes assort independently, write out the genotype of the parents in each of the crosses. Assume homozygosity unless there is evidence otherwise. *P* and *R* stand for the pointy and rounded phenotypes, respectively, and *A* and *D* stand for the attached and detached phenotypes. Use *P* and *p* for the pointy and round ear alleles and *A* and *a* for the attached and detached.

Parental Genotype	Parental Phenotype	Number of Progeny			
		PA	*PD*	*RA*	*RD*
	PA × PA	447	152	147	50
	PA × PD	107	98	0	0
	PA × RA	95	0	91	0
	RA × RA	0	0	153	48
	PD × PD	0	149	0	51
	PA × PA	140	46	0	0
	PA × PD	151	147	48	50

Problem 5

Geneticists working with the common fruit fly, *Drosophila melanogaster*, are studying the mode of inheritance of vestigial wings. Vestigial wings are a mutant phenotype in which wing development is curtailed. Wild-type or normal flies display normal wing development. Geneticists working with flies label genes and the alleles, and therefore the flies, after the mutant phenotype. Pure breeding vestigial flies have the genotype of *vgvg* (two copies of the mutant allele) and the phenotype is denoted as *vg* for vestigial. Pure breeding normal flies are denoted vg^+vg^+ for their genotype and their phenotype is described as vg^+.

After sorting flies for several hours, a freshman volunteering in the laboratory accidentally sneezes, mixing up all three piles of flies. Originally, one pile of flies displayed vestigial wings, a second pile of flies had normal wings but was heterozygous, and the third pile had normal wings and was homozygous. The principal investigator of the laboratory calls on you, as an expert geneticist, to resort the flies. By performing several crosses, you are able to determine from which pile each fly originated.

Which phenotype is recessive, vestigial wings or wild-type wings? (See the table below).

Using the symbols of *vg+* and *vg*, write the genotypes of the flies used in each cross in the table below.

Genotype	Phenotype	Number of Progeny	
		vg+	*vg*
	normal × normal	628	209
	vestigial × normal	333	340
	normal × vestigial	454	0
	vestigial × vestigial	0	121
	normal × normal	92	0

Problem 6

One day as you were traveling along a semi-deserted highway, your car has an electrical surge and suddenly stops. You can't get your car to run again so you decide to get out and walk to the nearest phone to call for a tow truck. While you are walking you see a dairy farm in the distance. As you get closer you notice that these are not ordinary cattle. Then, you see the signs indicating a government research facility. Because you are extremely interested in the genetics of these odd cows, you take a few with you as you run off. Amazingly, you and your new bovines have made it to your family's farm safely. You decide to breed the cattle to obtain a better understanding of the genetic basis for their coat color. The results from your crosses are found on the table below.

Parental Genotypes	Parental Phenotypes	Green Progeny	Purple Progeny
	purple × purple	0	98
	green × green	94	0
	purple × green	51	49
	purple × green	97	0
	green × green	71	27

Which color results from possessing a dominant allele?

Using G and g as allele designations in the form of $Gg \times Gg$, assign the most probable genotypes for the parents in each cross above. Assume homozygosity unless there is evidence otherwise.

Besides having either a purple or a green coat, your cattle have two other odd traits that you have observed to be inherited in a Mendelian manner. Your new bovines have either yellow eyes (Y) or blue eyes (B) and produce either regular white milk (W) or chocolate milk (C). Consider the table below. Which is the recessive eye color?

What type of milk is produced when a cow possesses a dominant allele?

Parental Phenotypes	Blue Chocolate	Yellow Chocolate	Blue White	Yellow White
$BC \times BC$	10	4	3	1
$BW \times YC$	7	8	0	0
$BC \times BW$	4	0	3	0
$BW \times BW$	0	0	15	5
$YC \times YC$	0	22	0	7
$BC \times BC$	12	4	0	0
$BC \times YC$	10	11	3	4

Using *C* to indicate the allele for chocolate milk production, *c* for white milk production, *B* for blue eyes, and *b* to indicate the allele for yellow eyes, write the parental genotypes for each of the crosses shown above. Assume homozygosity unless there is evidence otherwise. What proportion of the offspring of two parental cattle each of the genotype *Gg Bb Cc* (green with blue eyes and chocolate milk producing) will be *Gg bb cc*?

Consider the following cross: *Gg Bb Cc × gg Bb cc*

What fraction of the progeny do you expect to phenotypically resemble the first parent?

If 100 progeny resulted from crosses such as this, how many would you expect to exhibit new genotypes (i.e., do not genotypically resemble either parent)?

Pedigree and Probabilities
Problem 1

Part a. For the pedigree shown below, state whether the condition depicted by darkened symbols is dominant or recessive. Assume the trait is rare. Assign genotypes for all individuals using *A/a* designations.

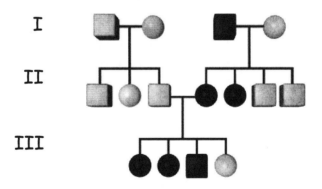

Part b. Assume the inherited trait depicted in the pedigree below is rare and state whether the condition is dominant or recessive. Assign genotypes for all individuals using *A/a* designations.

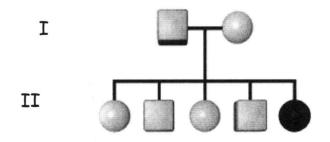

Part c. Is the pedigree shown below consistent with a dominant or recessive trait?

In analyzing this pedigree, would you conclude that this trait is caused by a rare or common allele?

Assign allele designations using *A/a*.

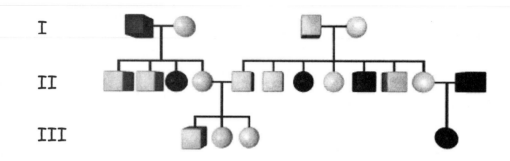

Problem 2

You have a single six-sided die. What is the chance of rolling a six?

What is the chance you will roll a five or a six? (HINT: Should you use the product rule or the sum rule to determine the probability of rolling one or the other?)

You are now given two additional dice. If you roll all three dice simultaneously, what is the chance of obtaining a five on all three dice?

Rolling your three dice again, what is the chance of obtaining no fives at all?

What is the chance of obtaining two fours and one three on any of the dice in a single roll?

What is the chance of obtaining the same number on all three dice?

What is the chance of rolling a different number on all three dice?

Problem 3

Preparing for a long night studying genetics, you open a big bag of candy-coated peanuts. Two friends are supposed to join you so, so you divide the bag into thirds. As you wait, you get hungry. But since you want to keep the bowls equal, you eat one from each bowl.

Bowl 1	Bowl 2	Bowl 3
30 blue candies	40 blue candies	30 blue candies
30 yellow candies	50 yellow candies	20 yellow candies
40 brown candies	10 green candies	50 red candies

What is the chance that you will select three blue candies?

What is the probability of selecting a brown, a green, and a red candy?

If you pick one candy from each bowl, what is the chance you will pick two blue candies and a yellow candy?

If you pick one candy from each bowl, what is the probability of obtaining no yellow candies?

If you pick one candy from each bowl, what is the chance of picking at least one yellow?

The probability of having two peanuts in a single candy is 2%. If you eat 300 candies, how many double peanuts do you expect to find?

Problem 4

The ability to taste the chemical phenylthiocarbamide (PTC) is an autosomal dominant phenotype. Lori, a taster woman, marries Russell, a taster man. Lori's father and her first child, Nicholas, are both nontasters.

What is the probability their second child will be a nontaster girl?

What is the probability their second child will be a taster boy?

What is the probability that their next two children will be nontaster girls?

Problem 5

A rare recessive allele inherited in a Mendelian manner causes phenylketonuria (PKU), which can lead to mental retardation if untreated. Fortunately, with a diet low in phenylalanine and tyrosine supplementation, normal development and lifespan are possible. Mike, a phenotypically normal man whose father had PKU marries Carol, a phenotypically normal woman whose brother had the disorder. The couple wants to have a large family and come to you for advice on the probability that their children will have PKU.

What is the probability that the couple's first child will have PKU?

Their first child, Greg, has PKU and the couple wants to have five additional children. What is the probability that out of their next five children only two will have PKU?

What is the possibility that they will have no more than one affected child in the five additional children they are planning?

Problem 6

John and Maggie are expecting a child. John's great grandmother (mother's lineage) and Maggie's brother have a rare autosomal recessive condition. What is the chance that their child will be affected?

John and Maggie have just discovered they are going to have twins. What is the chance that both twins will be affected if they are identical twins?

If the twins are dizygotic twins (non-identical or two-egg twins), what is the chance they will both be affected?

If they are dizygotic twins, what is the chance that at least one of them will be affected?

Chromosomal Theory of Inheritance
··

Goals for Chromosomal Inheritance:

1. Understand the classical evidence that genes are located on chromosomes:
    *   Similarity of behavior of chromosomes in divisions with the behavior of genes (alleles) in inheritance
    *   Identification of genes with sex-linked inheritance and chromosomes with sex-linked inheritance
2. Distinguish mitosis from meiosis figures, as well as the species diploid chromosome number, by inspection of simple diagrams with chromosome size, shape, and copy number as the cues.
3. Connect genetic inheritance with chromosome behavior during divisions:
    *   Assign alleles to chromosomes: sister chromatids have identical alleles, homologous chromosomes (in a heterozygote) have different alleles
    *   Homologous chromosomes pair and disjoin from each other in the first meiotic division, illustrating segregation of the alternate alleles
    *   Sister chromatids separate at mitosis, illustrating the constancy of the genotype in both daughter cells
    *   Genes that are on different chromosome pairs, i.e., non-homologous pairs, always assort independently
4. Be able to identify sex-linked inheritance in a pedigree.

Mitosis and Meiosis
Problem 1

Many cells undergo a continuous alternation between division and nondivision. The interval between each mitotic division is called interphase. Which stages (G1, S, G2 or M) make up interphase?

It was once thought that the biochemical activity during interphase was devoted solely to cell growth and specific functions the cell normally performs. However, it is now known that another biochemical step critical to the next mitosis occurs during interphase: the replication of the DNA of each chromosome. During which stage of the cell cycle is DNA replicated?

The concentration of DNA contained in a single set of chromosomes is frequently referred to as "c." Diploid somatic cells have two sets of chromosomes and alternate between 2c and 4c. A cell has 2c before DNA replication. Then, it has twice that amount, 4c, until it undergoes mitotic division. After mitosis, the two resulting cells will both have 2c.

The job of the mitotic division phase is to segregate the replicated chromosomes into two cells, each with the same chromosome and genetic complement as the parent cell. The four phases of mitosis are: prophase, metaphase, anaphase, and telophase.

DNA replication (in S phase) produces a second double helix. During prophase, the sister double helices condense, forming the X shaped chromosomes typically seen under the microscope.

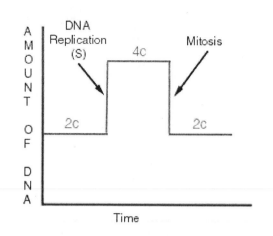

Draw a cell containing one pair of homologous chromosomes after it undergoes DNA replication (i.e., S phase) as seen in metaphase.

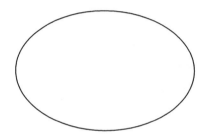

Suppose one homologue carries the *a* allele and the other homologue carries the *A* allele. Label each chromatid in your drawing.

Put each of the pictures below (containing two pairs of homologous chromosomes) in correct order and label with the appropriate phase. Note: Homologous chromosomes do not have to be next to each other during mitosis.

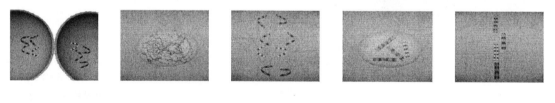

_____   _____   _____   _____   _____

Draw two daughter cells generated by a mitotic division with the appropriate homologues (labeled with *A* and *a*).

Problem 2

The results Mendel observed from his early crosses established the idea that alleles segregate. Unfortunately, Mendel couldn't figure out the biological mechanism behind this principle.

About 100 years ago, scientists realized that gametes (sex cells) were the result of a specialized cell division. This division process began with a cell that had two sets of chromosomes (a diploid cell) and ended with four cells with only one set of chromosomes (a haploid gamete). Today we know that the segregation of homologous chromosomes within this division process, meiosis, provides the mechanism for the segregation of alleles.

During sexual reproduction, two gametes (one from the mother, one from the father) combine in fertilization to form a diploid cell. A diploid fly has eight chromosomes. Without meiosis, how many chromosomes would the first generation progeny contain?

The purpose of meiosis is to convert one diploid cell into four haploid cells, each containing a single complete set of chromosomes. Meiosis can be divided into two cycles: Division I and Division II (or meiosis I and meiosis II). Two cells result from meiosis I. For a cell with a single pair of homologous chromosomes (labeled with *A/a*), draw the chromosomes found in the two cells formed by the first meiotic division. Then draw the four products of meiosis, with the appropriately labeled chromosome.

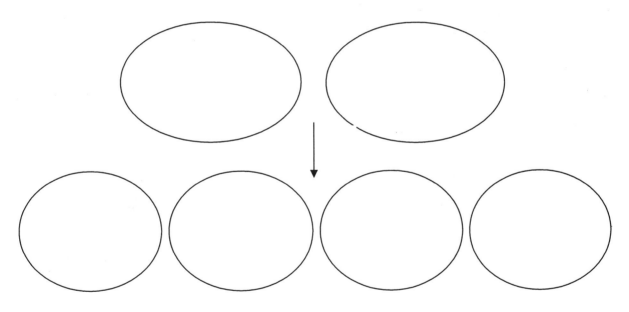

How many types of gametes with different alleles are formed in this example?

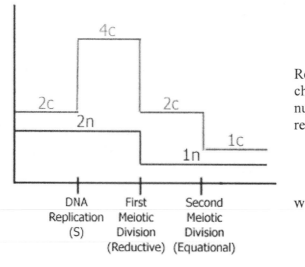

Regardless of whether there are one or two chromatids per chromosome, n indicates the number of chromosomes in a cell. Recall that c refers to the concentration of DNA in a cell.

w

a

What phase of meiosis best describes the cell in Figure a?

How many pairs of homologous chromosomes are there in this cell?

How many chromatids are present?

What is the value of c at this stage of meiosis?

b

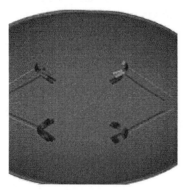

What phase of meiosis best describes the cell in Figure b?

Does the cell shown here have 1n or 2n chromosomes?

Is the amount of DNA best described as c, 2c, or 4c?

c

A karyotype (Figure c) is the ordered visualization of a complete set of chromosomes from metaphase. Karyotypes can be used to determine the number of chromosomes a species has, as well as any abnormalities an individual chromosome might have. This is the karotype of a male gorilla. How many chromosomes does a gorilla have? How many chromosomes are found in a gorilla gamete?

Autosomes are the same in both males and females. Sex chromosomes (bottom right corner) differ according to sex (XX in females, XY in males). X and Y chromosomes behave as homologues in males (the X and Y chromosomes pair and segregate at meiosis I). How many chromosomally different gamete types does a male produce?

Problem 3

Through the analysis of dihybrid crosses, Mendel was able to deduce that the genes he was studying were assorting independently, giving rise to gametic ratios of 1:1:1:1. However, in 1865, the physical nature of this independent assortment was not understood. After the discovery of meiosis in the 1880s, scientists recognized that genes located on different chromosomes should assort independently. A physical basis for Mendel's hypothesis was now possible!

The genetic material must be duplicated (or replicated) prior to its assortment. Draw two pairs of homologous chromosomes (labeled with *A/a* and *B/b*) in the cell below as they would appear after DNA replication (in prophase I).

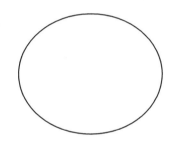

Two cells result from the first division of meiosis. Draw the chromosomes as they would appear in the cells after meiosis I. Remember that there are two different, equally probable ways for the chromosomes to assort (just draw one).

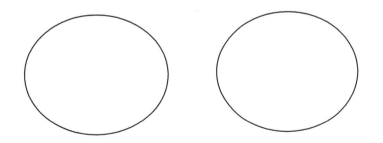

Write the possible gametes produced by a female with the genotype *AaBb*.

What are the possible gametes produced by a male with the genotype *AaBb?*

Notice that the gametes produced are identical to the meiotic products for an *AaBb* × *AaBb* cross. Recall that the Punnett square accurately predicts the genotypic frequencies when the genes involved assort independently of one another. In further testing Mendel's law of independent assortment, you cross an *Aa Bb Dd* female mouse with an *aa bb dd* male. How many possible gamete types can the female produce?

Suppose instead of seeing the number of classes of mice you expected, you find only four. You find 30 *abd*, 33 *aBd*, 29 *AbD*, and 28 *ABD* mice. In order to explain this enigma, we will look at one possible arrangement of genes on the mother's chromosomes.

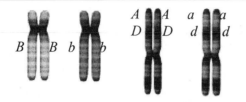

Two cells result from the first division of meiosis. Draw the chromosomes shown above to create the cells after meiosis I. Note that even though the two chromosomes are assorting randomly, the *A* and *D* alleles travel together since they are on the same chromosome.

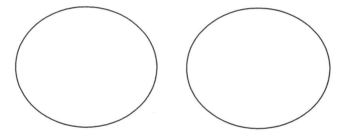

Completion of meiosis will produce four gamete types. What are the four possible genotypes (consider both possible assortments)?

This is an example of tightly linked genes on a chromosome where no crossing over can occur (see topic Linkage).

Problem 4

Problem 4 is a series of cartoons of cells at different stages of meiosis and mitosis and should be done on the CD-ROM.

X-Linked Inheritance
Problem 1

Is the pedigree below consistent with a dominant or a recessive trait? Assume the trait is rare.

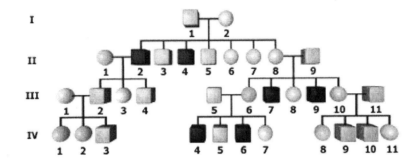

Is this pedigree consistent with X-linked or autosomal inheritance? Assume the trait is rare.

In the pedigree above, is there any evidence of father to son transmission of the trait?

What is the genotype of individual II-2? Write *A* or *a* for the X-linked alleles and Y for the Y chromosome.

What is the genotype of individual III-6? Write *A* or *a* to designate the X-linked dominant or recessive alleles.

Suppose in the pedigree above, individual II-8 was a male instead of a female and individual II-9 a female instead of a male. All other individuals are unchanged. Would this change the mode of inheritance deduced from this pedigree?

Is this pedigree still consistent with a recessive trait?

Can you tell whether the trait, in this hypothetical pedigree, is common or rare?

Now that the mode of inheritance is different, what is the genotype of individual II-2? Use *A* or *a* for the dominant and recessive alleles, respectively.

Problem 2

Hemophilia is a rare, recessive X-linked disease that usually affects only males. Consider the pedigree below of a family of normal parents with a son who has hemophilia.

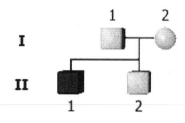

What is the father's genotype? Write H for X^H, h for X^h, or Y.

What is the mother's genotype? Write H for X^H and h for X^h.

They are going to have another child, whom they know is male (individual II-3). What is the chance that he will be affected?

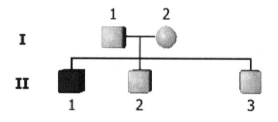

What is the chance that a daughter (individual II-5) will be a carrier?

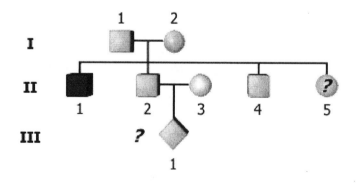

What is the probability that the Child III-1 will be affected?

What is the probability that the Child III-1 will be a carrier girl?

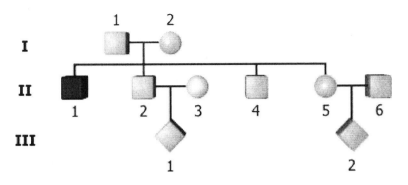

What is the probability that the Child III-2 will be an affected boy?

Problem 3

The pedigree below belongs to a family with a rare trait known as vitamin D resistant rickets. Is this pedigree consistent with a dominant or recessive mode of inheritance?

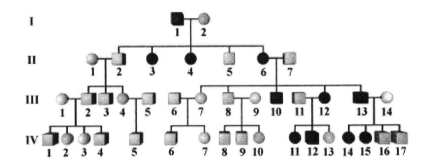

Is this pedigree consistent with an autosomal or an X-linked trait?

Assuming the trait is X-linked, what is II-6's genotype? Use *A* or *a* to designate the dominant or recessive X-linked alleles.

What is the genotype of individual II-5? Use *A* or *a* to designate the dominant or recessive X-linked alleles.

Suppose that III-14 has just found out she is going to have another girl. What is the probability that the child will be affected?

Individuals III-11 and III-12 are going to have another child but they don't know its sex. What is the probability they will have an affected son?

Problem 4

Red green colorblindness is a common X-linked recessive trait in humans that affects both males and females. Affected individuals are unable to distinguish red from green. Its prevalence accounts for the fact that you see some affected females as well as affected males.

Sickle cell anemia is an autosomal recessive disorder with a high incidence in the African American population (1/400). It is a structural hemoglobin abnormality that causes sickling of red blood cells with resulting complications.

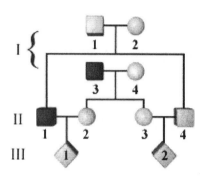

Consider this pedigree of two brothers marrying two sisters. All the individuals in generation II, namely II-1, II-2, II-3, and II-4 are carriers of the sickle cell allele. Individuals with filled blue squares are colorblind. Is there a greater chance that Child III-1 will have sickle cell anemia than Child III-2?

What is the chance that Child III-1 will be colorblind and have sickle cell anemia?

What is the probability that Child III-2 will be colorblind and affected with sickle cell anemia?

If the child (III-2) from the second marriage is colorblind, what is the child's gender?

Problem 5

Nondisjunction is a rare event that occurs in meiosis when paired chromosomes or sister chromatids do not disjoin properly. When this occurs in meiosis I, both homologues end up in one daughter cell. When nondisjunction occurs in meiosis II, one of the four meiotic cells ends up with both sister chromatids from one chromosome, and the other meiotic cell involved doesn't have any copies of that chromosome.

In humans, nondisjunction is generally lethal except when it involves chromosome 21 (leading to Down syndrome) or the sex chromosomes. In the following examples, nondisjunction of the sex chromosomes will be examined in conjunction with the X-linked gene that causes red green colorblindness. Affected individuals are unable to distinguish red color from green. In severely affected individuals, everything appears gray.

A man who is colorblind has a daughter with Turner syndrome who is also colorblind. You want to explain the origin of this daughter through nondisjunction in one parent.

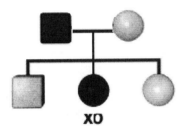

XO

Assuming the mother is not a carrier of the colorblindness allele, use the genetic marker of colorblindness in this family to determine which parent had the nondisjunction event.

Can you tell at what stage of meiosis (meiosis I or meiosis II) nondisjunction occurred?

A woman who is colorblind has a son with Klinefelter (XXY) syndrome who is not colorblind. You want to explain the origin of this son through nondisjunction in one parent. Start by assigning a genotype to each individual.

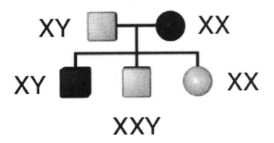

Can you use the genetic marker of colorblindness in this family to determine which parent had the nondisjunction event? Indicate that parent in the pedigree above.

Can you determine whether the nondisjunction event occurred in meiosis I or meiosis II of the parent's gametic line?

In the *Drosophila* fruit fly, the gene for white eyes is located on the X-chromosome. The white-eyed allele is recessive to the wild type. In crosses between white-eyed females and wild-type males almost all (regular) daughters have red eyes, except for about 1/2000 (0.05%) exceptional white-eyed daughters. Almost all (regular) sons have white eyes except for about 1/2000 exceptional red-eyed sons that are sterile.

Upon inspection of the flies' karyotypes, the white-eyed exceptional daughters were found to carry XXY. (Regular females are XX and males are XY.) Does this finding indicate that the Y chromosome determines sex in fruit flies?

Indicate the correct alleles of the exceptional females, one allele at a time. Follow the *Drosophila* nomenclature rules (see Fly Lab below) and choose between *w* and *W* for the white-eyed allele, and between *w+* and *W+* for the wild-type allele.

When the exceptional red-eyed sons are inspected, they are found to be XO, having a single X chromosome. Does this fact agree with the conclusion that the presence of Y does not determine sex in the *Drosophila* fruit flies?

Do these unusual findings of white-eyed exceptional daughters and red-eyed exceptional sons represent nondisjunction in the *Drosophila* father or *Drosophila* mother?

Can we tell whether the nondisjunction occurred in the first meiotic division or in the second meiotic division?

Fly Lab

Welcome to the fly lab! The following problems allow you to play the role of a geneticist working with the common fruit fly, *Drosophila melanogaster*. Since *Drosophila* nomenclature may be confusing at first, here is a brief review:

Genes are named after the mutant phenotype in *Drosophila*. Therefore, if the mutant gene results in the lack of eye formation, the gene could be named "eyeless." (Note that this differs from the Mendelian designations used for peas, for example, where the gene is typically named after the dominant trait.)

When allele symbols are assigned, lower case is used if the mutant allele is recessive (for example, *ro* for rough eyes). Capitalizing the first letter indicates that the mutant allele is inherited as a dominant trait (for example, *N* for notched wings).

Wild-type alleles are designated with a superscript + (eg., ro^+ or N^+). In the following problems simply indicate the + after the allele symbol (i.e., *ro+* or *N+*).

The problems below are set up as simulations on the CD-ROM. Here, the results are provided.

Problem 1

In *Drosophila*, the brown mutant is characterized by a brown eye color compared with the brick red, wild-type color. A wild-type fly is crossed with a brown-eyed fly to produce the F_1 generation and then the F_1 flies are crossed to produce the F_2 generation.

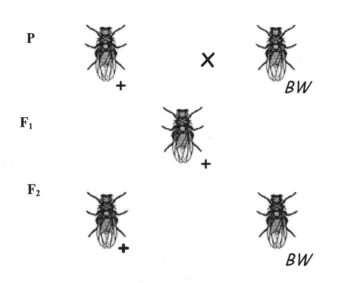

Is the brown mutation dominant or recessive?

Using the letters *BW* and proper Drosophila nomenclature (*bw* for recessive and *Bw* for dominant), indicate the proper mutant allele designation below.

What is the genotype of each of the flies above?

Problem 2

In *Drosophila*, the lobed mutation is characterized by smaller eyes compared to the wild type. A wild-type fly is crossed with a lobed fly to produce the F_1 generation (all lobed). Then the F_1 flies are crossed with each other to produce the F_2 generation (768 lobed and 259 wild type).

Is the lobed mutation dominant or recessive?

How would you indicate the mutant allele using the correct *Drosophila* nomenclature?

What is the genotype of each of the parental, F_1, and F_2 flies?

What fraction of the mutant F_2 flies is homozygous?

Problem 3

In *Drosophila*, the white mutant is characterized by white eyes compared to the brick red, wild-type color. A wild-type female fly is crossed with a white-eyed male to produce 974 wild-type F_1 flies. At this point what can you conclude about the mutation (is it dominant or recessive)?

The reciprocal cross is then performed: A wild-type male is mated with a white-eyed female to produce 436 white-eyed males and 421 wild-type females.

What can you now conclude about the mode of inheritance (autosomal or X-linked)?

Now go back to each of the crosses above and write the genotype of all the flies.

Problem 4

In *Drosophila*, the bar mutant is characterized by eyes that are restricted to a narrow, vertical bar. When a bar female is mated to a wild-type male, all the F_1 flies are bar. However, when a bar male is mated to a wild-type female, 857 bar females and 905 wild-type males are observed.

What is the mode of inheritance of the bar mutant?

What is the genotype of each of the flies in the two crosses above?

Problem 5

In *Drosophila*, the ebony mutant is characterized by an ebony body color and purple is characterized by purple eyes. Mating an ebony, purple female with a wild-type male yields all wild-type progeny. The reciprocal cross gives the same results.

What is the mode of inheritance for the ebony mutation?

What is the mode of inheritance for the purple mutation?

What is the genotype of the F_1 flies?

Mating the F_1 flies together yields 226 wild type, 74 ebony, 78 purple, and 25 ebony, purple flies. What is the ratio of progeny for each of the phenotypic classes?

Which F_2 fly should you use for a testcross of the F_1 flies?

How many different phenotypic classes, and in what ratios, do you expect from this cross?

Problem 6

In *Drosophila*, the sable mutant is characterized by a sable body color and dumpy is characterized by shorter, oblique wings. In a cross between a sable, dumpy female and a wild-type male, all the female progeny are wild type and the male progeny are sable. When the F_1 siblings are mated, the F_2 consists of 338 wild type, 336 sable, 114 dumpy, and 110 sable, dumpy (both male and female).

What is the mode of inheritance of the sable mutant?

What is the mode of inheritance of the dumpy mutation?

In order to understand the unusual ratios and the lack of apparent linkage to sex of the sable phenotype in the F_2, assign genotypes to the parental and F_1 generations. Write the genotypes above, using *s* or *s+* or Y to indicate sable alleles and *dp* or *dp+* to indicate dumpy alleles. Predict the proportion of each phenotype you would expect in the F_2.

Genotype/Phenotype

Goals for Genotype/Phenotype:

1. Understand the following phenomena that lead to variations on Mendelian phenotypic ratios:
   - Incomplete dominance
   - Codominance
   - Epistasis
   - Homozygous lethality
2. Recognize combinations of 9:3:3:1 phenotypic ratios, where the phenotype comes from two genes and involves epistasis.
3. Distinguish from phenotypic ratios whether multiple genes or a single gene with multiple alleles are involved in determining the phenotypes.

Problem 1

Gardeners at the Japanese Botanical Garden discovered that after planting only red and ivory snapdragons some plants with pink flowers appeared among the progeny. The gardeners decided to experiment with the snapdragon flowers and carried out a number of additional crosses. In this diagram, the parents are shown along the margins with progeny types inside the boxes.

Parents	Red	Ivory	Pink
Red	red	pink	red, pink
Ivory	pink	Ivory	ivory, pink
Pink	red, pink	pink, ivory	red, ivory, pink

Which of the parents are homozygous?

Which of the parents are heterozygous?

By crossing the pink plants we should be able to distinguish whether their color is caused by one or more gene differences between the red and ivory parents. Take a moment to predict what offspring you expect for one gene versus two genes.

From the original results, the pink × pink cross yielded three phenotypes among the progeny. The numbers of each of the phenotypes are listed below.

F_2: 261 red
489 pink
243 ivory

What is the ratio suggested by these results?

How many <u>genes</u> differ between the red and ivory parents in determining flower color?

Assume the gene differing between the red and ivory parents has two alleles, *P1*, associated with the presence of red pigment, and *P2*, associated with the absence of red pigment. What are the genotypes of the different progeny types in the F_2 above?

Which of the following concepts best explains the observed results of the F_1 and F_2 progeny: incomplete dominance, codominance, *P1* dominant, *P2* dominant, multiple alleles, or multiple genes?

Problem 2

The tools we use to infer genotypes are phenotypes and the results of crosses. A powerful advance is the ability to look at phenotypes that are closer to gene activity. Here we describe an example using gel electrophoresis of proteins.

Proteins can be separated based on size and/or charge using electrophoresis through a gel matrix. Migration distance through the gel serves as a phenotypic marker that could help determine an individual's genotype. Such is the case with the normal and the sickle cell hemoglobin molecules. With sickle cell, heterozygous individuals produce both normal and sickle cell hemoglobin molecules. When the normal and the abnormal hemoglobin molecules are run through a gel, they migrate at different rates and can be visually separated.

Part a. The family below has a history of sickle cell anemia. The hemoglobin electrophoresis pattern for each child is shown in the lane below that child.

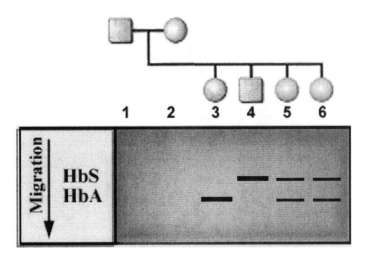

Which individuals in this family are homozygous for the sickle cell allele?

Which individuals from this family are heterozygous?

Imagine that this family is one out of a large number of families with heterozygous parents tested for their hemoglobin structure. Predict the ratios of children expected to be homozygous for *HbA*, homozygous for *HbS*, and heterozygous.

Which of the following concepts can best explain the expected phenotypic ratio: *HbA* dominant, *HbS* dominant, incomplete dominance, or codominance?

Part b. Gel electrophoresis can be used to screen individuals for the *HbS* allele.

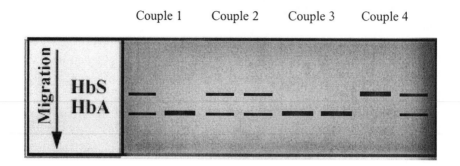

Which of the couples above are at risk of having an affected child?

What is the probability that Couple 2 will have a child with sickle cell anemia?

Part c. Use the Southern blot below to determine which of the three males (4, 5, or 6) could be the children's father.

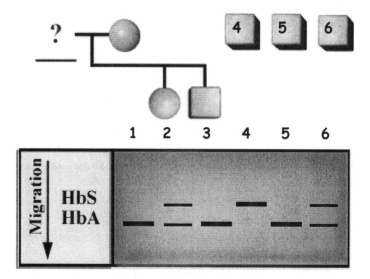

If this couple has another child, what is the chance the child will be anemic?

Problem 3

Karl Landsteiner was an Austrian-American physician who discovered that human blood differed in the capacity of serum to agglutinate red blood cells. By 1902, he and his group divided human blood into the groups A, B, AB, and O. Landsteiner concluded that two genes, A and B, control the ABO blood system he discovered. He proposed that each gene had two alleles, the presence and the absence of that allele.

Genotype	Phenotype
A– bb	A
aa B–	B
A– B–	AB
aa bb	O

Since you have been volunteering in Labor and Delivery for quite some time, you decide to compare the blood types you would expect using Landsteiner's two-gene hypothesis to the blood types you have observed. What genotype do you expect the children of two type O parents will have?

What is the expected phenotype of their children?

From your observations, 500 O × O parents have 763 type O children, whereas no type A, type B, or type AB children are observed. Is this consistent with the two-gene hypothesis?

One day you realize that in all 503 O × AB couples you have never seen any AB or O children. You have observed 600 type A and 650 type B children. Is this what you would predict based upon Landsteiner's hypothesis?

*Additional related questions are on the CD-ROM.

Problem 4

Suppose you are studying a novel bird species that displays a variation in feathers (blue, green, teal, and purple). Starting with pure breeding males and females from each phenotypic class, you perform the crosses diagrammed below.

Parents	F$_1$	F$_2$
teal × green	teal	¾ teal, ¼ green
green × blue	green	¾ green, ¼ blue
teal × purple	teal	¾ teal, ¼ purple
green × purple	green	¾ green, ¼ purple
blue × purple	purple	¾ purple, ¼ blue

Based on the data shown above, is feather color in this novel species segregating as if it were associated with multiple genes or multiple alleles?

Using the data shown above, which allele is dominant, blue or green?

Blue or purple?   Green or purple?   Green or teal?

What is the order of alleles corresponding to increasing dominance?

Cross	Parental Phenotypes	Phenotypes of Progeny			
		Blue	Green	Purple	Teal
1	blue × green	0	4	4	0
2	green × purple	0	3	3	0
3	teal × blue	0	4	0	4
4	teal × purple	0	4	0	3
5	green × green	0	6	2	0

For each of the crosses shown above, deduce the parental genotypes where f represents the feather gene and the following superscript designations represent the different alleles:

b - blue; g - green; p - purple; t - teal

*Assume homozygosity unless otherwise indicated.

Problem 5

A geneticist discovered two pure breeding lines of ducks. One line had white eyes and quacked with a "quack-quack." The other line had orange eyes and had a deeper quack, "rock-rock." In order to determine the mode of inheritance of these characteristics, she mated the two types of ducks and found that all F_1 ducks had yellow eyes, and uttered "quack-quack." When the F_1 ducks were interbred, the F_2 ducks were found in the following ratios:

24	yellow-eyed, quack-quack
12	white-eyed, quack-quack
12	orange-eyed, quack-quack
6	yellow-eyed, rock-rock
3	white-eyed, rock-rock
3	orange-eyed, rock-rock
2	yellow-eyed, squawk
1	white-eyed, squawk
1	orange-eyed, squawk

How many genes are involved in the inheritance of eye color?

P   orange-eyed × white-eyed

F_1 yellow-eyed

F_2 2   yellow-eyed
         1   orange-eyed
         1   white-eyed

Use $A1/A2$, $1/B$ $B2$, etc. to designate eye color alleles, assign genotypes to the individuals in the cross described above.

How many genes are involved in the inheritance of quacking?

P		quack-quack × rock-rock
F_1		quack-quack
F_2	12	quack-quack
	3	rock-rock
	1	squawk

Use *B/b*, *C/c*, etc. to designate quacking phenotype, assign genotypes to the individuals in the cross described above.

Problem 6

Recall from Problem 4 that blood types are classified by their surface antigens. Type A blood has antigen A on its surface, type B has antigen B, type AB has both antigen A and antigen B, and type O has neither antigen on its surface.

Ellen and Carl have just had a baby boy. Ellen's blood type is B, Carl's is AB and the baby's is O. Having a good knowledge of blood typing, you realize that something is not quite right with this story. You happen to know that Ellen's parents' blood types are B and O. From this, what is Ellen's genotype?

What is Carl's genotype?

Does it appear that Ellen and Carl can have a child with O blood?

Ellen's ex-boyfriend, Mark, has type O blood. Could Mark be the baby's father?

Ellen insists that Carl is the father, so you look into their family history (you may notice that Ellen and Carl are cousins). The results are seen in the pedigree below. Which parents have phenotypes that are incompatible with the blood types of their children?

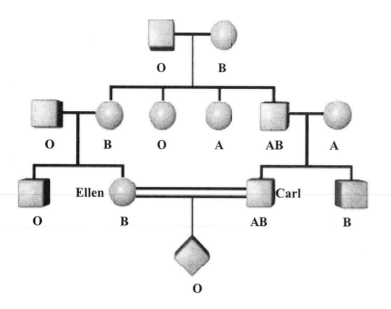

One of your friends suggests that this may be an example of lack of penetrance, i.e., the phenotype does not reflect the genotype. However, since Ellen and Carl share a grandfather who also has a suspicious O phenotype, you wonder if it's possible that they both inherited a recessive allele that is epistatic to the ABO blood antigens.

Upon further research you find there is such a rare recessive mutation, h, that is epistatic to the ABO system gene. Individuals who are homozygous for h cannot synthesize A or B antigens, so they have an O phenotype, referred to as the Bombay phenotype.

Assign genotypes for the H gene, assuming this is the reason for the unusual phenotypes (use H/h). With this information, determine the genotype for the ABO locus for individual I-1.

Does it now appear that Carl could be the father?

Does this information conclusively prove that Carl is the father?

In the hope of solving the paternity issue, you decide to run a Southern blot, probing for the alleles of the H gene.

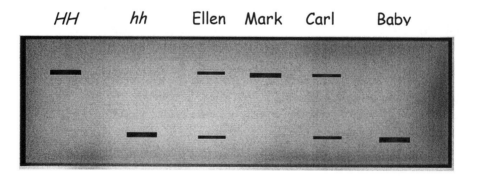

With this information, can you definitively say who is the baby's father?

Problem 7

Carolyn and Jeff are both cat lovers and neighbors in a local apartment complex. Recently, Carolyn's cat, Grace, escaped from her apartment and was seen mating with Jeff's ferocious feline, Chuck. Although Carolyn and Jeff were initially upset with the scandalous behavior of their pets, they felt reassured that Grace and Chuck would produce a beautiful and profitable litter of kittens. Weeks later, Carolyn and Jeff discovered the litter of kittens and were surprised to find one cat with curled ears.

As their close friend and local genetics expert, they turn to you to provide a genetic explanation for this strange occurrence. At this point, can you determine whether the curled-ear cat occurred as a result of a spontaneous dominant mutation or whether both parents were heterozygous for a recessive mutant allele?

In order to determine whether the curled-ear mutation is dominant or recessive to the normal condition, you mate the curled-ear cat with an unrelated normal cat. From this mating, *two curled-ear cats and two normal cats* are produced. Based on these results, is the curled-ear mutant allele dominant or recessive to the wild-type allele?

The allele responsible for curled ears in cats is located at the Ear locus. Using the allele symbols c and $c+$, assign genotypes to the cats in the cross described above. Be sure to follow the format shown below when assigning genotypes:

Homozygous individuals: cc or $c+c+$
Heterozygous individuals: $cc+$ or $c+c$

What ratio of curled-ear cats to normal cats do you expect from the mating of two heterozygous curled-ear cats?

After performing the mating between two heterozygous curled-ear cats, you wait several weeks and examine the litter of kittens. The mating produced four curled-ear cats and two normal cats. From these results, can you be certain about the mode of inheritance of curled ears?

What results do you expect for a mating between a homozygous curled-ear cat and a normal cat?

What results do you expect for a mating between a heterozygous curled-ear cat and a normal cat?

You decide to perform 170 matings between curled-eared cats to determine whether curled ears are inherited in a simple Mendelian manner. You obtain 1020 kittens from your 170 matings between heterozygous curled-ear cats. If curled ears are inherited according to our predicted 3 to 1 ratio, how many curled-ear cats do you expect to see out of the 1020 kittens?

You only observe 677 curled-ear cats and 343 normal cats from the 170 matings. After re-counting the kittens several times, you are certain that the number of curled-ear cats is accurate. What is our observed ratio of curled-ear cats to normal-ear cats?

What is the most likely genetic explanation for the results we have gathered?

Linkage

Goals for Linkage Analysis:
1. Recognize the difference between linkage and independent assortment in dihybrid crosses.
2. Relate parental chromosome input to linkage phase to calculate recombination frequencies.
3. Be able to make maps from recombination frequencies.
4. Use three-factor crosses to distinguish order and distances between genes.

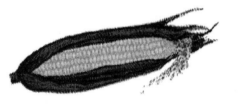

Problem 1

Consider two corn mutants: dwarf and glossy. Dwarf (gene symbol d) is a recessive trait characterized by short, compact plants. Glossy (gene symbol gl) is also recessive and is characterized by a bright leaf surface. F_1 progeny from a cross of pure breeding parents were back-crossed to the recessive parent. The results of this cross are shown below.

F_2:		
	286	wild type
	89	glossy
	97	dwarf
	277	glossy dwarf

Since the ratio of phenotypic classes is not the expected 1:1:1:1 for a test cross, what concept best explains the observed phenotypic ratio?

What phenomenon explains the four phenotypic classes of unequal ratios instead of two phenotypic classes as predicted by the tight linkage?

Complete the genotype of each F_2 progeny class. Which phenotypic classes carry the recombinant genotypes?

What approximate percentage of progeny is descended from distinguishable recombinant gametes?

What is the map distance between dwarf and glossy determined from these numbers?

Using the known map distance between glossy and dwarf determined above, what is the expected number of each F_2 phenotypic class out of a total of 1000 F_2 progeny?

Problem 2

In *Drosophila*, the forked phenotype is characterized by short bristles with split ends. The scalloped phenotype is characterized by scalloped wings at the margins and thicker wing veins. Both genes are X-linked and are marked with recessive mutant alleles. In the cross below, an F_1 is generated by mating a wild-type female with a scalloped, forked male to produce all wild-type progeny. The females are then mated to their fathers (a test cross) to generate an F_2.

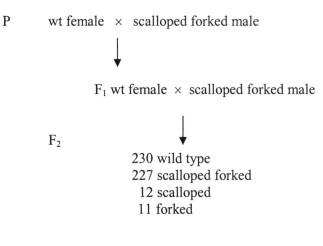

P   wt female × scalloped forked male

F_1 wt female × scalloped forked male

F_2

230 wild type
227 scalloped forked
12 scalloped
11 forked

What concept best explains the <u>observed</u> phenotypic ratio of the F_2 progeny?

Which of the phenotypes belongs to F_2 progeny that that are descended from distinguishable crossover bearing (recombinant) gametes?

What approximate fraction of progeny is descended from distinguishable crossover bearing gametes?

What is the map distance between forked and scalloped?

Problem 3

The *Drosophila* forked phenotype is X-linked recessive (as we saw in Problem 2) and is characterized by short bristles with split ends. The miniature phenotype, also X-linked recessive, is characterized by reduced wing size.

A miniature female is crossed with a forked male. The F_1 progeny consists of wild-type females and miniature males. These are then mated to yield the following F_2:

510 miniature females	418 miniature males
490 wild-type females	412 forked males
	87 miniature forked males
	83 wild-type males

What is the approximate distance in map units (mu) between forked and miniature that you calculate from this cross?

When a wild-type female is crossed with a miniature forked male, and the wild-type F_1 females are testcrossed, the F_2 are as shown below:

84 miniature
86 forked
417 miniature forked
413 wild type

Is the distance you find in mu the same in this cross as the distance you found for the markers in trans?

In Problem 2 you found that the distance between the scalloped gene and the forked gene is 5 mu. In this problem you found that the distance between miniature and forked is 17 mu. Is this information sufficient to find the gene order of the three genes on the X chromosome?

When a wild-type female is crossed with a miniature scalloped male, all the progeny are wild type. A testcross of the heterozygous females yields the following F_2:

51 miniature
45 scalloped
354 miniature scalloped
350 wild type

What is the approximate map unit (mu) distance between miniature and scalloped that you calculated from this cross?

Now that you know the distances between each pair of genes, construct a map of the three genes.

Problem 4

In *Drosophila*, the spineless mutant is characterized by shortened bristles compared to the longer bristles of the wild type. The radius incomplete mutant is characterized by an incomplete wing vein pattern. Both genes are autosomal and are marked by recessive mutant alleles.

A wild-type female is mated with a spineless, incomplete male. Then, an F_1 wild-type female is crossed to the parental male to generate the following F_2:

> 449 wild type
> 452 spineless incomplete
> 52 spineless
> 48 incomplete

In this cross you did not obtain the expected 1:1:1:1 ratio expected for a test cross of unlinked genes. This implies that the genes are linked. Use these data to calculate the distance between spineless and incomplete.

A wild-type male is mated with a spineless, incomplete female. Then, the F_1 wild-type male is mated with the parental female. The following F_2 progeny are observed:

> 361 wild-type flies
> 357 spineless incomplete flies

Note that very different results are observed in this F_2 compared to the first cross. To help explain these unusual results, let's first assign genotypes to the F_1 male and tester female.

Write the genotypes of the F_1 flies in the cross above. Use the symbols *sp, sp+* for spineless and *ri* or *ri+* for radius incomplete. Indicate linkage by writing the two linked alleles on one side of a / (e.g., *a+b+/ab*).

After examination of these genotypes, does it make sense that these two genes are on the same chromosome?

Geneticists made a novel discovery when working with *Drosophila melanogaster*. Scientists found that male flies do <u>not</u> undergo recombination. Therefore, test cross of the F_1 male yields only two gamete types. The distance between genes cannot be measured using F_1 males. What gametes are produced in the F_1 male fly?

The consequence of no recombination in males is that crosses to map linked genes must use F_1 females as the dihybrid parent. This is true only in *Drosophila melanogaster*. In other species there is frequently a difference in recombination between the sexes, but it is not usually 0 in males.

Problem 5

In *Drosophila*, the X-linked genes singed (*sn*), characterized by bent bristles, miniature (*m*), reduced wing size, and tan (*t*), a tan body color, are marked with recessive alleles. In this problem you will use three-factor crosses to determine the gene order and distance between these markers. (Mating the F_1 flies is essentially a test cross, allowing one to analyze all of the F_2 flies directly).

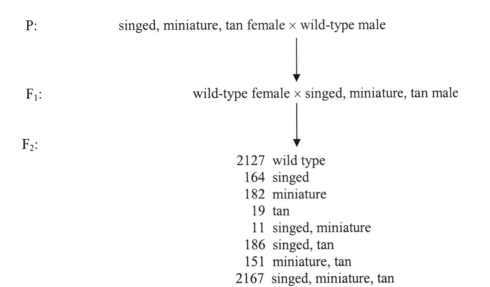

P: singed, miniature, tan female × wild-type male

F_1:   wild-type female × singed, miniature, tan male

F_2:

2127 wild type
164 singed
182 miniature
19 tan
11 singed, miniature
186 singed, tan
151 miniature, tan
2167 singed, miniature, tan

How many gamete types do you expect the F_1 female to produce?

Which two classes of gamete types are the produced most often by the F_1 female? (This is reflected in the phenotypic classes of the progeny.) These are the parental types.

Which of the reciprocal gamete types are the produced least often in the F_1 female? (This is reflected in the phenotypic classes of the progeny.) These are the double crossover types.

Compare the double crossover gametes with the parental gametes to determine which gene is in the middle. Write the correct gene order below.

Using your data, determine the distance between the singed gene and the tan gene, and the distance between the tan gene and the miniature gene.

Problem 6

Chromosome 13 of the mouse carries the locus for flexed tail with the alleles f and $f+$ and the locus for extra toes with the alleles Et and $Et+$. The flexed tail mutant is a recessive trait while the extra toes mutant is dominant. The next exercises will use 24 map units as the known distance between Et and f to predict the expected number of F_2 progeny from crosses.

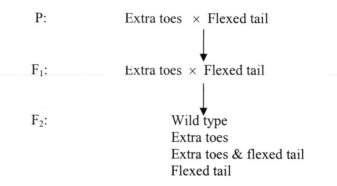

Find the genotype of each mouse above. Both parental mice are from pure breeding stocks. Use Et and $Et+$ for the extra toes gene, and f or $f+$ for the flexed tail gene.

Which F_2 progeny represent recombinants?

The distance between Et and f is 24 mu. From a total of a 1000 F_2 mice obtained from crosses such as this, how many are expected for each phenotypic class? (Start with the number of wild-type progeny.)

> Wild type
> Extra toes
> Extra toes & flexed tail
> Flexed tail

Let's now try to find the expected progeny for a three point cross. Chromosome 13 of mouse also carries the locus for satin fur texture with the alleles sa and $sa+$. The satin fur mutant is a recessive trait. Use the following map unit distances to map sa: $sa - Et$: 4 mu, $sa - f$: 20 mu. Draw a map of the three markers.

Use this map to determine the number of predicted progeny of each phenotypic class shown below with parents of genotypes $Et\ f/Et\ f$ and sa/sa. The F_2 progeny classes from these crosses are as follows:

> Wild type
> Flexed tail, satin with extra toes
> Flexed tail with extra toes
> Satin with extra toes
> Flexed tail, satin
> Extra toes
> Flexed tail
> Satin

Bacterial Genetics

Goals for Bacterial Genetics:
1. Be familiar with the mechanism of conjugation and transduction as methods of gene transfer.
2. Determine genotype from ability to grow on different media.
3. Determine type of media needed for selection of exconjugants.
4. Be familiar with replica plating and interpretation of resulting data.
5. Map genes relative to other genes by interrupted mating, natural gradient of transfer, and variations of these techniques.

Conjugation
Problem 1

Various strains of bacteria were incubated on plates containing minimal media plus several amino acids. From the pattern of growth, answer the following questions regarding the genotype of each strain.

What are the genotypes of colonies 1–4 with respect to *arg*, *met*, *asp*, and *ile*?

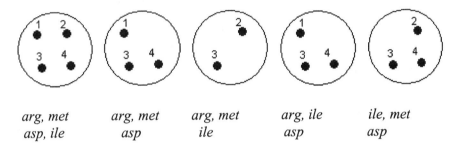

| *arg, met* | *arg, met* | *arg, met* | *arg, ile* | *ile, met* |
| *asp, ile* | *asp* | *ile* | *asp* | *asp* |

The following plates show bacterial strains that have been replica plated on minimal media plus the indicated mixture of amino acids and antibiotics.

Determine the genotypes of colonies 1–4 with respect to *arg, met, pen,* and *str*.

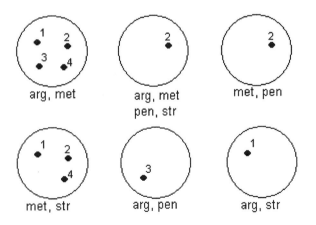

The plates shown below contain different sugars as carbon sources. Cells were first plated

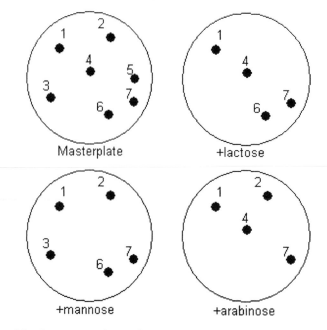

on minimal media with glucose as the carbon source (Masterplate) and then replica plated onto plates with lactose, mannose, or arabinose (no glucose).

Which colony has the genotype *lac− man+ ara+*?

Which colony has the genotype *lac+ man− ara+*?

Which colony has the genotype *lac+ man+ ara−*?

Which colony has the genotype *lac− man+ ara−*?

Which colony has the genotype *lac− man− ara−*?

Problem 2

Lederberg and Tatum made crosses by mixing pairs of strains. They used a C strain which was *arg−met+* and an E strain which was *arg+met−*. However, they did not know whether each of these strains was F−, F+ or Hfr. An example of the type of crosses they performed is shown below.

C strain: *arg−met+* × E strain: *arg+met−*

↓

arg+met+

What type of cross would give you many exconjugants?

Which of the following crosses will give you only a few exconjugants?

What kind of crosses will give no *arg+ met+* exconjugants? (select from the choices below)

F+ × F+   **F− × Hfr**   **F+ × Hfr**

F− × F−   **F+ × F−**   **Hfr × Hfr**

The table shown here contains the results of all possible crosses (0 = none; F = few; M = many). From the questions you just answered, you should now be able to identify which strains are F+, F−, and Hfr.

	E1	E2	E3	E4	E5
C1	F	0	M	F	0
C2	0	M	0	0	M
C3	0	F	0	0	F
C4	F	0	M	F	0
C5	0	F	0	0	F

Bonus Question

What we are looking for in our exconjugants are the recombinants of the E strains, *arg+met−*, with the C strains, *arg−met+*. What type of media should be used to select for the prototrophic exconjugants?

Problem 3

From the time that Hfr and F− cells are combined, the number of markers transferred for each mating pair depends on how long the two cells stay conjoined. The mating can be interrupted at specified times after the start of mating by vigorous shaking, frequently accomplished by blending. Without such mixing the disruption of the mating pairs occurs naturally generating a gradient of transfer.

In the following experiment, a prototrophic Hfr strain (streptomycin sensitive) is mated to a streptomycin resistant F− strain auxotrophic for methionine, leucine, and cysteine. A pipette is used to distribute cells to each of the plates shown below, each containing minimal media with streptomycin, methionine, and cysteine added in order to select only for leucine prototrophs.

Aliquots of culture A (Hfr cells only), culture B (Hfr cells + F− cells), and culture C (F− cells only) are added to the three plates containing streptomycin, methionine, and cysteine. The plates are then incubated. The results are shown in the following illustration. Genotypes: (Hfr: *str-s, met+, leu+, cys+*) (F−: *str-r, met−, leu−, cys−*).

A: *str, met, cys*
No colonies

B: *str, met cys*
375 colonies

C: *str, met, cys*
No colonies

Why were no colonies observed after plating the Hfr strain on plate A above?

Why were no colonies observed after plating the F- strain on plate C above?

Why were cells able to grow on plate B?

The mating mixture produced 375 exconjugant colonies that replaced the *leu–* allele of the F– strain with the *leu+* allele of the Hfr strain. Additional markers may have also been transferred and could be checked by replica plating.

To measure the time of entry of each of the markers, matings between the Hfr and F– strains can be interrupted at various time points and plated on selective media. Below is a graph of colonies versus time.

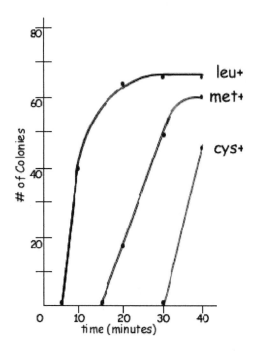

Draw a map below to indicate the order of the markers in the Hfr strain with respect to the origin.

Problem 4

An Hfr strain with genotype *met– leu+ his+ trp+* that transfers the *met* gene very late was mated with an F– *met+ leu– his– trp–* strain. After mating for 30 minutes, cells were plated on minimal media (MM) with the added nutrients listed below each plate. The number of colonies that grew on each plate is indicated.

Plate	1	2	3
Supplements	his, trp	his, leu	leu, trp
Colonies	250	50	500

What is the purpose of the methionine marker in this cross?

What markers are selected for on Plate 1?

What markers were selected on Plate 2?

What markers were selected on Plate 3?

Those markers closest to the origin are transferred first and yield the highest number of recombinant colonies. Based on the number of colonies on each plate above, determine the order of the markers in the Hfr strain with respect to the origin.

Problem 5

A small bacterial genome was mapped using three different Hfr strains and an F– *str* ^{r} *ala– ade– bio– his– ile– val–*. Each of the Hfr strains was *str* ^{s} and contained the wild-type alleles for all the markers. The matings were interrupted after 30 minutes and the exconjugants were selected on the following plates (the indicated nutrient is left out of the otherwise complete media + streptomycin).

Hfr: *str* ^{s} *ala+ ade+ bio+ his+ ile+ val+* × F–: *str* ^{r} *ala– ade– bio– his– ile– val–*

Mating	Missing Nutrient					
	Ala	Ade	Bio	His	Ile	Val
HfrA × F–	300	0	0	0	900	750
HfrB × F–	0	400	0	0	700	875
HfrC × F–	200	0	956	724	0	0

In the table above, some of the plates have no colonies. How can this be explained?

Consider the data from the Hfr A × F– cross. Which of the markers can be ordered using this cross?

What is the first marker to be transferred by Hfr A?

Order the three markers starting with *ile*.

What is the order of the markers indicated by the results of the Hfr B × F– cross?

What is the order of the markers indicated by the results of the third cross?

In every problem so far we have represented the bacterial chromosome as a circle. The original recognition of this fact comes from comparing the maps of different Hfr strains. As you can see from your results, the maps can only be reconciled as permutations from a common circular map, with the different Hfr strains having different origin points and directions of transfer. Use this information to construct a circular map.

Transduction
Problem 1

A generalized transducing phage is grown on a prototrophic strain and then used to transduce a recipient that is *arg– leu– gln– gua– thr–*. In this experiment *arg+* transductants will be selected by plating on media lacking arginine but containing all other supplements required by the recipient strain. This will be the master plate.

> DONOR: *arg+ leu+ gln+ gua+ thr+*
> RECIPIENT: *arg– leu– gln– gua– thr–*
>
> MASTERPLATE: leu, gln, gua, thr (supplements)

Fifty-nine colonies grew on this master plate. To test for co-transduction of the unselected markers, the colonies were replica plated to different media, omitting one supplement at a time. The results are shown in the following table.

Supplements to Minimal Media					
Plate	Leucine	Guanine	Glutamine	Threonine	Colonies
1	–	+	+	+	0
2	+	–	+	+	0
3	+	+	–	+	24
4	+	+	+	–	0

What does the absence of growth indicate on Plate 1?

What does the presence of growth indicate on Plate 3?

What is the co-transduction frequency (in percent) of *arg* and *gln*?

To develop a genetic map, let us now repeat the transduction experiment and make a new master by plating on media lacking leucine but containing all other supplements required by the recipient strain. This is our second master plate. Seventy-three colonies grew on this plate and were replica plated onto various plates containing the media indicated in the table below.

Supplements to Minimal Media					
Plate	Arginine	Guanine	Glutamine	Threonine	Colonies
1	–	+	+	+	0
2	+	–	+	+	22
3	+	+	–	+	0
4	+	+	+	–	45

What is the co-transduction frequency (in percent) of *leu* and *gua*?

What is the co-transduction frequency (in percent) of *leu* and *gln*?

To finish our experiment, we will do one more transduction. In this third master plate our media lacks guanine but contains all other supplements required by the recipient strain. Sixty-five colonies grow on this plate which is replica plated onto various plates containing the media indicated in the table below.

Plate	Arginine	Leucine	Glutamine	Threonine	Colonies
			Supplements to Minimal Media		
1	–	+	+	+	0
2	+	–	+	+	20
3	+	+	–	+	13
4	+	+	+	–	0

What is the co-transduction frequency of *gua* and *gln*?

Using these co-transduction frequencies you have determined, determine the order of the three markers: *arg, gua,* and *gln.*

Problem 2

In a transduction experiment, the donor strain is *kan^r lys+ arg+* and the recipient strain is *kan^s lys– arg–*. Transductants are plated on MM (minimal media) supplemented with kan (kanamycin), lys (lysine), and arg (arginine) and then replica plated on the plates shown below.

Plate	Kan	Lys	Arg	# of colonies
		Supplements to Minimal Media		
Master	+	+	+	500
Replica 1	+	–	–	20
Replica 2	+	+	–	21
Replica 3	+	–	+	200

In the table below, fill in the number of colonies for each of the four genotypes.

Genotype	# of colonies
kanr arg+ lys+	
kanr arg+ lys–	
kanr arg– lys+	
kanr arg– lys–	

What is the co-transduction frequency of *kan* and *arg*?

What is the co-transduction frequency of *kan* and *lys*?

Problem 3

This problem is a simulation and can only be done on the *Interactive Genetics* CD-ROM.

Problem 4

Frequently, mutations are so close together that it's impossible to determine their order with respect to a nearby marker by simply using conjugation or transduction.

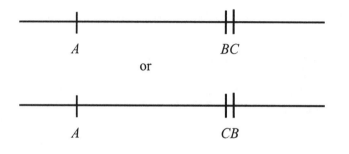

In the illustration above it is difficult to determine if *B* is closer to *A* or *C* is closer to *A*. The co-transduction frequency between *A* and *B* would be about the same as the co-transduction frequency between *A* and *C*. However, reciprocal three-factor transductions can be used to determine the order of these markers.

Unlike in plants and animals, reciprocal transductions in bacteria do not refer to switching markers between sexes. Instead, in bacteria we refer to switching the markers between the donor and recipient. That is, in Transduction I the donor may be *B+C−* and the recipient *B−C+*, while in Transduction II (the reciprocal) the donor may be *B−C+* and the recipient *B+C−*. Having the markers in trans is essential for this experiment since it forces a crossover between the two tightly linked markers (*B* and *C*) to generate a prototroph. The outside marker will be *A+* in the donor and *A−* in the recipient in all transductions.

In the two reciprocal transductions shown below the order is assumed to be *ABC*. Draw an X in the appropriate places to generate prototrophic (*A+B+C+*) transductants for each experiment.

Transduction I: Donor: *A+B+C−*   Recipient: *A−B−C+*

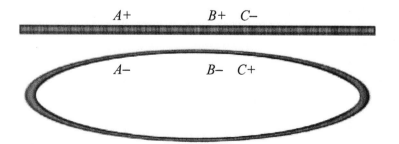

Transduction II: Donor: $A+B-C+$ Recipient: $A-B+C-$

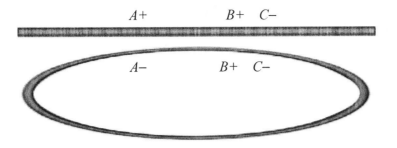

Notice that in Transduction I, prototrophs were generated with only two crossovers, whereas in Transduction II, four crossovers were required to generate $A+B+C+$. This is true only when the order is ABC.

Now we will see what happens if the order is ACB. Again, draw X's in the appropriate places to generate prototrophic ($A+B+C+$) transductants.

Transduction I: Donor: $A+B+C-$ Recipient: $A-B-C+$

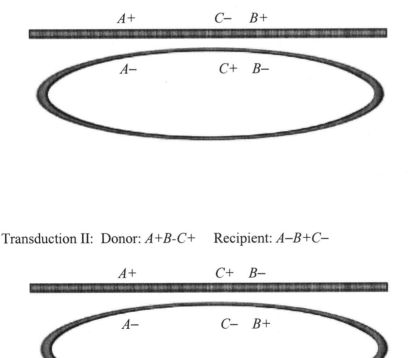

Transduction II: Donor: $A+B-C+$ Recipient: $A-B+C-$

Notice that Transduction I prototrophs were generated with four crossovers, whereas in Transduction II only two crossovers were required to generate $A+B+C+$.

A minimum of two crossovers is required to incorporate genes from the donor fragment into the circular recipient chromosome. This results in the frequencies of transduction we usually observe. However, a situation that requires four crossovers will yield many fewer transductants or none at all.

You need to compare reciprocal transductions in order to determine which experiment gives you normal transduction frequencies and which gives you reduced frequencies. This example asks you to determine the frequencies using two different orders, although there is only one order.

Reciprocal transductions were used to order three mutations (*trp1, trp2,* and *trp3*) required for metabolism of tryptophan with respect to a nearby marker tyrosine (*tyr+*). For each pair of mutants (*trp1* and *trp2; trp2* and *trp3; trp1* and *trp3*) a pair of reciprocal transductions will be made to order them with respect to *tyr*.

	Donor	Recipient
Transduction I	*tyr+ trpx– trpy+*	*tyr– trpx+ trpy–*
Transduction II	*tyr+ trpx+ trpy–*	*tyr– trpx– trpy+*

Experiment	x y	I	II
1	*trp1 trp2*	800	5
2	*trp2 trp3*	3	750
3	*trp1 trp3*	659	8

Which mutation, *trp1* or *trp2*, is closer to *tyr*?

Which mutation, *trp1* or *trp3*, is closer to *tyr*?

Which mutation, *trp2* or *trp3*, is closer to *tyr*?

Draw a map below to indicate the order of the three mutations with respect to *tyr*.

Biochemical Genetics

• •

Goals for Biochemical Genetics:
1. Determine the number of genes coding for steps in a biosynthetic pathway. Use auxotrophic mutant strains to determine the number of complementation groups for a pathway.
2. Determine a biosynthetic pathway from nutritional studies that supplement with compounds that may be intermediates in the pathway.
3. Determine whether any of the genes are linked to each other.
4. Distinguish the metabolism of amino acids between anabolic (biosynthetic) and catabolic (degradative) reactions. Most human biochemical disorders show up from the loss of a catabolic enzyme. Auxotrophic mutant strains of microorganisms result from the loss of an anabolic enzyme.

Problem 1

Six *Neurospora* mutants were isolated that require vitamin B1 to grow. To determine whether the six mutants have mutations in different genes, complementation studies were performed. Heterokaryons were formed and tested for growth on minimal media.

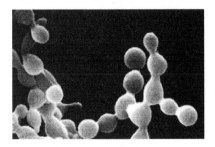

The following table shows the experimental results; + indicates heterokaryon growth and − indicates no heterokaryon growth in the absence of vitamin B1.

	A	B	C	D	E	F
A	−	+	+	−	+	+
B	+	−	+	+	−	−
C	+	+	−	+	+	+
D	−	+	+	−	+	+
E	+	−	+	+	−	−
F	+	−	+	+	−	−

Which mutants belong to the same complementation group as mutant A?

Which mutants belong to the same complementation group as mutant B?

Which mutants belong to the same complementation group as mutant C?

Which mutants belong to the same complementation group as mutant D?

How many complementation groups are there?

Would a heterokaryon formed from mutant A and the double mutant C,E grow on minimal media? The double mutant strain contains two gene mutations: the gene mutation in strain C and the gene mutation in strain E.

Would a heterokaryon formed from the double mutants A,B and C,F grow on minimal media?

Would a heterokaryon formed between the three double mutants A,E and B,C and C,D grow on minimal media?

Problem 2

You are a graduate student working in a *Neurospora* lab and have recently isolated five *Neurospora* methionine auxotrophs that each contain a single gene mutation. You set out to determine if any of the mutated genes in the *Neurospora* strains are linked. You decide to do this by crossing mutant strains and examining the phenotypes of the progeny.

You form a transient diploid by mating mutant strains (represented by the cross *ab+* × *a+b*). The diploids then undergo meiosis immediately by sporulation. A diploid cell with a crossover in the interval between the linked *a* and *b* genes (in this example) will yield four genetically distinct spores. Three of them are still auxotrophs and only the *a+b+* spore is phenotypically distinct as a prototroph. The frequency of prototrophs in a large collection of spores can be used to determine the map distance between two mutants.

The results of your crosses with the five *Neurospora* auxotrophic strains (A–E) in all combinations are shown below. For each cross, 1000 ascospores were plated on minimal media. The table shows the number of methionine prototrophs that you recovered from each cross.

	A	B	C	D	E
A	0	150	250	215	40
B		0	250	65	110
C			0	250	250
D				0	175
E					0

Since strains A–E are methionine auxotrophs, the prototrophic ascospores must have been produced from independent assortment or recombination between two mutated genes during meiosis. How many prototrophic ascospores would you expect from a cross if the recombinant progeny were produced by independent assortment?

For each of the linked genes in the table above, determine the map distance between them. Then, combine the information from each of the two factor crosses to assemble a map of the entire linkage group.

Problem 3

Wild-type *Neurospora* is orange when exposed to light during growth. You have isolated three albino mutants that are completely white even when grown in the presence of light. Using heterokaryon complementation studies, you find that these three strains have mutations in different genes, which you name *al-1*, *al-2*, and *al-3*.

You suspect that the albino mutants are white because they are unable to make the carotenoid pigment (known to cause the orange color). To test your idea, you seed the albino mutants on media supplemented with carotenoid pigment.

You are thrilled to find that supplementation with carotenoid pigment results in orange colored hyphae for all three albino mutants! Assuming carotenoid synthesis is affected in these mutant strains, you set out to determine which step in the biosynthetic pathway is blocked by each of the mutations.

Luckily, three precursors in the carotenoid pigment biosynthetic pathway had previously been discovered, although their order was unknown. To determine the order in which the precursors are converted into the carotenoid pigment, you grow the albino mutants on media supplemented with each of the three precursors.

You find that supplementing media with these different precursors restores wild-type color in some of the albino mutant. You compile your results into a simple table, where + indicates wild-type color and − indicates white color.

	Carotenoid pigment	GGPP	Phytoene	PPP
al-1	+	−	−	−
al-2	+	−	+	+
al-3	+	−	+	−

Use the data from your experiments to determine the order of the precursors as they appear in the biosynthetic pathway and which step in the biosynthetic pathway is blocked by each mutant.

Problem 4

Saccharomyces cerevisiae has played a fundamental role in human history and culture. For centuries, yeast has been used by humans for the rising of bread and for the fermentation of wines and beers. Today yeast is used as a model organism for both genetic and cellular biology studies. Yeast are eukaryotes that grow simply as either single cells or colonies.

Saccharomyces cerevisiae can exist as either a haploid or a diploid. Both haploids and diploids are able to grow and divide by mitosis, through a process called budding. This differs from *Neurospora* where the diploid zygote is transient and quickly undergoes meiosis to form ascospores. In the haploid state there are two mating types, a and α. When an a cell and an α cell are brought together, they fuse to form a diploid cell. First the cellular membranes fuse, then the nuclei (yeast do not normally form stable heterokaryons). The diploid cell can grow and divide indefinitely. However, when nutrients are depleted or when environmental conditions become unfavorable, the diploid cells will undergo meiosis to form four haploid spores (a process called sporulation). The spores can be separated manually and tested directly for phenotype. Since they are haploid, their phenotype directly reflects their genotype.

Ten mutant yeast strains have been isolated that cannot grow on medium with galactose as the sole carbon source (galactose medium). The ten strains are named A-J. Each mutant was separately crossed to the others to form a set of diploid strains. The ability of the diploids to grow on galactose medium was used as a measure of complementation. The results of this experiment are given in the chart below.

Gal Mutants

	A	B	C	D	E	F	G	H	I	J
A	–	+	+	+	+	–	–	+	–	+
B		–	+	+	+	+	+	–	+	+
C			–	+	+	+	+	+	+	+
D				–	–	+	+	+	+	–
E					–	+	+	+	+	–
F						–	–	+	–	+
G							–	+	–	+
H								–	+	+
I									–	+
J										–

Are mutants A and B in the same complementation group?

Are mutants A and F in the same complementation group?

How many complementation groups are present?

To determine whether the four genes identified were linked to one another, each diploid strain was sporulated and 1000 random spores were analyzed for growth on galactose medium. The number of spores that were able to grow on galactose medium is indicated in the table below.

	A	B	C	D	E	F	G	H	I	J
A	0	250	0	0	0	0	0	250	0	0
B		0	250	250	250	250	250	0	250	250
C			0	0	0	0	0	250	0	0
D				0	0	0	0	250	0	0
E					0	0	0	250	0	0
F						0	0	250	0	0
G							0	250	0	0
H								0	250	250
I									0	0
J										0

Since we know certain strains contain mutations in the same gene, we can simplify the chart above to reflect the same data by grouping together mutants from the same complementation group. We will arbitrarily give each complementation group a *gal* gene designation as follows: Mutants A, F, G, I—*gal1*; Mutants B, H—*gal4*; Mutant C—*gal7*; Mutants D, E, J—*gal10* (these designations are given based on actual *gal* genes in yeast).

	gal1	gal4	gal7	gal10
gal1	0	250	0	0
gal4		0	250	250
gal7			0	0
gal10				0

Is *gal1* linked to *gal4*?

gal1 shows no wild-type recombinants with either *gal7* or *gal10*. How can you explain this observation?

Analysis of 100,000 haploid spores produced from each of the crosses above were plated on galactose medium. Use this data to determine the map distance between *gal1* and *gal7*, *gal1* and *gal10*, and *gal7* and *gal10*.

	gal1	gal4	gal7	gal10
gal1	0	25000	35	16
gal4		0	25000	25000
gal7			0	18
gal10				0

Problem 5

This problem is a simulation and can only be done on the *Interactive Genetics* CD-ROM.

Population Genetics

Goals for Population Genetics:
1. Be able to relate allele frequencies at a gene to population homozygote and heterozygote frequencies.
2. Be able to use Chi-square test to determine if a population is in Hardy-Weinberg equilibrium.
3. Combine family studies with population frequencies to predict the chance a child will be homozygous for a disease allele.

Problem 1

Do you know your genotype with respect to your ABO or MN blood types? Will you consider these genotypes when you decide to marry? Most people are unaware of the alleles they carry for the majority of their genes. How do you study the genetics of animals in a natural environment, where family units are usually impossible to discern (e.g., fish). Because of the lack of pedigree data, individuals are regarded only as samples from the larger population.

A population, consisting of interbreeding individuals in a prescribed geographical area, contains a reservoir of all the gene copies (alleles) that will give rise to the individuals in the next generation. In these and similar examples, the population's allele frequencies are used to predict the genotype frequencies of the individuals.

The reservoir of alleles for a single gene is referred to as the gene pool for that gene. The genotypes of individuals can be considered a random (unbiased) sampling from the gene pool, with a gamete representing a single sample from the gene pool. In the following sets of questions we are going to use a bowl of ping-pong balls (whose different colors represent different alleles) to simulate random sampling.

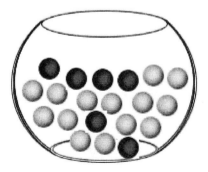

To represent diploid individuals subsequent questions will use pairs of samplings. What is the frequency of black ping-pong balls in the adjacent bowl?

Note that the probability of randomly obtaining a black ball, represented by p, is the frequency of black balls in the bowl.

What is the frequency of the white ping-pong balls in this bowl?

The frequency of white balls, represented by q, is the probability of randomly obtaining a white ball. Here black and white are the only colors of Ping-Pong balls in this bowl, and $p + q = 1$. To represent diploid individuals subsequent questions will use pairs of samplings.

What is the probability of picking two black balls (homozygous black)?

What is the probability of picking two white balls (homozygous)?

What is the probability of picking one white ball and one black ball (heterozygous) in either order?

As you have probably noticed, the sum of the probabilities found equals 1, that is $p^2 + 2pq + q^2 = 1$. This equation, which is derived from $(p + q)^2 = 1$ (when there are two alleles for a certain trait), describes the expected genotype frequencies of a population in Hardy-Weinberg equilibrium.

Problem 2

Consider this hypothetical population of 25 individuals of the FISH species. The sum total of all alleles present in the population is referred to as the population's gene pool. We will examine only one gene locus with two possible alleles: F for green color, and f for white color. In this population, the solid green fish are homozygous for F, the solid white fish are homozygous for f, and the spotted fish are heterozygous. Answer the following questions using the observed phenotypes.

Using the fish above, calculate p, the frequency of the green allele.

What is q, the frequency of the white allele? Recall, in the population above, ten fish are *FF*, five fish are *ff*, and ten fish are *Ff*.

To determine if this population is in equilibrium you need to determine the expected numbers using the Hardy-Weinberg equilibrium equation. To begin with, what is the expected frequency of *FF* individuals?

What is the expected <u>number</u> of FF individuals in this population?

What is the expected number of *ff* fish? What is the expected number of *Ff* individuals?

To find whether or not the population depicted above is in HWE (Hardy-Weinberg equilibrium), we will perform a chi-square (χ^2) test, using expected and observed values of each genotype. The χ^2 test is used to determine whether deviation from expected values are due to chance alone. If the deviations are too large we conclude that something besides chance is involved and the population is not in HWE. What is the value of χ^2 for this example? This χ^2 value can be converted into a probability value. To do this, we need the number of degrees of freedom (df) for the particular χ^2 test. The df equals the number of variables - 1. Although there are three genotypic classes, their numbers are determined by only two variables (p and q). Therefore, the degrees of freedom is 1. Using a χ^2 distribution chart, determine the probability (the p value) of obtaining these deviations due to chance alone. From your χ^2 test, is the population depicted above in Hardy-Weinberg equilibrium (HWE)?

Problem 3

Men are from Mars. Women are from Venus. When Mars and Venus collide, a new population arises. The allele frequencies on Mars and Venus differ. In the following problem, two alleles of a gene are represented by gray and blue ping-pong balls.

Mars   Venus

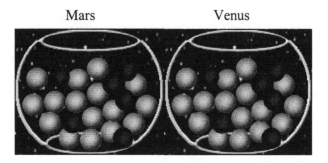

Predict the genotype frequency for homozygous gray in the new population.

What proportion of homozygous blue individuals do you expect in the new population?

What is the proportion of heterozygotes you expect the new population to have?

If the new population breeds only among themselves, will the next generation be in Hardy-Weinberg equilibrium?

Problem 4

The ability to taste the chemical phenylthiocarbamide (PTC) is an autosomal dominant phenotype. Tasters can detect it as extremely bitter, while nontasters cannot detect it at all. The genotype frequency of the homozygous recessive individuals (nontasters) in the U.S. population is 0.35, or 35%. Note that this population is in Hardy-Weinberg equilibrium (HWE) for this trait. What is the allele frequency for the nontaster allele?

q is commonly used to represent the allele frequency of the recessive (nontaster) allele. q^2, therefore, represents the genotype frequency of the homozygous recessive, nontaster individuals with a value of 0.35. What is the allele frequency for the dominant taster allele?

p is commonly used to represent the allele frequency of the dominant (taster) allele. After finding allele frequencies, we will use these values to find genotype frequencies. Answer the following questions using p and q. What is the frequency of heterozygotes in the general population?

What is the frequency of heterozygotes among the PTC taster population only?

A PTC taster man whose mother is a taster while his father is not, marries a taster woman. Recall that PTC tasting is an autosomal dominant trait. What is the probability that their child will be a nontaster?

Problem 5

Red green colorblindness is a recessive sex-linked trait. In a given population that exists in HWE, one in every eight males is colorblind. Using this information answer the following questions. What is the allele frequency for the red green colorblindness allele? (Or, what is q?)

What proportion of all women are colorblind?

In what proportion of marriages will all the males be colorblind but none of the females?

In what proportion of the marriages will all the males be normal and the females be heterozygous?

Molecular Markers

Goals for Molecular Markers:
1. Recognize the kinds of variation in DNA sequences between homologous chromosomes that can be used as codominant alleles, including restriction fragment site polymorphisms and variations in the number of repeat sequences.
2. Understand the techniques (Southern Analysis and PCR) used to detect these variations within a defined chromosome region.
3. Understand how molecular markers are used for linkage studies to locate a disease gene and to start the positional cloning of the gene.
4. Use molecular markers to determine the genotype of an individual for disease prediction, for relationship studies between individuals, and for forensic identification of unknown individuals.

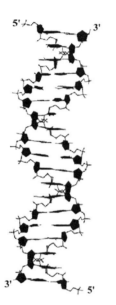

Problem 1

Restriction fragment length polymorphisms are identified by screening DNA isolated from members of families, using an array of different restriction enzymes. Random human genomic clones are then used as probes in a Southern blot. In the following problems you must identify the probe and restriction enzyme that gives rise to a polymorphism and then determine if the RFLP is linked to the inherited trait.

The DNA sample from each person is split into six tubes and digested with the following enzymes:

Apa I, Bam HI, Eco RI, Hind III, Sal I, Xma I

Gel electrophoresis separates the DNA fragments according to size. Denaturation of the DNA and transfer to nitrocellulose allows specific bands to be detected by hybridization with a radioactive probe (Southern blot). Choose the probe and restriction enzyme below that has an RFLP linked to the disease gene indicated by the pedigree.

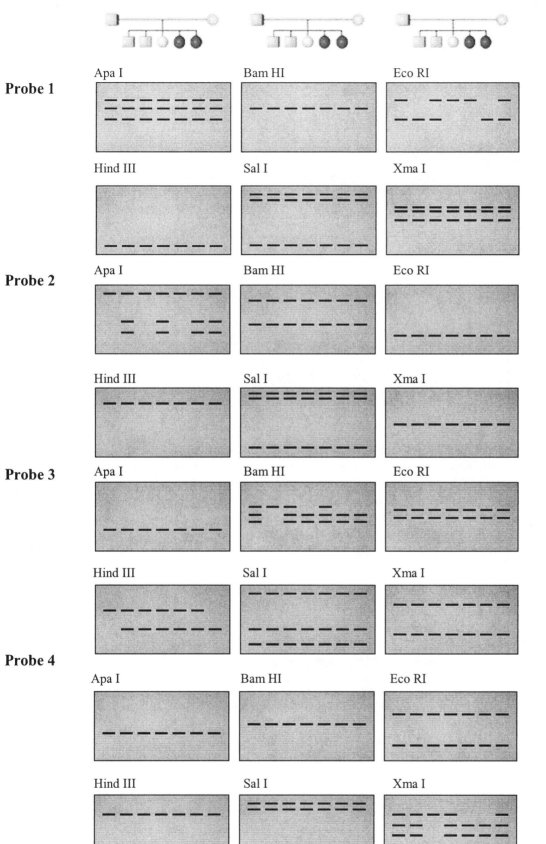

Problem 2

Part a

In the pedigree below, the presence of the rare recessive disease phenylketonuria (PKU) is indicated by filled symbols. Using the *A/a* above, assign genotypes based on the pedigree analysis.

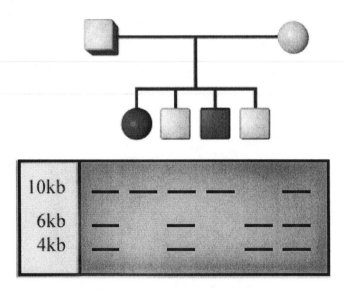

DNA was obtained from each of the individuals in the pedigree above, digested with a restriction enzyme and probed with a fragment known to hybridize to a linked RFLP. Analysis of the Southern blot above in conjunction with the pedigree allows determination of the linkage between the *A/a* gene and the RFLP shown. Focus first on the affected children. Use this information to determine the coupling (linkage) in each of the parents and then in the unaffected children. On the parental chromosomes shown below, fill in the galactosemia alleles (*A/a*) and the RFLPs (10 or 6-4) on the chromosomes to indicate linkage.

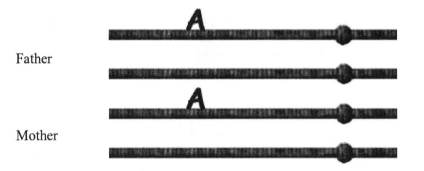

Part b

As a second example, let's look at linkage between the gene causing galactosemia and a nearby RFLP. In the pedigree above, the presence of the rare disease galactosemia is indicated by filled symbols. Using *A* and *a*, assign genotypes to each individual.

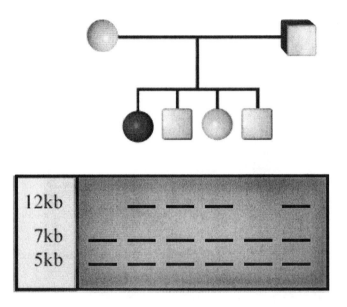

DNA was isolated from each of the individuals in the pedigree above and a probe from a linked RFLP was used in a Southern analysis to determine the genotype of each of the members of this family. On the parental chromosomes shown below indicate the linkage between the galactosemia alleles (*A/a*) and the RFLPs (12 or 7-5).

What is the probability that Child 3 is a carrier of galactosemia? (Assume the RFLP is tightly linked to the galactosemia gene and no crossing over occurs between them).

What is the probability that Child 4 is a carrier of galactosemia? (Again assume tight linkage and no crossing over.)

Problem 3

Part a

The pedigree shown below traces the inheritance of polydactyly, associated with a dominant, autosomal allele (*D*). Southern analysis reveals a closely linked RFLP. On the chromosomes shown below, indicate this linkage by typing in the appropriate *D* or *d* alleles and the linked 8 kb (8) or 6-2 kb (6-2) alleles.

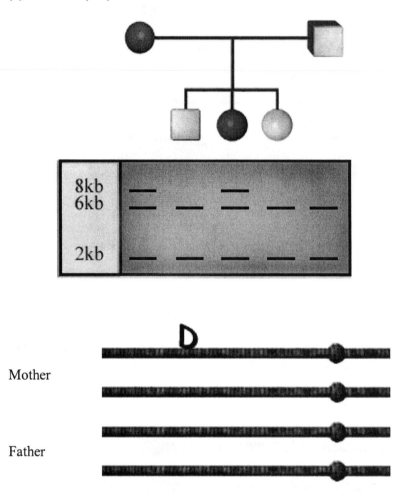

Use the Southern blot above to determine which children are heterozygous for polydactyly.

Child 1 marries a normal female and they are expecting a child. What is the probability the child will be affected?

Child 2 marries a normal male and they are expecting a child. What is the probability the child will be affected?

Would analysis of this RFLP in the fetus help them determine whether their child will have polydactyly?

Part b

Red-green colorblindness in humans is caused by an X-linked recessive allele (*c*). An STRP (small tandem repeat polymorphism) was found that is closely linked to the colorblindness gene. Assume no recombination in this problem. PCR was used to amplify this region of the genome for the individuals in the pedigree. Assign the appropriate alleles (*C*, *c* or *Y*) and linked STRPs (6 or 7) on the parental chromosomes below.

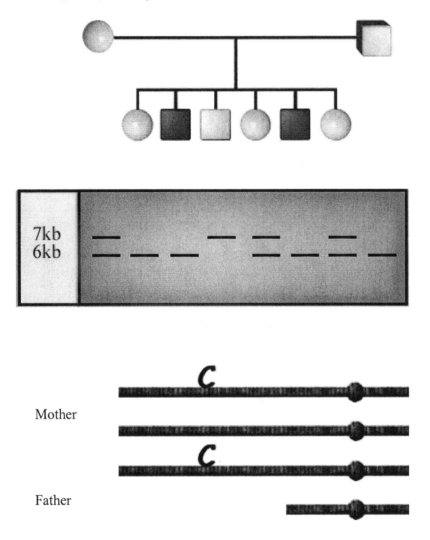

Use the gel above to determine which children are heterozygous.

Child 1 marries a normal male and they are expecting a child. What is the probability their child will be colorblind?

They find out that their baby will be a boy. Would analysis of this STRP in the baby help them determine whether their child will be colorblind?

Problem 4

You have recently identified two molecular probes (A and B) that hybridize to chromosome 12 in yeast. Although the loci are linked, you suspect they may be far enough apart to measure recombination. To test this, you mate two haploid strains to produce a diploid, which is then induced to undergo meiosis. You examine 100 meiotic haploid spores by Southern blotting and find the results shown on the next page.

Pattern	# spores	Probe A	Probe B
1	44	8 kb, 1kb	4 kb
2	7	9 kb	4 kb
3	42	9 kb	4.4 kb
4	7	8.1 kb	4.4 kb

You notice immediately that four patterns are discernable and that these are found at different frequencies. Using the information in the table, determine the genotype of the diploid cell formed from the two haploid parents. Write the linked alleles on the chromosomes below.

Probe A   Probe B

Using the data in the table, determine the distance between locus A and locus B.

Problem 5

Individuals with Duchenne's muscular dystrophy (DMD) are missing an important structural "glue" of skeletal muscle. This makes them susceptible to muscle tears and leads to the progressive death of muscle tissue. Many patients show pseudo-hypertrophy (false muscle enlargement), especially in the calves, as muscles die and are replaced by fat and connective tissue. Almost all are confined to a wheelchair by the age of 10, and most die in their twenties, as their respiratory muscles fail.

Although the gene leading to DMD has been cloned and sequenced (dystrophin located on the X chromosome) and many mutations can be detected directly, not all alleles have been identified. In these cases, a linked polymorphism can be used to determine probabilities. Suppose in the following family an STRP located 6 map units away has been identified.

Answer the following questions taking into account any recombination that may occur. Note that all known disease alleles are recessive, appearing primarily in males.

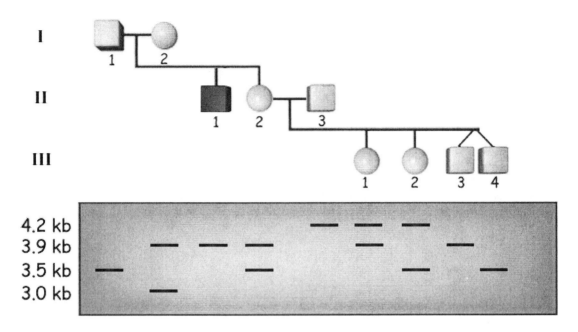

Roya (II-2) and Fardad (II-3) are in genetic counseling, after finding out she is pregnant with twin boys. Roya's brother has DMD, and it is believed to be inherited from their mother. What is the probability that Roya is a carrier? Remember, the polymorphism shown is known to be 6 map units from the DMD gene. Assume there is no crossing over in the mother I-2 leading to the child II-1.

Roya has two unaffected girls. What is the probability that her daughter Sonya (III-1) is a carrier for DMD?

What is the probability that her second daughter Kate (III-2) is a carrier for DMD?

Roya is concerned about her twin boys. The first question she asked the genetic counselor was if the twins were identical or fraternal. Based on the gel above, how would you answer this question?

What is the chance the first twin (III-3) has DMD?

What is the chance the second twin (III-4) has DMD?

Problem 6

Megan, a nineteen year old, was found severely beaten off the side of a small rural road by a patrol officer. She was taken immediately to the nearest hospital, where she was treated for her injuries and it was discovered that she had also been raped. A semen sample was recovered and stored for DNA analysis.

Two men seen in the area at the time of the crime were brought in for questioning. They claimed to have been together all through the night and knew nothing about the young woman. Blood samples were taken from both men and four STRPs were examined and compared with the semen sample.

Southern analysis of both the suspects' blood samples (1 and 2) and the semen sample (S) was performed using well characterized STRPs. Allele frequencies determined using the FBI databases for the U.S. population were then used for probability determinations (only alleles A, B, and C shown). It is assumed these alleles are present in Hardy-Weinberg equilibrium. From the data below, can either suspect be excluded?

STRP 1

Frequency in US Population

Allele	Frequency
A	1/3
B	1/9
C	1/15

What is the probability that an individual would have the genotype consisting of STRP alleles A and B? Does this prove that Suspect 1 committed the rape?

STRP	Allele	US Population
1	A	1/3
	B	1/9
	C	1/15
2	D	1/53
	E	1/78
	F	1/5
	G	1/23
3	H	1/7
	I	1/90
	J	1/43
	K	1/28
4	L	1/86
	M	1/35
	N	1/4
	O	1/13

An additional three independent STRP loci were typed (as shown above). Looking at this additional data, did Suspect 1 rape the victim?

What is the probability of an individual in the United States having the set of alleles found in the semen sample for all four STRPs?

In a population of six billion (the current population of the earth), how many individuals will have this pattern?

Medical Genetics

Goals for Medical Genetics:
1. Understand differences in classification of genetic diseases as chromosomal, single gene, or multifactorial in origin.
2. Understand how to use pedigree information to determine the mode of inheritance.
3. Interpret karyotypes to identify genetic diseases associated with chromosomal disorders.
4. Be able to understand and use in diagnosis molecular genetic techniques (ASO, FISH, and PCR) that recognize mutant DNA causing genetic diseases.

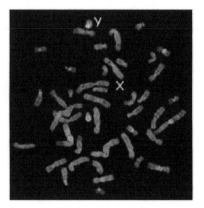

In this section, you will use case studies to see how the genetic techniques you have studied are used in medicine to help doctors form a diagnosis. To aid you in your diagnosis, the patient's clinical description and family history will be available to you. You will be able to perform various chromosomal and molecular genetic tests. At any point in your investigation, you can use the genetic reference database, and whenever you are ready, you may submit a diagnosis. You may want to take notes along the way to keep track of the information you have gathered.

Case I

Frank and Mary Smith asked that their six-year-old son George be seen by a pediatrician. His parents report that that he was slower than their other children to make his developmental milestones. They had attributed his delays to a heart defect that he had when he was born that had required surgery in infancy. He has now started school and is having trouble because he is acting out and having some behavioral problems. They are concerned that he is not ready for school.

The pediatrician is concerned about the developmental delays but notes that George's language skills appear normal or even advanced for a six year old. He notes facial dysmorphisms (unusual facial features) that seem different from his family, including full lips and puffy eyes. The pediatrician suggests that genetic tests be performed to rule out the following conditions for which further information is given:

> Angelman syndrome
> Di George syndrome
> Down syndrome
> Fragile X
> PKU
> Prader-Willi syndrome
> William syndrome

Case II

Han Chen and Yuh Nung Lee are referred to the Westside Fertility Clinic after a series of miscarriages. Han Chen is 43 and her husband is 38. Neither reports any major health problems. Han Chen was born in Taiwan. Four brothers and sisters survive in her immediate family. Two siblings died shortly after birth and one was stillborn. Yuh Nung is an only child with no family history of infant mortality.

The clinic first evaluates Yuh Nung's sperm count and finds it is normal. During this evaluation Han Chen became pregnant, but loses the fetus after three months. Fetal tissue was obtained and examined for chromosomal abnormalities in the cytogenetics laboratory. Refer to the chromosomal section under disease reference for the list of chromosomal disorders that are considered.

Case III

Sara and David Goldenstein have brought their daughter Rebecca in for her six-month checkup. They are concerned because she has stopped gaining weight, though she seems to have a healthy appetite. She has had frequent and numerous colds and suffers from diarrhea. Sara and David are first cousins, but there is no other family data that is useful.

The physician finds no evidence of mental retardation. Based on the relatedness of the parents, a recessive single Mendelian gene is suspected of causing the symptoms. Refer to the single gene section under disease reference for the list of disorders that may be considered in this case.

Case IV

Jan de Broek (age 60) was referred for medical examination by social services after his arrest for vagrancy and disorderly behavior. Social services reports that he has lost his job and apartment and is currently homeless. Jan's family was contacted and states that his behavior seems similar to both his father's and uncle's at this age (they are both deceased).

The physician records several of the symptoms associated with senile dementia. These include episodes of severe irritability, forgetfulness, anxieties, ataxia, and alcohol abuse.

Behavioral changes are among the hardest to diagnose; age-related causes may include alcoholism, Alzheimer disease, and Huntington disease. Schizophrenia is usually an early onset disorder, but should be considered if the patient history is uncertain. Further information is given in the disease reference list.

Molecular Biology

....

Gene Expression

Goals for Gene Expression:
1. Understand the molecular mechanisms by which the genetic information in a DNA sequence is converted to an amino acid sequence, or protein.
2. Be able to compare the similarities and differences between prokaryotic and eukaryotic transcription and the proteins involved in this process.
3. Understand the mechanism of splicing and how introns and exons are mapped in eukaryotic genes.
4. Understand how RNA is translated into a protein and how changes in a gene's sequence can give rise to defective proteins.

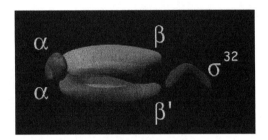

Problem 1

RNA Polymerase holoenzyme in the cell contains a sigma factor that is 70 kD (σ^{70}). However, there are several other sigma factors that recognize different promoter sequences and respond to specific cellular signals. One such sigma factor (σ^{32}) is activated in response to heat shock and recognizes promoters encoding chaperone proteins and proteases. Expression of these genes helps the cell deal with the heat denatured proteins to prevent further damage.

Comparison of promoter sequences found near genes that are transcribed under heat shock conditions revealed the recognition sequence of sigma-32. Shown below are ten such promoter sequences. Determine the best consensus sequence for this promoter.

GCCTATATA
GCCCAACTT
CCCCATGTA
CCGCATTGA
CGCCACGTA
CCCGCTATT
CGCCATCTA
ACTCTTTTT
CCCTAGATA
CGCCATGTA

Promoters that have sequences that are a good match with the consensus bind RNA polymerase more often and lead to an increased amount of transcription. These are considered strong promoters. Which promoter sequence above has the best match to the consensus sequence? Which promoter would be considered the weakest promoter?

> 5' AGCCTAGCTCCATATAGAACGATCATCTAAG 3'
> 3' TCGGATCGAGGTATATCTTGCTAGTAGATTC 5'

You have recently found a new heat shock gene above and decide to use the consensus sequence to give a first approximation of the transcription start site (+1). Which nucleotide in the sequence above represents +1?

Write the appropriate substrates in a chemical equation below to describe formation of the first phosphodiester bond from this promoter.

Which end of the mRNA has the triphosphate group?

Only one of the two DNA strands is used as a template for RNA synthesis. Which strand is used in this example?

Problem 2

Exon-intron structure of genes can be determined in a number of ways. One method involves the comparison of cDNA with genomic DNA. This can be done either by DNA sequencing or Southern analysis. In this problem, the gene structure of calcitonin will be examined by Southern analysis.

Calcitonin is a peptide hormone synthesized by the thyroid gland and serves to decrease circulating levels of calcium and phosphate. This is achieved primarily by inhibiting bone decalcification (or resorption).

mRNA was isolated from human thyroid cells and converted to cDNA using reverse transcriptase. Isolation of the calcitonin cDNA was done by hybridization with the calcitonin genomic DNA (previously cloned). Ten subclones from the regions of the genomic DNA shown below were isolated and labeled for use as probes (1–10).

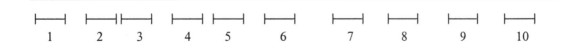

In order to determine where the calcitonin gene is located in the original genomic clone, Southern blots of the thyroid cDNA clone were hybridized with each of the ten probes (lane t, vector plus cDNA insert). A control of total genomic DNA digested with Not1 is included in lane g. See CD-ROM for the hybridization results.

The Poly-A site was found to be on the right side of the genomic DNA above, orienting the gene in a left to right direction. Use this information and the results from the Southern analysis to create an exon-intron map for the calcitonin gene.

Surprising results were observed when the same probes were used in conjunction with a cDNA clone isolated from neurons. Go to the CD-ROM to observe the results of Southern analysis on the neuronal cDNA clone (n) versus genomic DNA (g). Additional questions are available on the CD-ROM.

Problem 3

Hemoglobin is a tetrameric protein ($\alpha_2\beta_2$) that functions to carry oxygen in all vertebrates and some invertebrates. Each subunit of hemoglobin is associated with a prosthetic group, iron containing heme (in white above), which provides the ability to bind oxygen. Disorders associated with alterations to, or the disrupted synthesis of, hemoglobin are the most common genetic diseases in the world.

The β-globin gene is a relatively small gene, composed of three exons and two introns. One major class of hereditary disorders associated with hemoglobin is characterized by alterations in the amino acid sequence of β-globin but not the amount (structural variants). A second major class is characterized by a reduced level of hemoglobin (thalassemias), due to either mutations affecting transcription or splicing. In the following problem, β-globins were obtained from mutant hemoglobins. Use your knowledge of gene expression to help determine the cause of each of the mutations described.

Blood samples were collected from individuals homozygous for several distinct hemoglobin disorders. Protein from each sample was electrophoresed through either a native polyacrylamide gel or an SDS gel (shown below) and the separated proteins transferred to nitrocellulose. β-hemoglobin was then detected using antibodies (Western analysis).

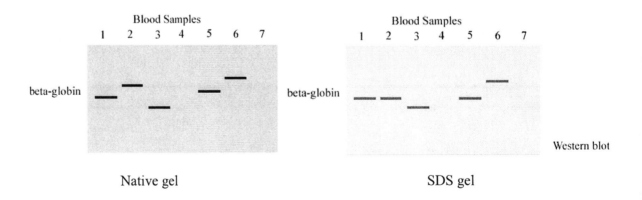

The first lane (blood sample 1) contains wild-type β-globin. Which lanes from 2–7 above represent structural changes in β-globin?

Compare the sample in lane 2 on a native versus SDS gel. Do these results suggest a change in the length of the polypeptide or the charge of the polypeptide?

Compare the sample in lane 3 on a native versus SDS gel. Do these results suggest a change in the length of the polypeptide or the charge of the polypeptide?

DNA and RNA samples were then examined by <u>Southern</u> and <u>Northern blotting</u> to detect changes in the DNA or RNA.

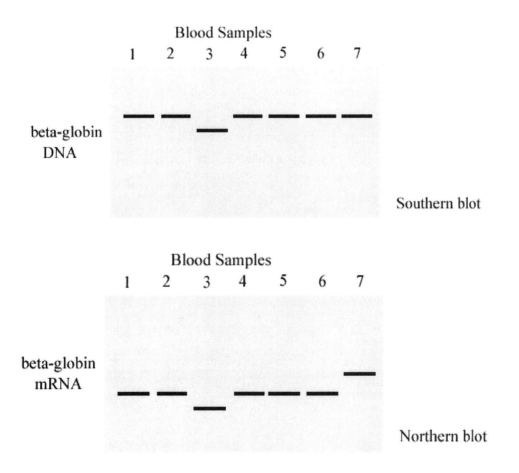

Again, the first lane contains a wild-type sample. To begin, let us examine sample 2 in detail. You have already determined the β-globin polypeptide differs from the wild type by a charge change. Does the Southern blot detect any difference in sample 2 from wild type?

Does the Northern blot detect any changes in mRNA length in sample 2?

DNA from sample 2 was analyzed by dideoxy sequencing and compared to the wild-type gene. A single mutation was found in exon 1, as shown below (only the nontemplate strand is shown). Use the genetic code to determine whether this mutation represents a nonsense mutation (stop codon), missense mutation (amino acid substitution) or a frameshift mutation (insertion or deletion).

Wild type   ATG GTG CAC CTG ACT CCT G̲AG GAG AAG TCT GCC

Sample 2   ATG GTG CAC CTG ACT CCT A̲AG GAG AAG TCT GCC

Now look at samples 3–7 and determine which fit the following descriptions. Which of the samples above could be explained by a nonsense mutation?

Which of the samples can be explained by a deletion of more than a few nucleotides?

Sickle-cell anemia is known to be caused by a mutation that changes a glutamate codon at amino acid 6 to a valine. Which sample above is consistent with the sickle cell allele (*HbS*)?

Many thalassemias are caused by mutations that effect splice sites. Which sample above would be consistent with a splicing mutation?

The only sample that has not yet been identified is sample 6. What explanation is consistent with the results obtained for sample 6 above?

Molecular Biology

Gene Cloning

Goals for Gene Cloning:
1. Review the use of restriction enzymes and be able to use restriction enzymes to make a physical map of a clone.
2. Review basic cloning techniques, including vectors, library construction, and screening techniques.
3. Review the DNA sequencing, Southern analysis, and PCR amplification.
4. Be able to design a cloning strategy, beginning with understanding what is to be cloned, why it is to be cloned, and what will be done with the clone. The cloning strategy consists of the origin of DNA (genomic or cDNA library), choice of cloning vector, and choice of screening technique to recognize clone.

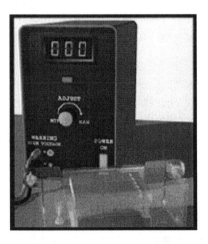

(-) electrode

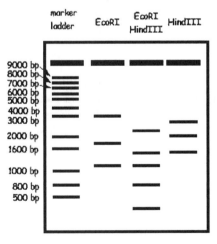

(+) electrode

Problem 1

A restriction map provides information as to the presence and position of one or more restriction sites. In many ways it is like the road map to a DNA molecule, and it is used in many recombinant DNA techniques.

Digestion with a single enzyme can provide information about the number of restriction sites and the distance between them, but the order of the sites cannot be discerned. However, if a second enzyme is used both alone and in combination with the first enzyme, a map can be constructed.

In this problem, you will determine the restriction map for a linear DNA sample using the enzymes Eco RI and Hind III. Three samples of DNA are first digested with either Eco RI, Hind III, or both Eco RI and Hind III. The resulting fragments (restriction fragments) are separated by size using gel electrophoresis.

The results are shown here.

Since the DNA fragments in each lane are derived from the same linear DNA fragment, the sum of the size of fragments in each lane should be the same and reflect the size of the original DNA fragment. How many base pairs are in the original DNA molecule?

Focus first on the Eco RI single digest. Since we know our sample DNA is linear, how many EcoRI sites are present in the DNA?

When the same DNA sample is digested with Eco RI and Hind III, five fragments are generated. Let's look at this one band at a time. Since the 3000 bp Eco RI fragment is not present in the double digest, it must contain at least one Hind III site. Which fragments in the double digest add up to 3000? Which bands in the double digest add up to 1800? Now there is only one fragment left in the double digest. This is the same size as one of the Eco RI fragments in the single digest. It must correspond to the band in the Eco RI lane.

Now, let's focus on the Hind III digest. The largest band, 2500 bp, disappears in the double digest. Which two bands in the double digest add up to 2500 bp? Which two bands in the double digest add up to 2000 bp? That leaves just the 1500 bp Hind III fragment, which is also present in the double digest. This means that there are no Eco RI sites inside this fragment and may indicate the end fragment on a linear DNA.

Since this is a linear DNA fragment, let's start with one of the fragments that is the same in both the single and double digest. This could be an end fragment. Although we could start with either the 1200 or 1500 bp fragment, let's begin with placing the 1200 bp fragment on the left side and build the map from there. Create a map using the information gained from the single and double digests above.

Problem 2

You have just identified a patient who appears to have an altered hemoglobin protein (β-subunit). Analysis of the patient's blood by Southern, Northern, and Western blotting reveals no detectable change of the DNA, a larger RNA, and smaller protein. You surmise that the mutation affects hemoglobin splicing. With a clone of the wild-type hemoglobin in hand, and cultured skin cells from the patient (as a source of DNA or RNA), select a type of insert, a type of vector, and a method in which to identify and study the mutant hemoglobin gene.

Problem 3

After many months in the laboratory, you have finally obtained a pure preparation of DNA polymerase from a novel fungus. However, you have a very limited amount and cannot do all of the biochemical experiments you would like. Your advisor suggests that you use your purified polymerase to clone the gene encoding the polymerase. After consulting several other graduate students, you come up with a plan. Select a type of insert, a type of vector, and a method to identify the DNA polymerase gene.

Problem 4

Polydactyly is known to be caused by a dominant allele and is characterized by extra fingers and/or toes. Recently, an RFLP has been identified which is linked to polydactyly and you decide to use this information to clone the gene. Determine what type of insert you would use in making your library, what type of cloning vector would be optimal, and how you would go about identifying the correct clone.

Exploring Genomes

TABLE OF CONTENTS

Introduction

Computer science has had a major impact on the biological sciences, and this is particularly true in the field of sequence analysis. As sequencing technology improved and became highly automated in the 1990s, researchers around the world accumulated a wealth of information that rapidly grew beyond the scope of what scientists could analyze independently. Scientists and government officials with considerable foresight lobbied for centralized institutions that could store this information and make these resources available to researchers worldwide through the internet. At the same time, programmers were developing potent analysis tools that could mine the information in the databases for comparisons at a very detailed level. With the advent of easy and near universal internet web access, researchers have come to rely on these institutions to an ever increasing extent. *Genomics*, the study of whole genomes, and *bioinformatics*, the development and use of computer tools to analyze them, now contribute to virtually all areas of biological science. It is safe to say that understanding how to make use of these resources is an essential skill for anyone in the fields of biology and biomedical sciences.

The resources available are located on thousands of different web sites and are quite varied in size and nature. Some are very narrowly focused, such as databases created for a particular organism or sequencing project. Alternatively, a site might be set up simply to allow the use of a particular type of analysis or a new software tool. Some sites may represent the efforts of a single laboratory. Others, such as the National Center for Biotechnology Information (NCBI) in Washington, D.C., have as their mandate the collection of all publicly available sequence and related information and the development of software tools for its use.

The curators and computer programmers at comprehensive centers such as NCBI compile sequence and related data as it becomes available from sequencing centers around the world and present it in a form that is easily accessible to researchers. Their efforts are coordinated with similar genome centers in many countries. NCBI and its counterparts, such as the European Molecular Biology Laboratory (EMBL) in Germany, and the DNA Database of Japan (DDBJ), are the primary centers for the collection and archiving of sequence data.

The software tools that are used for access and analysis differ at the various web sites. First and foremost, an archival site such as NCBI allows researchers to retrieve the original sequence record and all subsequent updates and modifications for a gene, protein, or genome. The sites also provide tools for gene alignment that are essential in answering questions regarding the conservation of genes among organisms. This often allows information from major genetic systems such as yeast, *Drosophila,* or *Arabidopsis* to provide clues as to the biological function of a human gene, or vice versa. At another level, the programmers actively participate in designing new software tools for problems as diverse as identifying the coding regions within newly sequenced genomes, for linking gene discoveries to human diseases, and for examining and comparing the three-dimensional structures of proteins.

Students trying to find their way through one of these web sites are often intimidated by the unfamiliarity of the material and terms, as well as a sense that their computer skills are not as advanced as they would like. They needn't worry on either count. Web-based access at its best is designed to allow very sophisticated analyses to be carried out with little or no detailed knowledge of the underlying programming. All the student needs is a little guidance.

The tutorials in this book and on its accompanying web site at **www.whfreeman.com/young** focus almost entirely on a single site, the National Center for Biotechnology Information. It is one of the world's largest resources and it has a wide range of useful tools and presentations available. The tutorials here are designed to introduce the basics of genomic analysis at a level appropriate for college students. These tutorials give students initial guidance and practice in exploring the breadth and depth of the resources by helping them understand the structure and content of the databases, and facilitating direct practice in the use of search and alignment tools.

Each tutorial is written as an interactive guide to an aspect of the NCBI site. The user's browser window is split into two parts. Guidance is provided as text on one side of the screen, while the main window is directly connected to NCBI. This latter window runs live on the NCBI computers. Using real examples from the research literature, the tutorial walks students through the exercise and tells them precisely how to run the programs and interpret NCBI results.

These tutorials are not intended to be comprehensive. In many ways, they simply scratch the surface of the resources. They do, however, allow a student to gain practical skills and to do independent research very quickly. A student that becomes familiar with this material will have no problem going forward to more advanced use of the resources. NCBI itself has a broad array of tutorials associated with their web site. Many of these are at a more advanced level and the introductory ones provided here serve as a bridge to them.

The tutorials are designed to display some of the major public resources of the NCBI, keeping in mind the level and interests of the beginning genetics student. For each tutorial the searches are run with known genes or sequences and for the most part examples from our major genetic systems such as *E. coli, Drosophila*, yeast, *Arabidopsis,* and human. Usually a fairly common protein or gene is chosen so that it might be familiar to a student in genetics. The choice of example is sometimes dictated by a desire for a relatively simple result (not always possible).

The tutorials include an overview of the NCBI site as well as examples of how to use the information resources such as Entrez, PubMed, OMIM, and the Cancer Genome Anatomy Project. There are several tutorials dealing with gene alignments using BLAST, as well as searches for conserved domains using pattern and profile searching. The student can explore the 3D structure of nucleic acids and proteins and even search the database for proteins with similar 3D structure. Lastly, taxonomy resources are examined and a multiple alignment of a gene family is used (ClustalW) in order to draw a phylogenetic tree. In all cases, students are asked to go back to the program using an input of their choice (with some suggestions made) so that they can reinforce their knowledge.

Database resources such as NCBI and the European Molecular Biology Laboratory (EMBL) are dynamic and constantly being improved. They increase in size daily and frequently add to the services that they offer. The inevitable consequence is that occasionally hyperlinks are altered. Although these tutorials are frequently monitored to ensure that they work, sometimes a glitch occurs. We would appreciate hearing of any problems so that they can be corrected. Please email us at **techsupport@bfwpub.com**.

Detailed Table of Contents

examine the domain structure of proteins and the way in which this is conserved through evolution.

1. Introduction to Genomic Databases

Starting link: **http://www.ncbi.nlm.nih.gov.**

Everyone has a difficult time keeping up with the flow of new information. This is particularly true in biology now as the pace of discovery accelerates. Databases have become an essential tool for accumulating and archiving raw data. They also play a major role in analyzing and presenting information to researchers and the public in an easily accessible form. In this *Exploring Genomes* tutorial we will survey one of these resources: The National Center for Biotechnology Information (NCBI) located in Washington, D.C.

One of the roles of NCBI is to archive raw DNA sequence data. The sequence information comes from research efforts in laboratories around the world as well as from large-scale, dedicated genome sequencing centers. The resulting database is referred to as GenBank. GenBank shares its resources with the European and Japanese equivalents so that there are three primary public repositories of such information in the world. Because of the automation of DNA sequencing over the past decade, these databases are increasing in size exponentially. GenBank includes the sequences of the *E. coli*, *Drosophila*, and human genome, as well as data from thousands of other species. At present it comprises some 13 million DNA sequences with a cumulative length of 31 billion base pairs. You can see the rate of growth by going to **http://www.ncbi.nlm.nih.gov/Genbank/genbankstats.html.**

Now let's go back to the NCBI home page, the public access point to its many resources (**http://www.ncbi.nlm.nih.gov/**). A quick glance at this page shows that NCBI contains far more than just the sequence repository. It is a rich source of information on all aspects of genetics and genomics. All of the divisions are searchable and information ranging from gene sequences, to the position of a locus on a human chromosome, to direct access to the scientific literature dealing with a particular gene, is immediately accessible.

There are six major categories of service listed across the top of the NCBI homepage. These are PubMed, Entrez, BLAST, OMIM, Taxbrowser, and Structure. In addition there are specialized projects and databases listed on the right hand side of the page. In this *Exploring Genomes* tutorial we will take a quick look at some of these resources. Subsequent tutorials will explore a particular resource in much greater depth. Let's click on each of the six resources at the top in turn. Click on '**PubMed**' first.

PubMed is the NCBI gateway to the biomedical research literature. It is a searchable database and information can be retrieved based on combinations of parameters such as author, subject key words, or organism. A complex query can be entered and a list of publications matching it will be returned. For instance, we could enter a simple search by author. If we entered **Hartwell LH** (one of the winners of the 2001 Nobel Prize in Medicine) and pressed '**Go**', PubMed would return a list of his current publications. Give it a try.

The list of Hartwell's publications is 5 pages long. Only the top and therefore most recent papers are displayed on the first page. Each paper is linked to its abstract and sometimes the full text of the articles. We will explore PubMed much more fully in a later tutorial. Let's go back and try another NCBI division by clicking the **NCBI icon** on the top left to get to the NCBI homepage again, and then clicking the button for '**Entrez**'.

The Entrez button opens a search engine that links all of the NCBI databases together. This allows access to everything from sequence to structure. The options are in the drop-down window at the left. The PubMed link that we looked at initially was only one of these options. Let's try searching for the human keratin protein sequence in the Protein database. Choose **'Protein'** from the drop-down menu, and then type **'keratin AND human'** in the text box. Now press **'Go'**.

The search returns a list of database entries including keratin-associated proteins and keratins themselves. Note that only the first 20 entries of around 500 are displayed on the first page! For each entry, clicking on the associated links will display the sequence, related information, and links to other parts of the databases. We can explore these later. Let's go back to the **NCBI** homepage and try the **'BLAST'** button.

BLAST is a powerful nucleic acid or protein alignment tool. It allows us to dynamically search the sequence databases (all 31 billion base pairs!) to find similar sequences in different organisms. It is extremely versatile and comes in many different forms for doing different types of searches. The underlying method is the same in each case, however. This is our standard software tool for doing such searches. It is very important and we will devote an entire tutorial to its use later. For now, let's move back to the **NCBI** homepage and click on to **'OMIM'**.
OMIM is the Online Mendelian Inheritance in Man database. Note from the overview on the OMIM homepage below that it integrates the known Mendelian genetics of human disease with the resources made available through Entrez at NCBI. We will explore this in detail later but for now let's open the **OMIM Statistics** link, under OMIM Facts on the left-hand menu.

The OMIM statistics page gives an overview of the breadth of the resource. Note the categorization by Mendelian inheritance pattern. Note also that over 13,000 entries are available with at least some information. Keep in mind that we believe there are more than 35,000 human genes! All genes will not necessarily be associated with a disease or visible phenotype and therefore for many genes little information is yet available. The Human Genome Project has provided a glimpse at large numbers of new genes of unknown function, reminding us of how much remains to be done.

Next, click back to the **NCBI** homepage and on to the **TaxBrowser** database.

The Taxonomy section groups all data by taxonomic classification. You may type in a species name to find out if any sequence information is available; for the major genetic systems, simply click on the links on the Taxonomy Browser home page. Try *Caenorhabditis elegans*, a nematode worm, one of our most powerful model genetic systems whose full genome sequence was recently determined.

This organism-specific page gives important information regarding phylogenetic lineage as well as the number of sequences of various types that have been deposited. These groupings may be called up at will and each of the genes listed are linked to the Entrez system. One important use of the groupings is to restrict other types of searches. For instance a BLAST search can be launched from the BLAST page and the database searched restricted to *Caenorhabditis elegans* only (or any other species).

Now, let's go to the last major subdivision, **'Structure'**.

The structure database contains the 3D structure for all nucleic acids and proteins whose shape has been determined by X-ray crystallography or nuclear magnetic resonance. We can call up these 3D models at will. The structure database is associated with the VAST program that allows for 3D structural comparisons among different proteins. It also will search the database on the basis of structure. These are very powerful tools and we will give them a try later in the term. Now, let's go back to **NCBI** home page to see some of its other resources.

Apart from the basic databases and access software, NCBI has a wide variety of highly specialized databases and analyses. These focus on particular problems or interest groups. Some are listed on the right hand side of the NCBI homepage below. We will just touch on a couple of them for now to get a sense of their capabilities.

The first is the Human map viewer. This allows us to visualize the various human chromosomes and the genetic loci on them. Let's take a look by clicking **'Map Viewer'** under the Hot Spots list. From the resulting page, choose **'Homo sapiens (human)'** from the Mammals category.

The human genome view is a visualization of the full chromosome set. Clicking on any chromosome number will expand the view of that particular chromosome. Let's try the **Y** chromosome.

The graphic of the Y chromosome is expanded so that we can see its banding pattern as seen by a cytologist. Aligned along it are the blocks of DNA that have been sequenced and represented here (contigs) followed by the genes located on these pieces of DNA. The genes are represented as blue diamonds in the second column labeled "Uni…", which stands for Unigene, the NCBI database that contains the compiled summary data for each gene. Roll over each diamond to get the identification information for each gene. Subsequent columns provide links to the various genes that have been annotated on the DNA sequence. Click on any of the **entries** in the Unigene list.

The UniGene summary is an annotated view. It is compiled by a curator and takes into account all known information regarding the sequence. UniGene also includes links to OMIM as well as a comparison to the most similar genes in other organisms. It is a very rich source of information. All of the data is cross-referenced by links into Entrez and various other databases that provide the gene sequence and research literature references. Ultimately the entire human genome will be available in finished form from this database.

Now let's go back to the **NCBI homepage** and click on the last resource under Hot Spots, NCI-CGAP, the **'Cancer genome anatomy project'**.

The Cancer Genome Anatomy Project is the last topic we'll briefly explore. This database focuses only on tumor tissue and strives to provide cross-referenced information for all genes thought to be involved in cancer. Each of the subsections enters the database with a different type of approach. Ultimately you can explore the database to great depth, searching out the genes involved, chromosomal locations and aberrations, and biochemical pathways. All of this information is related to the tissues and cells where the tumor that you are interested in originates.

Over the course of these *Exploring Genomes* tutorials, we will look at parts of these databases in much greater detail. A facility for handling them is an essential skill for a modern biologist. Ultimately, the only way to familiarize yourself with a resource of this type is to go to the web site (**http://www.ncbi.nlm.nih.gov/**) and start exploring some of the links. You might start by searching for the answer to the following simple questions:

- What is the publication history of your Biology and in particular your Genetics instructors?
- What organisms do they work with?
- What types of question are they trying to answer?

2. Learning to Use Entrez

Starting Link: **http://www.ncbi.nlm.nih.gov**.

The Entrez retrieval system is the entry point for searching most of the material at NCBI. Its strength is that it provides links between related types of information. For instance, in storing a DNA sequence file, the file is associated with the protein translation of the sequence, with the literature reference, and with links to similar genes or proteins in other organisms. It also provides a characterization of any notable features, such as conserved regions, and ultimately chromosomal location and 3D structure of the gene product. The retrieval system moves relatively effortlessly among these various types of data. The links are updated as new data are added to the databases (or new databases are developed) and the resource becomes richer and richer over time.

We looked very briefly at Entrez in the introductory tutorial. Now let's learn to use it by walking through an analysis of the dystrophin gene in humans, which is responsible for causing Duchenne muscular dystrophy (DMD).

Click on **'Entrez'** on the top blue bar to access the Entrez home page from the NCBI home page.

Remember that in the drop-down window on the left (in the 'Search' box) there are several options indicating different databases. All of these databases are ultimately linked within the framework of the Entrez retrieval system.

Let's start with a research literature search for dystrophin on PubMed. Since it is a well-characterized protein involved in an important human disease, expect a long list of retrievals.

Choose **'PubMed'** from the Search drop-down menu, type **'dystrophin'** into the text box, and then click **'Go'**.

The search returns a rather daunting several thousand entries, obviously a very active research area. They list in reverse chronological order and only the last few are displayed here. Notice the titles of some of the papers. Clicking on the authors' names will link you to the publication information and abstracts. Note that the reference numbers change as new publications are added to the database.
For this exercise, we are going to take a look at a paper written in 1987, when the molecular genetics of the dystrophin locus and the Mendelian genetics of Duchenne muscular dystrophy were converging.

Let's call up the Hoffman, Brown and Kunkel (1987) reference. Since it is so old, and the articles are listed in reverse chronological order, it will take a lot of clicking through the pages to get to it. A faster way to reach the article is to do a search with limits. Click **'Limits'** on the light blue bar above the Display button. In the Publication Date field From, type in **'1987 01 01'**. In the field To, type in **'1987 12 31'**. Now press **'Go'**.

When the search is completed, you should see the paper by Hoffman, Brown and Kunkel (1987) near the top of your screen. Now click on the **authors' names** to link to the information related to this paper.

The abstract of the paper is displayed as well as a link to the full text article. For most articles there is a registration and charge to go beyond this point. Also included is a set of links on the right (in blue). These point to closely related research articles, to the gene and inheritance pattern, as well as books and other resources. Clearly we could pursue an extensive library research project related to this gene simply by following a number of these links. For the moment we will find the chromosome map of the gene responsible for DMD.

From this page, click on 'OMIM', from the menu that appears when you click 'Links' near the title of the paper.

This link will connect us to all the literature related to DMD and dystrophin. For now, though, we just want to see the gene map locus. Click on 'Xp21.2'.

Now you should see a chart with all the gene locations related to DMD and dystrophin. Click on the first location, 'Xp21.2'. This will take us to Entrez Genome page.

The graphic that appears is a view of the human X-chromosome. The DMD (Duchenne Muscular Dystrophy) gene is highlighted. Clicking on the DMD link itself will take us to the LocusLink page where there is an extensive compilation of information about the gene and its expression. Click on 'DMD'.

LocusLink is a very rich resource of data (scroll down through the pages and examine it) that links back to the other databases at NCBI. By now you should be getting a sense of the extent of the interconnection built into the Entrez structure. Clearly we could pursue this gene in many directions from this page. However, let's go back to the Entrez home page and initiate our search for dystrophin by another route.

Click 'Entrez' on the black menu bar. From the Entrez page we could start our search a different way with a search for the Protein record for dystrophin. Choose 'Protein' from the Search drop-down menu, type 'dystrophin' into the text box, and then press 'Go'.

The output is very long with hundreds of entries, including various alleles of dystrophin and many dystrophin-related proteins. The sequence we are interested in has the identifier P11532. This identifier number is the **accession number** and was permanently attached to the sequence file when it was first archived in the database.

You can find P11532 by clicking through the pages of results until you see it, or you can do another search in the Protein database for 'P11532'.

When you find the sequence, you will see that on its right are a number of links. These include BLink, which compares the sequence to others in the database, a link to conserved domains in the proteins, and a drop-down menu for several other links. Keep in mind that many of these links are displaying different aspects of the NCBI information in different ways. The underlying databases are the same. Let's click on 'BLink'.

The BLink program displays a graphic of an alignment of our gene with all related sequences in the database. Note that there are many dystrophin entries and the graphic on the left shows the regions of similarity in sequence. Some of these are different isoforms and alleles; some are from other species. Some are older records. NCBI is an archival database, and once generated, all records are kept, although they can be updated. All of these records are available from this page simply by clicking the links. A researcher would use this resource to examine the various records for the gene, and to see, in a simple graphical way, different forms of the protein in humans, or the extent to which dystrophin from other species is similar to the human form.

From here, let's look at the dystrophin protein record itself by clicking on the link **'A27605'**, the first entry under the Accession number column.

The page that comes up is the protein record for our accession number A27605. This record is a computer translation of the DNA sequence determined by sequencing the mRNA for the protein. Computers require standard formats to store data, and humans need accessible formats to view it. This file (in a human readable form) is a **GenBank flatfile.** The display you see is the standard output format for GenBank. It has a specific list of sections, each with a particular type of information. It also includes highlighted links to other aspects of the databases. We don't have to worry about them all, but let's look at a few.

The locus is identified by the accession number, at the top of the record. The DEFINITION line shows it as "dystrophin, muscle – human".

Scroll down to see the REFERENCE section, which gives us all of the literature references related to the sequencing and characterization of this protein.

Further down, the FEATURES list tells us about various predicted or biochemically verified conserved domains, regulatory sites, and binding sites for other proteins. The features are arranged in order starting at the amino terminus of the protein (designated 1) and their location is identified by the amino acid position along this very large protein (3685 amino acids).

Lastly, at the bottom of the record is the protein sequence itself in single letter code.

The format of the GenBank flatfile is a rich source of information and you should familiarize yourself with it. There are other file formats, however, that are much simpler and sometimes more appropriate to use, particularly if all you want is the sequence for insertion into another program. Some analytical programs will accept the accession number as input, but many require that you cut and paste in the sequence. One of the simplest and most widely used formats is the **FASTA** format, which was developed for one of our important search and alignment programs (FASTA).

We can see the FASTA version of our dystrophin protein by toggling the display drop-down menu (at the top left side of the page) to **'FASTA'** and pushing the **'Display'** button.

The format has one line of information beginning with the > sign and followed by the accession number and species information. Starting on the next line is the sequence of the dystrophin protein. Notice that there are no line numbers or punctuation – it is perfect for cutting and pasting.

Now let's go back and learn how to execute more specific searches in Entrez. Close the window that we've been working in, and then go back to the Entrez home page at **http://www.ncbi.nlm.nih.gov/Entrez/**.

Oftentimes in searching for information on a protein such as dystrophin, we can be overwhelmed with the hundreds or thousands of responses. If you have a more narrow interest it is possible to phrase searches using Boolean operators (AND, OR, NOT). For example if we searched the protein database for "dystrophin AND chicken" then our output is sharply focused. We only get a small number of entries including dystrophin and some dystrophin-related proteins from chickens.

See how many results you get when you select the **Protein** database, write **'dystrophin AND chicken'** in the text box, and then press **'Go'**. (Note: The Boolean operators must be written in ALL CAPS.)

There are other ways to limit searches as well. For example, since there are so many human records for such an important human genetic disease gene locus, we might want to exclude human data from the list. Search the protein database for **"dystrophin NOT human"** to simplify the output. Now we get everything but human and the list shrinks from thousands to a few hundred entries.

There is really only one way to become conversant with such an important resource. Go to the NCBI Web site at **http://www.ncbi.nlm.nih.gov/** and continue to explore the Entrez program. This is one of the most useful tools for any biologist and could be considered an essential skill.

You have already examined the publications produced by your professor. Choose one of those publications related to a particular gene, protein, or organism and search through Entrez to find the associated links. Start your search from the publication you've chosen in PubMed. Using the links on the records, go to Entrez and input a gene or protein name that you have gleaned from one of the abstracts of the papers. This will initiate your search – see how far you can go!

3. Learning to Use BLAST

Starting link: **http://www.ncbi.nlm.nih.gov**.

This tutorial is a more detailed look at the BLAST (Basic Local Alignment Search Tool) program at NCBI, the most important single software tool for searching sequence databases. BLAST can be used to search databases using nucleic acid or protein query sequences.

We will walk through two protein examples in this tutorial. The first is the protein insulin, which should be familiar to you for its role in the regulation of blood sugar and in diabetes. The second is dystrophin, which we examined in the Entrez tutorial, a large and complicated protein.

To get to the BLAST homepage from NCBI, click on '**BLAST**' at the top menu. This page is the starting point for several BLAST programs available from the BLAST homepage. Since insulin is a protein sequence, we will search using BLASTP. Click on '**Protein-protein BLAST [blastp]**' under the Protein BLAST heading.

The BLASTp page that appears contains a window for pasting in the query sequence as well as several optional parameters that can be set.

For our first database search we are going to use the insulin protein from the Zebra fish (*D. rerio*). The sequence of the insulin precursor protein in FASTA format is written as:

>gi|12053668|emb|CAC20109.1|insulin[Danio rerio]
MAVWIQAGALLVLLVVSSVSTNPGTPQHLCGSHLVDALYLVCGPTGFFYNPKR
DVEPLLGFLPPKSAQETEVADFAFKDHAELIRKRGIVEQCCHKPCSIFELQNYCN

You should copy and paste the sequence above into the **Search** text box. Alternatively, you could do this search by typing the accession number into the window. Blast will recognize it and retrieve the sequence from the database.

The FASTA format has an identification line followed by the primary amino acid sequence in single letter code. The mature insulin molecule is a processed version of this primary sequence. It is derived by limited proteolysis from the initial translation product shown here.

The remainder of the search form allows you to set various options.

Set subsequence allows you to search with a particular portion of the sequence. Leave it blank so that the entire sequence will be used in the search.

Choose database has a drop-down menu that allows you to choose which part of the database you want to search – the entire database (nr for non-redundant), or sections of it such as *Drosophila* proteins only. Leave it on 'nr', the complete protein database.

Do CD-Search allows a comparison of the query sequence to a database of conserved domain patterns. This is a powerful tool for finding functional domains in genes. Leave it toggled on.

Last, click the **BLAST!** button to start the search.

Pushing the BLAST! button sends the file to the NCBI computer for the search . The page that appears next tells you that the sequence has been submitted and gives you a Search ID number.

At the top is the result of the **Conserved Domains** search. Conserved domains are the functional modules of proteins. They might include a pattern of amino acids typical of a particular catalytic site, or perhaps the binding site for a regulator of a protein.

Click on **Insulin** to see the full list of conserved domains. These will appear in a pop up window. Our search identifies two known domain patterns: insulin and insulin-like growth factor. The colored bars can be activated by rolling your mouse over them, so that the identification of the pattern shows up in the window. **Roll over** the bar for insulin. You should see the text in the box above the bars change to give the identification number of the pattern match, the fact that it is for insulin, an alignment score (S) indicating how strong the match was (higher is better), and a statistical measure of the significance of the match (E). The E value is the expectation that the match would have been found in the database by chance alone (lower is better). In this case, E is very small indicating that our query sequence is likely insulin (but we already knew that!). **Close** the pop up window.

Now, click the **FORMAT!** button to tell the computer at NCBI to send the results of the actual search and close the pop up window.

The output of such a search is many pages long and is made up of several components.

The first major section is a graphical display of the strongest matches to the query sequence. Notice that some hits are to the full-length insulin precursor and some are to the shorter processed form. They are color-coded according to the alignment score. If you roll your **mouse over** the various lines, the identification information, alignment score (S), and E value appear in the text box above the chart.

The second section of the results is a detailed list of hits ordered by their alignment scores. They correspond to the ones displayed graphically. Note that each line gives the identification information for the protein followed by the alignment score and the E value. The entries are ranked from the lowest to the highest E value, which can be interpreted as from most similar to more distant. The top line, not surprisingly, is the record for Zebra fish insulin itself. If you click on the gene identifier link, it will call up the sequence from Entrez. When you click on the **Score** link, it will show you the particular alignment from lower down in the output file. You can return to this page by using the **back** button on your browser.

Further down the list, the output gives the actual gene alignments for the various proteins in the list we just looked at. The first of these represents the top alignment. It is a perfect match to the subject, the Zebra fish insulin record. Note however, that early in the sequence is a masked region (shown by XXXX). This is a low complexity, or repetitive, region that is masked out in the query and ignored in the database search. Such regions may interfere with the alignment. You can see the actual masked sequence, since it is the same protein, in the subject line (LLVLLVVSSVS); it is a mostly hydrophobic, repetitive sequence.

Now let's look at two examples with a less than perfect match.

Scroll down the list of sequences producing significant alignments until you find the *Xenopus* sequence (gi|124513|sp|P12706|INS1_XENLA), and click on its score, **'90'**.

The alignment with *Xenopus* insulin (South African clawed toad) is a less than perfect match. Many regions are still strongly conserved however (notice the low E value). Empty spaces indicate mismatches and a + sign indicates similarity between the two different amino acids compared. For instance, at the beginning of the sequence, a hydrophobic leucine is scored as similar to a hydrophobic valine. Notice also that a gap had to be inserted in the *Xenopus* sequence to give a good alignment with the query. This would be the site of an insertion or deletion event during evolution.

The last example that we will look at is the alignment with human insulin. **Scroll** back up to the list of sequences producing significant alignments, and look for gi|4557671|ref|NP_000198.1|, and click on its score **'79'**.

Notice that there are many records for human insulin in GenBank. The alignment itself is still very strong (note the E value). But, compared to the *Xenopus* alignment, there are several more gaps placed in both the query and subject, not surprising for more distantly-related sequences. Note that the full length of the protein is not shown. BLAST is a local alignment tool and only displays the most strongly matching regions of the overall comparison.

We are finished looking at this sequence, and we're going to search for another sequence. So, **close** the window that the Zebra fish results appeared in, and you should just have the tutorial window open.

For our second BLAST search, we will use the dystrophin protein that we examined in the Entrez tutorial. Dystrophin is a very large protein (several thousand amino acids) with multiple conserved domains. It is also very highly conserved within the vertebrate group.

To run the search let's go back to the BLASTP page by clicking on the **'Protein'** link at the top of this part of the page.

Now, we could call up our dystrophin accession number (P11532), choose the FASTA output and cut-and-paste the sequence into the search window. It is much simpler, however, to simply type in the accession number since in this case we know it and the NCBI BLAST program will accept it.

Try entering **'P11532'** in the Search box and pressing **'BLAST!'**.

We see the output from the Conserved Domains database. . This is a database of evolutionarily conserved patterns of amino acids. These small patterns identify the amino acid residues that are almost always conserved through evolution in particular functional domains. There are two types of conserved domains in dystrophin as shown in the graphical representation. The domain at the far left is the calponin homology domain.

Open the list of conserved domain results by clicking on it. Calponin is an actin binding protein in the cell. Note the E value in the window by rolling over the blue blocks that say 'CH'. The red and blue tags simply denote hits in two different Conserved Domain databases. Each database uses a somewhat different pattern, hence the different E values. The remainder of the domains marked are spectrin repeat domains.

Spectrin is an important protein of the cell's cytoskeleton. Dystrophin has 6 domains showing similarity to a motif in Spectrin. These are smaller and/or have weaker similarity than the calponin domain. Compare the E values yourself. The various domains are allowing you to see the modular nature of proteins.

Now, let's go **back** to the original results page and click **'Format!'**

Once we receive the search result, it is obvious that there are many similar proteins in the database. Moving the **mouse over** the graphic will display the information line for dystrophin for a variety of vertebrates, a large number of different human dystrophin alleles, and sometimes gene fragments or other proteins aligning with specific domains. Now let's look at the subject hit list below the chart.

Note the E values for many of these subject hits on the right. The expectation is not exactly 0, of course, but rather is very close to a 0 chance of it being by chance alone, and has been rounded off. Note that the first real score is a rather small number, so you can imagine how small the other expectations must be! If we click on one of the scores, we can see the alignment. Let's try the mouse gene by clicking the score of **'1513'** for 'gi|192972|gb|AAA37530.1| (M18025) dystrophin [Mus musculus]'.

The mouse sequence is obviously very similar to that of the human but there are lots of positions where they have diverged. Given longer periods of evolutionary separation, we would expect to see more divergence.

We are finished looking at this sequence, and now we're going to try a more targeted search for another sequence. So, **close** the window that the dystrophin results appeared in, and you should just have the tutorial window open.

Let's now ask whether *Drosophila* has a dystrophin sequence. One way of focusing the search is to specify a smaller database. In this case we can search the *Drosophila* genome alone.

Go back to the BLASTP home page by clicking on **'Protein'** at the top of page. Paste the dystrophin accession number, **P11532**, into the Search text box. Then, in the 'Choose database' drop-down menu, select **'*Drosophila* genome'** if that option is available. Alternatively, under the Options for advanced BLAST, choose **Drosophila melanogaster** from the 'select from' drop-down menu (top window on the right of the Advanced BLAST section). Now click **'BLAST!'**

When the results are in, click the **'Format!'** button to get the hit report.

Here the result is not as complicated as for the entire GenBank database. Clearly there are some very good scores, as well as various shorter alignments. Perhaps some of the latter represent calponin or spectrin repeat domains. Let's have a look at the top score by clicking on the **first number** underneath the 'Score (bits)' column.

The *Drosophila* dystrophin gene product is obviously quite strongly diverged from the human version. Here our overall score is very good but the sequence comparison shows only about 40% identities. Nevertheless we would expect, if we examined the conserved regions carefully, that they would represent the important functional domains of the protein. At this level of analysis, we are perhaps not so different from the fruit fly.

You have been investigating the publication record of your genetics professors and searching for information on a gene of interest to them. Go to the NCBI homepage (**http://www.ncbi.nlm.nih.gov/**), find the BLASTP program and do a BLAST search on one of the genes that you found.

4. Using BLAST to Compare Nucleic Acid Sequences

Starting link: **http://www.ncbi.nlm.nih.gov**.

In the previous tutorial, "Learning to use BLAST", you learned how to use BLAST to search the genomic databases and to compare protein sequences. Proteins have a high complexity and information content since there are 20 different possible amino acids at each position in the sequence. As we have seen, there are also groups of amino acids with related properties that are scored as being similar. Nucleic acids, on the other hand, have only four possible choices at each position (AGCT[U]). For the most part, we only look to match identical residues at a particular position in a DNA or RNA molecule. In this tutorial we will start using BLASTN to compare the sequences of different transfer RNA molecules.

From the NCBI home page, choose '**BLAST**' from the top menu, and then choose '**Nucleotide-nucleotide BLAST [blastn]**' from the list underneath the heading "Nucleotide BLAST".

Transfer RNA molecules have a complex tertiary structure that is essential for their function and is dependent on intramolecular complementary base pairing. The requirement that tRNAs maintain this structure in order to interact with the ribosome and with aminoacyl tRNA synthetases results in strong selection in favor of retaining the primary sequences through evolution.

The following figure shows the cloverleaf stem-loop secondary structure of a tRNA molecule. This conserved structure is common to all tRNA molecules. In the diagram, we can more clearly see the complementary base pairing in the stems. Ribosomal RNAs that serve a structural role in the ribosome also show strong conservation of some regions. As we will see later, however, the constraints on the structure of messenger RNA sequences are less and there is correspondingly less conservation of sequence.

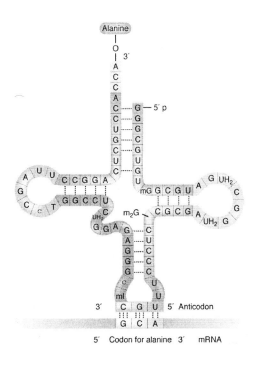

Figure 3-21 from *Modern Genetic Analysis, Second Edition* by Anthony J. F. Griffiths et al, © 2002 W.H. Freeman and Company.

BLASTN is very similar to BLASTP but obviously will only accept nucleic acids as input sequence. Cut and paste the phenylalanyl tRNA from *Drosophila melanogaster* into the **'Search'** text box.

>gi|174308|gb|K00349.1|DROTRF2 D.melanogaster phe-trna-2
GCCGAAATAGCTCAGTTGGGAGAGCGTTAGACTGAAGATCTAAAGGTCCCCGGT
TCAATCCCGGGTTTCGGCACCA

Just as for BLASTP, the program is run by pressing the **'BLAST!'** button and then the **'Format!'** button when it appears.

A search against the entire database at GenBank yields many hits, since tRNA molecules are highly conserved. The output list from this search shows many strong hits with low E values. Not surprisingly, many are from *Drosophila*—but also note other species, including human, in the list.

Look at the alignment for one of the human subject sequences by clicking on the Score **'137'**. You should see that it is almost identical to the query sequence.

Now let's look at the source for many of these sequences by clicking on the GI or accession number of any of the sequences. You will find that they are mostly in very large genome sequence files containing multiple genes; the tRNA subject sequence is somewhere internally in the DNA fragment.

Let's look at more distantly related hits, which would be much further down the list on the previous page. To illustrate a hit with low identity, first you should **close** the window that opened with the first search results in it. Then, click **'Nucleotide'** on the nucleotide-nucleotide BLAST page, so that you can start a new search.

Now, **search** again using our *Drosophila* tRNA, but choose the ***E. coli* database** option.

>gi|174308|gb|K00349.1|DROTRF2 D.melanogaster phe-trna-2
GCCGAAATAGCTCAGTTGGGAGAGCGTTAGACTGAAGATCTAAAGGTCCCCGGT
TCAATCCCGGGTTTCGGCACCA

Then, press **'BLAST!'** and then **'Format!'**

The output of our search is similar to the previous one—but note the E values. At this evolutionary distance (*Drosophila* to *E. coli*), there are only a few tRNAs retaining substantial similarity to the query sequence. The graphic display shows that most of these hits are towards the 5' end of our query sequence.

Click on the **'Score'** for some of the top subject sequences to see the alignments. Note that parts of our query sequence sometimes align with sequences within subjects that are thousands of nucleotides long. Remember that BLAST is a local alignment tool that finds high-scoring regions within a comparison. It does not usually produce an alignment from one end of the gene to the other.

Scroll down to look at the alignment for accession AE016756.1 (usually number 2 in the list) and remember the subject nucleotide position
numbers for the alignment. Then click on the **GI** or **accession number** to access the Entrez data for this large genome sequence in *E. coli*. The long subject file is for a long genomic sequence. Note in the information line that there are 18 such segments in the database representing the entire *E. coli* genome.

Scroll down the data until you see subject nucleotide position for your alignment. Note that it is described as a threonine tRNA. Now click on the word **'gene'** on the left of the position.

This link takes us to the Sequence feature view of the region. Scroll down to the bottom of the data, and you will see the actual sequence for this particular region.

BLAST does not attempt to do a global alignment from one end of the sequence to the other unless there is an uninterrupted high score throughout the alignment. Our hit reported a match of 23/25 for the 5'end of the molecule. If we align the complete sequence, we see that there are blocks of identity throughout the molecule with a match of 51/76 but only the block from 1-25 gives a high enough score to be reported by the BLAST program.

By examining the full sequence as shown in the following figure, it is obvious that the subject has significant similarity over its full length.

Remember the E value for this hit (0.015). This is the probability of finding a match of 23/25 in the database (*E. coli* genome) by chance alone. It is very high because the complexity of nucleic acids is low, having only the four possible base choices at each position. An alignment of two proteins with 23/25 identities would have a much lower E value because there are 20 potential choices at each position and the chance of finding the corresponding matches would be low.

5'
gccga**a**atagctcagttggg**a**gagagcgttagact**gaag**atctaaaggt**cccc**gg
gccga**t**atagctcagttgg**t**agag**c**agcgcattcgt**aatg**cgaaggt**cgta**gg

ttcaatcccgggttt**cggcacca**
ttcgactcctattat**cggcacca**
3'

Now that we have had a look at highly conserved tRNA sequences, let's try mRNA. Messenger RNA sequences for the same conserved protein drift substantially through evolutionary time. In addition to the drift of the amino acid sequence over time, the degeneracy of the genetic code may have caused the same amino acid to be encoded in multiple ways. Moreover, since the tertiary structural constraints are not as large on mRNA as they are on tRNA or ribosomal RNA, the sequences can diverge quite rapidly.

The p34 cyclin-dependent protein kinase is essential for mitosis and is strongly conserved at the amino acid level throughout the eukaryotic world.

Close the window with the *E. coli* results in it, and then click **'Nucleotide'** from the top of the previous search page. Now enter the accession number **'M12912'** (for the sequence for the p34 gene from *S. pombe*, the fission yeast) into the Search box, and click **'BLAST!'** and then **'Format!'**

When the results are in, we get a few strong subject hits. The graphic and hit list show strong alignments to several GenBank records for the fission yeast p34 gene itself. Not surprising, they are identical to each other. After that, however, the hits are only for very small regions, scattered throughout the gene and with significant scores in only a few other organisms, mostly other fungi. They do however show clustering in some particular regions presumably associated with conserved domains essential for function of protein kinases.

For comparison, if we run a BLASTP search using the protein product of the fission yeast p34 gene (accession P04551), we would get a very different result.

Close the window with the *S. pombe* nucleotide results in it, and then click **'Protein'** from the top of the previous search page.

Now enter the accession number **'P04551'** into the Search box, and click **'BLAST!'** and then **'Format!'**

Here the hit list is extensive with very low E values and clearly we have found a large number of homologues of this gene.

If we look at the third match (Click the score **'408'**), we see a highly similar protein in *Homo sapiens*. Here you can see that the protein in the human genome is 66% identical and 80% positive (identities and similarities) to that of the yeast. Furthermore, the similarity extends over the full length of the protein.

There is a lesson in this. If you are searching for matches to a protein encoding sequence, then always search with the translated product.

Now that you have experience doing nucleotide and protein searches, do a search yourself using a gene that is mentioned somewhere in your textbook. Find the accession numbers for the DNA sequence and for the protein sequence in Entrez (make sure both are for the same species and that neither is for a multiple sequence file). Run BLASTN and then run BLASTP and analyze the result.

5. Learning to Use PubMed

Starting link: **http://www.ncbi.nlm.nih.gov.**
Note: For this tutorial Cookies must be allowed on your web browser. If you are unsure how to enable these, turn to your browser's help menu.

It is the responsibility of scientists to communicate their findings with their peers and also with the world at large. The major means of doing so is the publication of scientific papers and based upon these, textbooks. One of the most difficult challenges for the modern researcher is to keep up with the current research literature.

Researchers must relate appropriate publications to the problem at hand and use the findings of others to help direct their own progress. It is not an easy task. Even if someone had all day every day to sit and read, it would be a formidable accomplishment to say that they had read all relevant material within their own field, let alone that of others. The scientific literature is growing in size so rapidly that computer searches are indispensable. A second aspect of this is that access through the computer increasingly allows direct access to electronic publications.

You have probably already become familiar with searching the databases at your institution's library. In this tutorial we will take a closer look at the research literature database at NCBI. Within the Entrez program at NCBI is the PubMed database, which is produced in collaboration with the National Library of Medicine MEDLINE database. Most importantly, all of the publications at PubMed are cross-indexed to gene and protein sequences and all of the other databases at NCBI. They are also cross-referenced to textbooks in the form of key word searches. Notable in the list of textbooks is Griffiths *et al., Modern Genetic Analysis* (MGA).

Click **PubMed** on the top menu. We have already looked briefly at this page in previous tutorials. In our introductory tutorial we did a simple author name search for L. H. Hartwell.

Let's try a number of other searches to demonstrate the potential of this system. Let's start with the words in the title of Chapter 6 of MGA: 'Genetic Recombination in Eukaryotes.' First type **recombination** in the search window and press **Go**. The result is a daunting 130,000 plus research articles that mention the word in title, abstract, or key words. It is a popular and important area, but this is far too general a search term to be useful.
How about **genetic recombination**? Try it. We still get more than 110,000 returns. We are retrieving articles actually studying the process of recombination but also many that are of more peripheral interest to us, for instance, simply using recombination as a method to insert genes into chromosomes.

Try the whole title **genetic recombination in eukaryotes**. Now the output is cut to less than one thousand. That number would certainly be far fewer publications than are actually concerned with recombination in eukaryotes. The problem is probably with the term eukaryotes. The key words of publications might include phylogenetic group, but not be as broad as all eukaryotes.

Try **genetic recombination in mammals**. We get almost 50,000 entries. Many of these clearly do not have recombination as the primary subject of the paper.

Let's have a look at how our terms might be related in the formal subject headings within the National Library of Medicine. These Medical Subject Headings (MeSH) are used to classify publications. Click on **MeSH Database** under PubMed Services on the left panel. At the new screen, type **recombination** in the search window and press **Go**.

Note that Recombination, Genetic appears in the search window as well as below with subheadings. This is more like the list of topics relevant to recombination that you would learn in a genetics course. Having a look at these terms sometimes helps narrow your search field by presenting you with some alternative (and narrower) terms. If we searched using the MeSH term *Recombination, Genetic*, we would get more than 100,000 returns. When we look at the subheadings covered we can see why this might be so.

Click on **Recombination, Genetic**. Note that **transfection** is included in the tree below. Click on it. The definition shows that it includes all papers that use transfection of DNA or plasmids into cells for experimental purposes. Notice that it is included in the techniques tree as well as the recombination tree. Many of the papers would be using it solely as a technique and the publication would not be about the process of recombination itself.

Click **Recombination, Genetic**, in the lower tree to go back to our original search.

Now let's try another term by clicking **Crossing Over (Genetics)**. This is narrower in scope and focuses on the chromosomal recombination process. Notice that it is listed under Cytogenetics in one tree, and under Recombination, Genetic in the other. Press the **'Clear'** button under the upper search window. Then check the box beside Crossing Over and press **Send to: Search Box with And**. Press **PubMed Search**. This yields a few thousand entries and a glance at the titles shows us that most are directly concerned with chromosomal recombination.

We could also focus our search by excluding some of the subheadings in the Recombination tree. Let's go back. Open the **MeSH Database** again (from the left-hand menu) and search **Genetic Recombination**. Click on **Recombination, Genetic**. Now click **Gene Transfer, Horizontal** in the tree, check the box beside Gene Transfer and then toggle the Send to: drop down menu to **Search Box with NOT** and click **Send to:**. Notice that the term appears in the search window with the NOT operator.

Next press **PubMed Search**. This returns an extensive list of papers dealing with various aspects of chromosomal recombination.

Now click the **Preview/Index** link in the bar under the search text box. This feature shows you a list of the searches you have run and the number of results for each. We could add more terms at this point. Toggle the drop-down menu at the bottom of the page to **Author** and type **Holliday** in the text box. Press **'AND'**, then **'Preview'**. A new result appears in your list of searches with only a few entries. If you click on this link for the number of results, you will see a subset of Holliday papers dealing with recombination processes.

There is a lesson here. It is relatively easy to define a narrowly focused search. It is often difficult however, to define a broad topic that does not include a substantial amount of irrelevant material (at least to you). If the search is narrowed to keep the amount of irrelevant material low, then it will almost invariably exclude material that you would want included in your list.

This means that you must use a somewhat broader search and then make a judicious choice of material to keep. Let's do a new PubMed search for **chromosomal recombination AND**

yeast. If you wish to see the formal MeSH search terms for your search, press Details at the top. (To return to your search results, click on the number under the Result line.)

Our hit list now is down to less than a thousand articles, with the titles showing. To assess the relevance of each article, a simple click on each list of authors will give us the Abstract of the paper.
Move down the list and choose a title that you think might be relevant and click on its **authors** to display the Abstract. Go back and forth to find several that meet the criteria of being about chromosomal recombination in the yeast system. For each of these **check** the little box to the left of the reference. When you have found a set, click **Send to: Clipboard** on the top menu. This will transfer these references to the Clipboard.

When you have finished, click **Clipboard**, on the gray menu bar. This will display only your selections and allow you to collect them by printing or saving to your computer. You could save it as a web page and retain all of the links to the abstracts. Alternatively, you could click Text to convert it to a text file that you can save to your disk and open with programs such as Notepad.

Now let's examine the links to textbooks. **Open** one of the abstracts on your list, and then click on the **Links** then **Books** link at the upper right. This searches the keyword lists of the textbooks against the abstract. All words or phrases found will light up as a link. **Open** one. You are presented with a list of textbooks where the topic is covered.

Click on the link to the number of **items** in a specific book to go directly to the list of locations where your keyword is discussed.

Alternatively, you could open the books themselves by clicking on the **book cover**. (You might have to click the very small **Books** link on the top black bar.) Try opening the first edition of *Modern Genetic Analysis* to see the relevant table of contents. From here, you can you can search for a term in the search box to see the paragraphs that will help you define terms and understand the concepts. This will allow you to read many scientific abstracts in areas that where you have little experience as yet.

From the book's table of contents you can also **open** the chapters directly. You will see various terms underlined in the text. **Click** on one of these. A glossary appears to provide you with a precise definition.

Initiate your own search by investigating Holliday structures in recombination. You might narrow the search by including a term such as **'sequence'** or perhaps **'genes'**. Use the resources of Entrez to find the sequence of at least one of the proteins involved in resolving the complex.

6. OMIM and Huntington Disease

Starting link: **http://www.ncbi.nlm.nih.gov.**

We looked briefly at the Online Mendelian Inheritance in Man (OMIM) resource in the introductory tutorial. Now we will explore it more fully. Note that if you are searching for information concerning a particular gene or disease, an OMIM search can be run directly from the initial drop-down menu on the NCBI home page. This is done by choosing OMIM and entering a search term.

However, in this tutorial we want to look at some of the structure of the OMIM database and this information is available from the OMIM home page. Click on the **OMIM** link on the NCBI home page to get to the OMIM home page.

Now let's start with the OMIM statistics page that we saw earlier. Choose **'Statistics'** from under the heading 'OMIM Facts', on the list on the left.

The statistics page shows the number of loci included in the OMIM database, as well as a breakdown of their distribution by inheritance pattern. Remember that our most recent estimate of the number of human genes is about 35,000. Now, notice that barely one third of all human loci are currently included in this database. There is still a long way to go before they are all categorized.

The various links in the table will call up a list of the genes fulfilling a particular mode of inheritance. Click the first **number** in the X-linked column.

The link takes us to a list of genes in the format of an Entrez search result. The individual records, however, are to annotations within the OMIM database. The Entrez links to sequence and related genes are over on the right side of the page. Let's look at a typical OMIM record by clicking on the first record **'*314998'**.

This record is a description of the discovery of the gene. It includes the literature references and information about linkage and location. This is all within a brief description written by a curator. This is an actively annotated database that changes with time. As research progresses, records of this type must be updated to keep them current.
Now, click on the green **OMIM logo** at the top of the page to get back to the OMIM homepage.

From here, click on **'Update Log'** from under the heading 'OMIM Facts', on the list on the left to call up a table of the frequency of updates and new additions of records to the database. Note that approximately 100 new records are added each month. Many times that number of existing records are updated to reflect our increasing knowledge of particular genes. Now, click on the green **OMIM logo** at the top of the page to get back to the OMIM homepage.

Now let's look at a particular case, Huntington disease. Type **'Huntington'** into the OMIM search window, and press **'Go'** to see the display for the OMIM record for this gene. The complete record for such an important and well-studied locus is tens of pages long. We will examine a few of the features.

First, click on the 'Links' button on the right of the first entry (OMIM number *143100), and then choose 'Protein' to take us to the Entrez protein record for Huntington.

Next, find the entry for accession number NP_002102 by doing a **Protein search** (it actually appears on the second page of the Huntington Protein list). Then, click on the accession number 'NP_002102' to see the record.

Scroll down through the human huntingtin protein record until you reach the Comment section. Notice that here is a description of the protein and its expression, the phenotype of mutants and even the nature of the common mutation in Huntington disease. The disease is associated with the expansion of a trinucleotide repeat sequence leading to an enlarged polyglutamine stretch in the protein. Notice also in the text how the NCBI staff indicate precisely the history of this record relative to the initial submissions to GenBank.

From here, scroll to the top of the page and click on the blue word 'Links' in the cluster of links to the right of the heading for huntingtin. From the drop-down menu that appears, choose 'OMIM'. Next, click on '*143100' to reach the OMIM entry for this disease.

The OMIM record for Huntington disease is a very rich source of information. The research literature is summarized under various headings including Clinical and Biochemical Features, Inheritance

Pattern, Population Genetics of the disease, and lots more. Clicking on the various headings on the left will take you to the subsections of the file. This is an immense resource. Entire term papers could be written based on this page alone! Clicking on the Mini-MIM button in the left panel at the bottom will take you to a much simpler version of the page. Clicking on any of the light bulb icons gives you the relevant literature sources.

Now let's have a look at the gene map. Click on the chromosome location **4p16.3** just under the page title and then click on **4p16.3** again at the top of the table that appears.

Now we have the graphical display of chromosome 4 with the Huntington disease locus highlighted. We can see from the graphic of the entire chromosome on the very left that the locus is very near the telomere on the *p*-arm of the chromosome. (Remember that the two arms of a human chromosome are named *p* and *q*.)

By zooming in on the locus (use the zoom in feature just above the chromosome graphic on the very left), we can display the genes that are located close to the Huntington locus. The list on the right is compiled from sequence data. Under the Morbid column are various loci associated with disease phenotypes.

Now, click on 'HD' under the "symb" column to go to the LocusLink record for the Huntington disease gene.

LocusLink is another annotated database and a very rich resource. Scroll through it. As with all of the other resources we are looking at, it is richly decorated with links to databases and information resources of all types. In the colored boxes are links to PubMed, OMIM, and other databases we have touched upon. However, some are new. Let's click on 'HGMD', the human genetic diseases database.

The HGMD database is an outside resource specializing in the nature of characterized mutations for various loci. Note that only the trinucleotide repeat is listed as present for the HD gene.

Now, let's go back to the Huntington disease LocusLink page at **http://www.ncbi.nlm.nih.gov/LocusLink/LocRpt.cgi?l=3064.**

This page will take us back to the gene map page if we click on the yellow **'MAP'** box in the menu.

Now we are back at the map view of the Huntington region. We can go to greater and greater depth from this map, which is largely a linkage map, to the actual nucleotide level. There we can examine the structure of the gene itself. Click on **'sv'** in the highlighted huntingtin row (the row with 'HD' in the 'symb' column – you may have to scroll down the page slightly).

'SV' stands for sequence viewer. In this case, it links to a graphical display of the annotated nucleotide sequence for the huntingtin region on chromosome 4. If you clicked on it for other genes in the display, you would activate different regions of the chromosome for viewing. The display shows about a megabase of DNA with several open reading frames indicated graphically. An expanded view of a particular region can be seen by clicking on a particular region of the chromosome and zooming in or out using the tool on the right.

The expanded region displays the start of the huntingtin coding region, indicated by HD at the left, and the three parallel lines under the sequence. Below this is the DNA sequence itself.

Scroll further down along in the sequence and you will see the transcription start site for the huntingtin mRNA. A bit further along, you can see the start of the coding region for the huntingtin protein with the amino acid sequence indicated below. Note the QQQQQ polyglutamine region — this is the region that expands and interferes with function, and is the mutational event that causes Huntington disease. The forward and backward blue arrows at the top of the sequence allow you to walk in both directions along the DNA strand at will.

We have explored OMIM to a sufficient depth for the moment. It is clear that there is much more here. As with all of the other sections of these tutorials you must go to NCBI and examine the links yourself.

Choose a well-defined human genetic disease based on one discussed in your textbook or class. Use OMIM to identify its chromosomal location and determine the nature of the gene product. What is the biochemical lesion involved? Use sequence viewer to find the start and stop sites for the gene. Also find the 5'-end of the first intron in the gene.

7. Finding Conserved Domains

Starting link: **http://www.ncbi.nlm.nih.gov**.

In the Protein BLAST tutorial we saw how the BLAST site at NCBI first compares the query sequence to a database of conserved patterns or motifs, the Conserved Domains Database. The patterns in the database are designed to detect structural motifs in proteins. By focusing on these conserved domains, such searches help us assign functions to unknown genes and see the domain architecture of complex proteins. We have already seen output from such a search for the dystrophin protein. Let's start there to investigate this feature more thoroughly.

Instead of starting with the BLAST page (where a Conserved Domains search is run as part of the BLAST search), let's go directly to the Conserved Domain Search page.

Click on **Structure** on the top black menu bar, then on **CDD** (for Conserved Domains Database) on the left-hand side bar. (You will have to scroll down to see it.)

Our basic evolutionary model is that sequences will drift further and further apart over time unless selection is at work. When we do a BLAST search, we compare proteins looking for regions of sequence conservation. When we find such evolutionarily conserved regions, we assume that they represent amino acid sequences that are responsible for the compact functional structural features of the proteins.

The order of the various conserved features that might be present in a protein is referred to as its 'domain architecture.' Complex domain architecture is very common among proteins in the human genome and other similarly complex organisms.

The CDD page contains a search entry window as well as information about the program and its underlying databases. Note also that it is closely integrated with the structural databases MMDB, Cn3D, and VAST, which are accessible from the side panel. We will learn to use these in a subsequent tutorial. Keep in mind for the moment that a conserved domain is really a compact structural feature, perhaps a binding site for another protein, or a catalytic pocket.

First, let's look at the nature of a search pattern for a conserved domain. Click on **Pfam**, in the paragraph beneath the search window. This takes us to Washington University in St. Louis, the home of the Pfam (Protein Families) database. Each Pfam search pattern is derived from a multiple alignment of a family of related proteins. The pattern is designed to recognize members of that family when an unknown sequence is searched against the Pfam database.

In the list of options, click on **BROWSE PFAM**.

Pfam is a large database with over 3000 conserved search patterns for all sorts of different conserved domains. The page that first appears shows the top twenty families, a sort of popularity list. Other lists are organized alphabetically. Each line in the top twenty families list gives its name, an accession number within the database, some statistical information, a link to its 3D structure, and a description. This database is an extremely rich source of information about proteins.

Let's try one. Click on the **LRR** motif, the Leucine Rich Repeat. An information page should appear telling you about the motif and what its function is thought to be—a protein-protein interaction domain. If you scroll down the page, you can retrieve the protein alignment that gives rise to the conserved pattern. Click on **Retrieve alignment**.

This list is hundreds of entries long. Note that sequences from the various genes (accession numbers on the left) are all a little different, but similar to each other. Also note that the pattern is not a continuous stretch of amino acids but instead is most highly conserved in three distinct clusters. In generating the Pfam search profile, the database combines all of this information to generate a pattern that will recognize all of these entries and others like them. Click the **back** button on your browser to go back to the information page.

Now try the **Retrieve domain structures** button to the right of the Retrieve alignment button. Here we can see the domain architecture for a number of proteins with leucine rich repeats (in green). Notice that different proteins have different numbers of domains and they are located in different locations along the primary sequence. What they all have in common is a conserved primary amino acid sequence in that region, which reflects a compact conserved structural domain with similar 3D structure in each case.

This is a diverse group of proteins. They also have a variety of other functional domains. If you roll your mouse over the domains you can see what they are.

Let's go back to the NCBI CDD page and run a search. Go to **http://www.ncbi.nlm.nih.gov/Structure/cdd/cdd.shtml**.

When we run a CD-search, our query sequence is compared to the thousands of stored patterns located in the Pfam database for various protein families (and also the Smart database at EMBL, a related resource).

Let's run a familiar search. As we have already seen, dystrophin has a complex modular structure containing a number of conserved functional domains. Type the accession number for *Drosophila* dystrophin (**NP_732427**) into the Search window in the Run CD-Search box in the middle of the page. Leave the Search Database drop-down menu toggled on **CDD**. (If you open this menu you will see that you can restrict the search to just Pfam or just Smart.) Now press **Submit Query**.

The graphic displays the location of a variety of conserved domains located along the length of the dystrophin protein. It is similar to the result we saw when we ran human dystrophin in BLAST.

Move your **mouse over** the colored domains and watch the display window provide information about the different regions.

Below the graphic is a list of the domains found. It tells you the Pfam or Smart accession number, a verbal description, and the E value for the match. Lastly, there is a detailed alignment for each of the domains found as well as considerable descriptive information about them.

As you can see, with a single search, researchers are able to glean an enormous amount of information regarding the structure and potential function of various regions in the protein.

Let's go back to the CDD search page by clicking the **back** button on your browser.

By analyzing dystrophin for conserved domains, we have discovered its domain architecture, that is, the order of different domains along its primary sequence. This was visualized as the graphic display with the various features arrayed along it. In principle, functional homologs of our protein would have a similar set of domains arrayed along them. When we did our BLAST searches for similarity, we found proteins with similar primary sequences to our query. In the CDART program (Conserved Domain Architecture Retrieval Tool) we are able to search for a pattern of conserved domains rather than for the primary sequence itself. Let's try it.

Click on **CDART** at the left side of the screen. Enter our dystrophin accession number (**NP_732427**) in the search window and press **Search**.

In the result, the graphic of our *Drosophila* dystrophin domain architecture is at the top. Below it are proteins with similar domain architectures. If you click on any of the icons on the graphics, it will open a window with the Pfam or Smart pattern information page.

Notice in the graphics aligned below our gene that some of the domains are in similar positions to those in our protein, notably the two calponin domains to the left, and the WW and ZZ domains on the right. The spectrin repeats are more variable, but present. In addition there are other domains that are not present in *Drosophila* dystrophin. This program is scoring similar domain architecture patterns and is a very rich source of information.

Note that the overall output of our search is about 20 pages long. **Click** on some of the later pages at the bottom and see some of the more divergent comparisons.

Remember that the Pfam and Smart motifs are identified by running programs such as BLAST to find families of related sequences. The patterns are then distilled from the conserved regions of these gene families. These patterns can now be used to search and identify new sequences and to place them in functional families.

Choose a gene that interests you, either from your text or related to your earlier searches, and run a CD and CDART search.

8. Determining Protein Structure

Starting link: **http://www.ncbi.nlm.nih.gov**.

WARNING: Depending on how fast your computer is and how much memory you have available, your computer may be very slow in responding to parts of this tutorial. To speed it up, we suggest closing any other programs you may have open before you start the tutorial.

Genome sequencing has given us an enormous quantity of data regarding the genetic breadth and potential of living organisms. As we have seen, tools such as BLAST make it relatively easy to search the genome databases using the primary sequence of hypothetical proteins. This allows us to ask questions about similar proteins, homologs, and gene families. The primary sequence determines the three-dimensional shape of a nucleic acid or protein. Structural databases such as Entrez Structure at NCBI contain details of protein structure based on X-ray crystallographic and NMR studies. From this database it is possible to retrieve and examine a particular three-dimensional structure in a viewer called Cn3D. In this tutorial, we will examine some structures and learn to use the Cn3D structural viewer.

First, let's download the Cn3D viewer plug-in for your browser. Click on **Structure** at the top black bar of the NCBI homepage. Then, click on **Cn3D v4.1** on the left-hand side menu.

Cn3D provides a graphical view of the three-dimensional structure for a molecule. Follow the directions to download the program. First click on the appropriate operating system for the computer you are using (**PC, Mac, Unix**) on the blue bar at the top of the page. Read the system requirements and directions carefully. In some cases (for PCs) there is more than one version of the program. Next, click on **here** to start the process. Either allow the file to open automatically, or save it to your disk and then click on it to run. The program will install itself and will recognize structure files automatically when we later download them.

Now let's go back to the Structure page by clicking the **Structure** heading at the top of the page.

The structure database can be searched through Entrez. In the search text window, type **1JM6**. This is the identification number for rat pyruvate dehydrogenase in the PDB protein database. Click **Go**. When the result appears, click **1JM6** to see the NCBI Molecular Modeling Database (MMDB) Structure Summary page.

This page provides a description of the file including the MMDB ID number, PubMed link, species, and citation information. Use the back button on your browser to go back to the MMB summary page. Make sure that you downloaded the viewer, then click on **View 3D Structure**.

You will see the file download and then two popup windows for Cn3D will appear. Move them to the left if need be so that you can still follow this text.

The upper pop-up window (Cn3D 4.1) shows a representation of the 3D structure of the protein. This is drawn from the 3D crystal coordinates for each of the atoms in the molecules. The representation is in structural mode: alpha helices are shown as green cylinders and beta pleated sheets as gold flat arrows. The connecting amino acid chain is in blue.

The display is interactive. The image can be rotated in three dimensions by placing your mouse on the image, hoding your left mouse button down and moving the mouse. Try it. The calculations are done on your computer and the rotation speed will be affected by how fast your computer is. As it rotates, note that it consists of two mirror image complexes consisting of the A and B chains. Any view of the molecule may be obtained.

As the structure rotated, did you notice the bound ADP molecule in each complex (partly colored in red and complexed with a Mg ion)? Can you identify the adenine ring structure in the ADP molecule? You may enlarge this window to full screen to make it easier to see, but when finished, reduce it again. Alternatively there is a Zoom In feature under View in the top menu.

The lower pop-up window (Sequence/Alignment Viewer) shows the primary sequence of the two amino acid chains. Marked along its length is a representation of the helical and pleated sheet domains. Notice that the primary sequence is color-coded to match the representations in the 3D viewer.

Now go back to the Cn3D 4.1 viewer window. The representation is of secondary structure. On the menu at the top, open the **Style** drop-down menu, open **Rendering Shortcuts**, and then click on **WireFrame** to see a different representation of the structure of the amino acids chain. It is still color-coded for the different structural features. Lastly, try **Spacefill** under Style. (This model requires much more computation for each view and may take several minutes to download. It will also be slow when you rotate it. If your computer is very slow, you should skip this step.) This is a more realistic view of what the surface might really look like, if you were another molecule! Try rotating it.

Close the pop-up windows and let's look at another example.   Click **Structure** at the top of the page to go back to the homepage.

The Structure database is linked to Entrez. Therefore we can enter terms and search in the same manner that we have done in other tutorials. Enter '**GCN4 AND yeast**'. This protein is a transcription factor that binds DNA. Click **Go**.

The list of hits is two pages long and represents all of the entries in PDB for this protein, its fragments and various mutants. As with all of the databases that we have looked at, it takes a certain amount of sorting to get what we want. Go to the bottom of the list. (Select page: **2** in upper right corner then scroll down the page). Choose the last entry, **1YSA**, by clicking on it. The MMDB structural summary for 1YSA is shown. Click on the **View 3D Structure** button. As before, the two Cn3D pop-up windows should appear.

This view shows us the interaction of chains of GCN4 transcription factor (in green) bound to a promoter site on DNA (in red). We are initially looking down the length of the DNA helix. First, let's switch the view to **WireFrame** in the Style menu at top. This will let us see the DNA base pairing. Now **rotate** the image through about 90 degrees. Stop when the red/blue nucleic acid chain is vertical. As it rotates we can plainly see the DNA double helix with its planar base pairs (in blue). We can also see the GCN4 chains contacting the DNA strands on each side. This is the sequence-specific recognition of the DNA molecule by the transcription factor. We can also see that the protein chains interact with each other at their opposite ends.

Now take a look at the Sequence/Alignment Viewer. It can be toggled (under File, sequences) to show either the protein or nucleic acid sequence. If you use your mouse to highlight a part of the nucleic acid or protein sequence (whichever you have showing in the panel) then it will color the corresponding sequence in the 3D graphic. **Try it.**

Lastly, place the viewer in **Spacefill** mode to have a more realistic sense of what the molecules are really like. (Again, depending on the speed of your computer, this may take some time, though not as much as the first protein, as this is a much smaller molecule.) **Close** the viewer windows, and go back the Structure homepage by clicking **Structure** at the top of the page.

In addition to allowing an investigator to search for the known structure of a particular protein, NCBI also allows an investigator to search the database for structures similar to the one of interest. This allows a crystallographer with a newly acquired sequence to search for structurally related proteins utilizing the VAST program (Vector Alignment Search Tool).

Within VAST, all of the structures in the MMDB database have already been compared to all other structures in the database. Since we don't have a new crystal structure to try, let's explore the structural comparison function using a previously known sequence. Type the PDB identification number **1G5S** in the search window and click **Go**. This displays a record for human cyclin dependent protein kinase, one of the proteins that controls the cell cycle.

First let's click on **View 3D Structure** to see what this protein looks like. **Close** the OneD-Viewer window, and minimize the Cn3D viewer window for the moment.

The Structure Summary page includes a line labeled 3D domains 1, 2, 3. These symbols refer to the subdomains of the protein. These links take us to the most structurally similar files in the database. Click on **2**.

A VAST Structure Neighbors summary page appears. Notice that it has a table of records with varying degrees of amino acid identity. Click on the small **square** to the left of 1IA8, near the top. This record is for another protein kinase found in humans. Now click on **View 3D Alignments**. A new Cn3D window will open. This contains a comparison of 1IA8 to our original 1G5S file.

Enlarge the original Cn3D window containing 1G5S and arrange it along side of the new window containing 1IA8. Keep the sequence alignment window for 1IA8 open below it. Toggle both windows to **Worm** under **Style**, Rendering Shortcuts. Notice the overall similarity of shape and organization to our first protein kinase. Note that in the new window the full structure of 1G5S is shown, but only the second domain of 1IA8. This is not surprising since they are both protein kinases and we have found the second one by structural comparison to the first. **Close** the 1G5S window.

A careful look shows that the new window contains both the 1G5S and the 1IA8 structures superimposed on each other. You can highlight the two sequences by using your mouse to highlight portions of the sequence in the alignment window DDV at the bottom. For instance, **highlight** amino acids 110 to 130 on the 1G5S (top) sequence. (There is a position counter in the lower left corner of the window.) You can see that one of the alpha helices in the 3D model turns yellow. If you look carefully you will see that close to the yellow is a red and blue chain directly paralleling the yellow one. This is the 1IA8 sequence. If you now use your mouse to **highlight** 110 to 130 on the 1IA8 sequence, you will see the corresponding change on the 3D image. The VAST program statistically compares how closely the two models fit to each other and aligns them in three dimensions.

Programs such as VAST are capable of finding very distant homologs, even where we do not see statistically significant sequence identity using alignment programs such as BLAST. Ultimately, it is the 3D structure of a protein that is selected during evolution. It is possible to have similar 3D structures, but quite different amino acid sequences. **Close** all of the pop up windows.

This is just a brief look at the richness of the 3D information that is available. Such information is critical to researchers wanting to understand how proteins interact or how a drug inhibits a protein.

Go back to the NCBI Structure page and search in Entrez Structure for a structure that interests you. It could be one of the sequences used in earlier exercises or something new. Keep in mind that only a small proportion of all sequences in GenBank have had their structure determined. As a start, you might try looking up the structure of a transfer RNA.

9. Exploring the Cancer Genome Anatomy Project

Starting link: **http://www.ncbi.nlm.nih.gov**.

A number of specialized databases are available through NCBI. These often represent collaborative efforts between outside resources and NCBI. We have already made use of several of these, for instance in the Conserved Domains tutorial and in using OMIM.

Many others are listed on the right hand side of the NCBI homepage. The Cancer Genome Anatomy Project (NCI-CGAP) is organized by the National Cancer Institute. NCBI provides a number of database and bioinformatic resources to the project. This allows cross-linking among all of the related databases.

Cancer is a very complex disease. In general, it is characterized by inappropriately controlled cell proliferation. The resulting cells interfere with other tissues and they characteristically spread in the body. There are many ways in which a cell's machinery can go awry and still have this overall end result. Cancer, therefore, can be thought of as hundreds of different diseases.

Most cancers arise from spontaneous somatic mutations accumulating over your lifetime. In this sense it is a disease of aging, although it too frequently occurs in younger people. Somatic mutations, of course, are not passed on to the next generation. There are also cancers caused by inheritable germ line mutations and they can be followed through families by pedigree analysis. Characterizing all of the genes and mutations that directly cause or predispose us to cancer is the goal of the Cancer Genome Anatomy Project.

Click on **NCI-CGAP** at the bottom in the right-hand menu. This page contains a description of various aspects of the NCBI-CGAP collaboration.

The Cancer Genome Anatomy Project (CGAP) pulls together a large amount of public data on genes altered in cancer cells. This ranges from gene sequence, mutational analysis, cytogenetic localization, chromosomal aberration, biochemistry, and clinical manifestation for cancers. By making all of this information easily accessible, CGAP helps the researcher analyze new information and relate it to what is already known. Since cancer ultimately is a problem of expression of a mutated gene product(s) and consequent inappropriate expression of other genes to give the cancer phenotype, analysis of gene expression has been one of the major efforts of CGAP.

Click on the **Cancer Genome Anatomy Project** link.

This is the homepage for CGAP (**http://cgap.nci.nih.gov/**). Notice that we can enter the databases from several directions and that they are extensively linked with NCBI at various levels. We could ask about gene or tissue expression patterns, chromosome aberrations, biochemical pathways and various other resources.

Let's try several. First, click on **Genes**. The GENES page provides access to a number of bioinformatics tools. Click **Gene Finder** in the top section.

Gene Finder allows us to search the database by gene name, accession number, or tissue where a gene is expressed. The key words and identifiers are from the UniGene classification at NCBI. Let's try an example. In the lower search window, toggle the Tissue Type to **Breast/Mammary Gland** and then type **BRCA1** in the Keyword in Gene Name text box. BRCA1 (breast cancer associated) is known to be involved in some forms of early onset breast cancer and occurs in families as a germ line mutation. Press **Submit Query**.

Gene Finder returns the BRCA1 gene and a number of records for it. It also returns records for a number of genes whose gene products are known to interact with the BRCA1 protein. The **Gene Info** link associated with BRCA1 takes us to a summary page containing the various GenBank accession numbers as well as links into various databases. Click on it.

The Gene Info page is a compilation of information and links concerning our gene. Scroll down the page and look at the scope. Near the top are a number of database links. Click on **UniGene**. The pop-up window that appears gives a summary of information about the gene and about the various GenBank records for it. Note the related proteins in a variety of eukaryotes, even plants. Remember that cancer is a disease that affects the basic cell proliferation machinery and much of this is conserved in all eukaryotes.

Close the pop-up window.

Clicking **LocusLink** takes us to a wealth of information about BRCA1 in the LocusLink database at NCBI. Try it. Scroll down through the page. **Close** the window when you are finished.

Similarly, clicking **OMIM** takes us to a long list of entries on the nature of the gene and its inheritance. We have looked at some of these resources before. It would take many hours to follow all of the links and text. **Close** the window.

Remember that these databases could also have been found by entering BRCA1 in Entrez.

Now let's go back to CGAP and see what other types of information it might offer. Click on the **CGAP icon** at the top left of the page.

As we unravel the biochemical function of various genes and gene products, we can place them in functional pathways using all of the information available to us. Click on the **Pathways** link.

Click on the **BioCarta Pathways** link. BioCarta has an extensive list of biochemical pathway charts. Under A, click on **ATM Signaling Pathway**.

Notice the schematic of the cell cycle across the top of the map. The ATM pathway is important for delaying cell cycle progression to allow time for repair if there has been damage to DNA. This helps to prevent permanent damage or chromosome breakage and the consequent genomic instability. Maps of this type help the researcher place a particular gene product into a broader context. Our BRCA1 gene product is shown just to the left of center. Its precise role, since mutations in BRCA1 lead to a high risk of breast cancer, is the focus of much research around the world. Clicking on any of the icons in the map takes us back to the Gene Info page for that gene. You might have noticed, when we examined it, that the bottom of the BRCA1 Gene Info page contained a link to this map.

Let's go back to the CGAP Genes page and try another link. Click on **Genes** in the green menu near the top of the page.

One of the problems facing all researchers (and students) is the development of a formal language describing the function of genes. This is essential if software is going to be developed that will allow us to move transparently among a variety of different databases. An 'ontology' defines function in a hierarchical fashion, moving from the general to the specific. Click on the **Gene Ontology** (GO) browser.

The pop-up page that appears contains a very general opening page with three descriptors: Biological Process, Cellular Component, and Molecular Function. The + sign in the boxes indicates that there are hidden trees for each.

Click on each of the **+ signs** next to the three terms in turn and open up the next level of the hierarchy. Each line starts with a symbol.

Try clicking on the + sign next to 'binding', and then 'nucleic acid binding' under Molecular Function. This opens up a further four categories including whether gene products bind DNA or RNA. (You may have to scroll down the page to find it.)

Click on the **+ sign** next to 'DNA binding' and a further 20 categories open with increasing levels of specificity.

If you click on one of the numbers it will call up a list of the genes in the main browser window. This pop-up window will go to the background and you will have to click on it in your tool bar to bring it to the foreground again.

In some ways working through this hierarchy is the opposite of what we have done in previous tutorials. We have started with a gene and built our information tree from the gene up, until we get clues to function. In the hierarchical structure of the Gene Ontology we narrow the functional fields and then ask "What genes belong to this group?". These terms and the hierarchical structure play a major role in categorizing the genes of organisms as new genomes are sequenced. The terms provide a common language for comparison. Keep in mind that BLAST scores are a major criterion in deciding whether a newly discovered sequence has a similar function (and ontology) to a protein whose function has been previously investigated.

A gene product may belong to more than one category. This is especially true of human genes where very complex domain architecture often correlates with multiple functions.

Let's go back to the CGAP/Gene home page. **Close** the Gene Ontology Browser, and click on **Genes** in the green menu near the top of the page.

We now know of several million different single nucleotide polymorphisms (SNPs) in the human genome. In the case of hereditable disease, such as BRCA1 associated breast cancer, we can track the alleles responsible. Most importantly we can use the information for diagnostic tests to assess the risk to individuals. Click on the **SNP Gene Viewer** in the second section.

Again, a pop-up window will appear. Type **BRCA1** into the search window and press **Submit**. A list of SNPs for the BRCA1 gene and some BRCA1 associated proteins appears. Notice in the column at the right that many SNPs do not cause changes in amino acid sequence in the gene product. Click on **view** under the accession number of one of the

BRCA1 mutations that does cause a change. A graphic appears at the bottom of the page along with a list of SNPs and the amino acid change (e.g. Pro871Leu is a change from proline to leucine at amino acid position 871). The graphic of the gene also displays the conserved domains for the protein to orient the viewer. Each point mutation is a different allele.

Let's **close** the pop-up window and go back to the CGAP main page and examine other resources by clicking the **CGAP icon** at the top left of the page.

Next, let's try clicking the **Chromosomes** link. Several resources are listed. Click on **GeneMap99** in the blue menu at the left. A new pop-up window will appear.

The GeneMap project compiles mapping data for the human genome. It maps known genes as well as ESTs. ESTs, or expressed sequence tags, are fragments of cDNAs derived from mRNAs in the cell and thus are a part of a gene sequence. ESTs represent a major portion of the sequences submitted to archival databases and at NCBI they are held in a database called dbEST. Let's look at what has been mapped.

Click on **Chromosome 1** on the upper menu. This yields a graphic of chromosome 1 as well as information on what has been mapped. Notice the red tag on the left hand vertical line. If you move your mouse over this line and click on various regions, you will see the red marker move. As this moves to new regions, the list at the bottom of the page (scroll down) shows you the physical mapping information for all known genes in the region as well as various ESTs that have been mapped. Such detailed mapping information is instrumental in compiling and interpreting the human genome sequence. It is of tremendous value in detailing the nature of chromosomal rearrangements that are often common in cancer cells.

Now let's **close** this pop-up window and go back to the CGAP homepage by clicking the **CGAP icon** on the top left of the page.

CGAP aids the researcher both by being an archive of information as well as being actively engaged in the compilation of new information and resources. Start your own search through the system. Perhaps begin at the BioCarta map for Cell Cycle: G1/S checkpoint. Checkpoints delay cell cycle progress while DNA repair occurs and many of the proteins in these pathways are associated with cancer. Find the p53 gene and click on it. This gene is mutated at high frequency in human cancers and you will find a wealth of information about its function.

10. Measuring Phylogenetic Distance

Starting link: **http://www.ncbi.nih.gov**.

Natural selection depends upon heritable variation within a population. The variation is a result of random mutation as well as the random choice of gametes. The net result is phenotypic differences upon which natural selection can work. Differential reproductive rates based on these phenotypic differences result in changes in the genotypes represented within the population over time.

It follows that over long periods of time, random change and selection will lead to increasing DNA sequence divergence among organisms. With certain underlying assumptions, we can use this information to estimate evolutionary distance. We have already looked at this to some extent in the BLAST tutorial and also when considering the nature of Conserved Domains. In this tutorial, we want to explore the relationship in a bit more detail and to examine some of the available resources.

NCBI is the repository for vast amounts of sequence data. It currently has at least one bit of sequence from each of more than 100,000 different organisms. By comparing homologous genes among different organisms we can get a measure of sequence divergence and from that, relative estimates of evolutionary relationship. Different genes, however, may diverge at different rates, so evolutionary relationships inferred from limited data must always be treated with a certain amount of skepticism. Clearly these data must also be combined with insight gained from comparisons of phenotypic characters and the geological record to reach a consensus within the scientific community regarding the relationship of one organism to another.

First let's look at the taxonomy resources at NCBI. NCBI has undertaken to generate a consensus taxonomy linked to all of the sequences in the database. Obviously sequence data plays a major role in defining such relationships. First let's look at the NCBI taxonomy resource. Click **TaxBrowser** on the upper menu.

This page is the gateway to a variety of resources. One of the goals of the Taxonomy section at NCBI is to generate a consensus Taxonomy for all of the organisms represented in the database. Let's have a look at one of our most commonly used organisms. Click on the **Drosophila melanogaster** link in the organism list.
This opens a summary page of taxonomic position as well as a list of the number of records of various types in the database. The taxonomic tree for *Drosophila melanogaster* is shown. Click on the word **Drosophila** to open the Genus level of the taxonomic tree.

There are a lot of fruit fly species in the world! On the upper gray menu, **toggle** the **proteins** to on and press **Display**. This will display the number of sequence records available for each item. Notice that for some, there is only a single entry.

Search down the list and find the melanogaster group (under Sophophora), melanogaster subgroup and eventually *D. melanogaster* itself. Each of these species has a unique genome. Our premise in looking at sequence data is that it would diverge more and more relative to *D. melanogaster* itself as we went further up this taxonomic tree. The quality of our comparison is dictated by the amount of sequence available for each species and the extent of our comparisons. Although we have the entire genome for *D. melanogaster*, that is not the case for the others.

To use sequences in order to help build a phylogenetic tree, we must align the sequences with one another to score their divergence.

In the BLAST tutorial, we saw that more distantly related organisms had more divergent sequences. When we examined the output from a BLAST search, the hits were ordered from most similar to most divergent and ranked by the BLAST score.

BLAST results are a complicated list of accession numbers and formal species names. It is often difficult to understand the relationship of the various organisms to each other simply by looking at the list. Many times, without expanding each entry, we cannot tell what organism we are looking at.

BLAST results, however, are linked to the Taxonomy browser. Let's run a BLAST search and explore it. Click on the **NCBI icon**, then **BLAST** and **Protein BLAST**. We will search using an *Arabidopsis* PCNA sequence. This protein is part of the DNA synthesis machinery and is strongly conserved in all organisms. Enter the accession number **NP_180517** in the window, press **BLAST**! and then press **Format!.**

As we have seen before, the BLAST results page is a detailed list of hits, but with limited visible information about the organisms in the list. Clearly we can open each entry and examine them one by one. However, BLAST results are linked to the Taxonomy Browser. Go to the left of the page (above the graphic display of results) and click on **Taxonomy Reports**.

The BLAST result is now formatted in the new window with the lineage of *Arabidopsis*, our query organism at the top, and then each hit on the list arrayed according to phylogenetic position.

Clicking on the **Organism Report** link at the top takes us down the page to list the hits from each organism.

Similarly clicking **Taxonomy Report** gives us the consensus taxonomy with all of the hits listed in it.

This is a very useful tool for orienting yourself to the strange organisms that we deal with.

Close the pop-up windows.

It is possible to use the quantitative measure of sequence divergence to construct phylogenetic trees. The underlying assumption is that the sequence divergence is related to the evolutionary distance.

We will build a small phylogenetic tree based on a sequence comparison of a single protein that is strongly conserved in all organisms. The protein is PCNA, which we used above, and which is part of the DNA replication machinery. We will use the ClustalW program to generate a multiple sequence alignment and then the Phylip program to form our tree. These two programs are not available at NCBI, so we will run them at the European Molecular Biology Laboratory (EMBL) site, one of the archival sequence databases centers in Europe.

We have looked at sequence file formats in earlier tutorials. For this exercise we need to construct an input file for the ClustalW program. We will do this as a multiple FASTA file of PCNA sequences chosen from a number of different organisms. This can be done by collecting several different PCNA sequences on the Clipboard at NCBI, displaying them in FASTA format and saving the result as a text file (.txt) that can be edited in Notepad or a similar text editor.

Let's choose some sequences. Go to the **NCBI** homepage by clicking the icon at the top left of the page. Then, go to **ENTREZ**, toggle the search to **Protein** and enter **pcna AND human** in the search window. Press **Go**. In the output list, **check** the box in front of accession number P12004, and then at the bottom of the page, toggle 'Send to' to **Clipboard** and press **Send**. **Repeat** the search for *Arabidopsis thaliana* (AAC95182), *Drosophila melanogaster* (NP_476905), *Danio rerio* (NP_571479) and *Gallus gallus* (BAB20424), and after each search add these sequences to your clipboard. At the end you should have five sequences on your clipboard.

Click **Clipboard** on the blue menu at the top of the page to see your sequences. It should display the summary for each of the five records. Toggle the Display window to **FASTA** and Click **Display**. Now you should see each of the files in FASTA format. When converted from its current HTML format to text format it will be a suitable multiple FASTA input file for a multiple alignment program.

We have to export this back to our own computer as a text file. At the top of the page, toggle the 'Send to' window to **Text**, and then click **Send**. Now the file consists of identification lines (starting with the > symbol) followed by the sequence in each case. There are no spaces between the entries. **Highlight, copy,** and **paste** the sequence file to a Notepad page on your computer, and then **save** it.

Open the file with Notepad or a similar text editor program on your computer. Towards the end of the identification line for each sequence is the name of the organism in brackets. We want to keep this information and eliminate all of the accession numbers, etc. For each of the identification lines edit the information behind the > symbol until only the name of the organism remains. This information will be printed out on our phylogenetic tree and shorter is better.

To run our multiple alignment, go to the ClustalW program at EMBL at **http://www.ebi.ac.uk/clustalw/#**.

This is a somewhat complicated web page, but we do not have to concern ourselves with all of it. Most of the windows can be left on their default settings. We do have to fill in several of them, however, and load our sequences.

First, create a title for your alignment in the upper left-hand window. Toggle Color Alignment to **Yes** in the top row. In the fourth row, toggle the Tree Type to **Phylogram**. Then push the **Browse** button at the bottom and locate your PCNA text file on your disk. This will upload your file to EMBL.

Last, push **Run**.

Scroll down the output page and you will see the multiple alignment of your files. The most similar sequences are on the top and the most divergent on the bottom. Not surprisingly, the chicken and human are closely related and on top and the plant sequence is at the bottom.

The color scheme identifies residues by properties and allows you to easily see the regions that are most conserved. Below each column the (*) identifies residues that are identical in all five sequences and the other symbols give degrees of similarity.

An alignment of this type is the starting point for doing the phylogenetic tree. To build a tree, a program such as Phylip quantitatively scores the similarity between each pair of sequences based on this alignment. In this case, the tree building program has already been run.

If you **scroll** further down the page, you will see a simple diagram displaying this information in graphical form. The length of the lines between organisms and branch points are related to the degree of similarity between the sequences in the alignment. Based on our assumption of constant mutation rate over time, these distances are then proportional to evolutionary distance.

You should **save** this output page to your own disk or print it for your use.

This has been a simple exercise to look at Taxonomy resources and the basics of multiple alignments and their use in tree building. Keep in mind that there are many assumptions underlying such exercises and many alternative ways of doing this type of analysis. Ultimately we satisfy ourselves of the validity of the result when multiple methods give us a similar result. For instance, we might expect to get a similar tree if we used a different protein from the same set of organisms.

Try running an analysis on your own. Choose a protein and make a new multiple FASTA file. You might choose a different protein for the same set of species that we just analyzed. Do you get an identical tree?